Carbohydrate Chemistry

Proven Synthetic Methods

Volume 2

Carbohydrate Chemistry: Proven Synthetic Methods

Series Editor: Pavol Kováč

National Institutes of Health, Bethesda, Maryland, USA

Carbohydrate Chemistry: Proven Synthetic Methods, Volume 1
by Pavol Kováč

Carbohydrate Chemistry: Proven Synthetic Methods, Volume 2
by Gijsbert van der Marel, Jeroen Codee

Carbohydrate Chemistry

Proven Synthetic Methods

Volume 2

Edited by
Gijsbert van der Marel
Jeroen Codee

CRC Press
Taylor & Francis Group
Boca Raton London New York

CRC Press is an imprint of the
Taylor & Francis Group, an **informa** business

CRC Press
Taylor & Francis Group
6000 Broken Sound Parkway NW, Suite 300
Boca Raton, FL 33487-2742

First issued in paperback 2019

© 2014 by Taylor & Francis Group, LLC
CRC Press is an imprint of Taylor & Francis Group, an Informa business

No claim to original U.S. Government works

ISBN-13: 978-1-4398-7594-0 (hbk)
ISBN-13: 978-0-367-24684-6 (pbk)

Visit the Taylor & Francis Web site at
http://www.taylorandfrancis.com

and the CRC Press Web site at
http://www.crcpress.com

Contents

SECTION I   Synthetic Methods

SECTION II   Synthetic Intermediates

In Lieu of an Introduction

FOR YOUNG READERS STARTING THEIR CAREERS IN SYNTHETIC CARBOHYDRATE CHEMISTRY

TIPS AND TRICKS FROM UP MY SLEEVE

This section consists of excerpts from a document every new associate in my laboratory receives upon joining the group. The pamphlet, which my young associates refer to jokingly as "The Bible" (no blasphemy intended), has been updated periodically at irregular intervals and is still in use. Over the years it has developed into a compilation of useful bits and pieces of information not typically found in textbooks or manuals and has been highly valued by my young colleagues. They tell me that most of the information found in the pamphlet would be hard to come across other than by personal interaction with more senior colleagues who are willing to share.

Here, I offer to young readers some tricks and tips that I have collected during many years of working at the bench in several carbohydrate laboratories. Usefulness and reliability of these have been verified by many generations of undergraduate and graduate students as well as post-doctoral fellows, and I hope that many readers will also find them useful. I should like to ask the more experienced readers not to get offended if they find some basic material. Experience with recently degreed researchers tells me that many of them came out of school as self-made lab workers, often with bad lab habits, because nobody told them that things could be done better, easier, more appropriately, faster, or more efficiently than the way they routinely do.

Before going into detail I would like to convey a separate piece of advice: Do not get misled by what you often read even in the so-called high-impact journals. Contrary to the impression you may get when reading the experimental parts of many currently published synthetic works, there is more to organic/carbohydrate chemistry than putting two things together and taking mass and nuclear magnetic resonance (NMR) spectra of the product. Do not ignore the classics; try to learn from them. It would be a mistake to ignore, or consider inferior and boring, work done by the founders of carbohydrate chemistry because they did not use fancy reagents, chromatography, and sophisticated spectroscopy equipment. Instead, they had to be patient; they crystallized, sublimed, and distilled (yes, distilled) carbohydrates to purify them. They *routinely* confirmed structures of newly synthesized substances painstakingly by chemical means, and they should be admired for it. The accomplishments of Fischer, Zemplén, Helferich, Levene, Wolfrom, Isbell, Hudson, Tipson, Haworth, Purves, Smith, and Stacey, to mention just a few, are truly stunning considering that they did it all without those modern tools available to the present generation of carbohydrate chemists. Works and biographies of these and other great carbohydrate chemists of the past are readily available through the Advances in Carbohydrate Chemistry and Biochemistry series, and other sources. I urge every person determined to become a carbohydrate chemist to look these up, to fully appreciate accomplishments of these great scientists, and get inspired by them the same way as many in the more recent past have been.

GENERAL

Safety first! Even though nobody may have told you this explicitly and you may not even realize it, some years back, when you decided that you wanted to become a chemist, you chose to accept certain risks. That should not discourage you from taking safety precautions. Remember, there are no harmless chemicals: Research in chemotherapy has shown that anything will kill a rat if you give him enough of it. Nevertheless, with a little common sense, street smarts, knowledge, and chemical intuition, every chemical reaction or operation can be performed in a safe manner. Most of the time, your common sense and chemical instinct will tell you that the hood in your lab is the best place to run a particular reaction, but not all reactions have to be performed in the hood. Be reasonable.

Most experiments in a chemical laboratory require undivided attention. This will tremendously diminish the likelihood of accidents. As an example of an accident that did not have to happen, let me briefly mention circumstances that led to a serious accident that resulted in a fire and burned down the whole laboratory. Fortunately, nobody was hurt. When properly executed, the involved experiment would not be considered dangerous. The same person had carried out the same experiment successfully a short time before. The fire started as a result of an exothermic reaction when the only mistake the person who carried out the experiment made was that the order of addition of requisite chemicals was reversed. The fire started not because of the person's inexperience, lack of training, or the dangerous nature of the experiment, but solely because of the worker's lack of concentration and attention to the work. While repercussions can be vastly different, lack of concentration can lead to disasters not only in a chemical but also in a biological or biochemical laboratory. When working in any life science laboratory, full concentration is as important as when driving a motor vehicle, because the worker's life or future quality of his or her co-worker's life can depend on it. Even though listening to classical or "elevator" music at low volume without headphones would probably not distract one very much, listening to news or talk shows may do so. Also, the chemical laboratory is not the right place to make a fashion statement. Long hair (regardless of sex of the individual) should be worn tied away from the face, and use of protective wear and goggles should be the norm when working in the lab.

It is virtually impossible to do good work in a dirty, untidy laboratory. There are better ways to show that you are busy than having the bench cluttered with multiple marked or unmarked round-bottomed flasks or vials containing variously colored concoctions or residues. Equipment should be maintained, taken good care of, and used as it is meant to be; for example, magnetic stirrers, a heating bath, or other equipment should not be left turned on overnight when it is not necessary. Similarly, oil in vacuum pumps should be changed or filled up *not* periodically but as required by use, rather than waiting until poor performance of the equipment reminds users that maintenance is required.

MORE CONCRETE TIPS AND PROCEDURES

A good practice in a synthetic laboratory is to have a stock of saturated, aqueous solutions of sodium chloride and sodium hydrogen carbonate ready, as these are used often when processing reaction mixtures. The same holds for ready-to-be-used, recycled ion-exchange resins. Recycle a *newly purchased* batch applying protocol recommended by the manufacturer, if you do not find a good amount of used material to recycle. Wash and air-dry the resin after recycling, and store it in a tightly closed container. Replenish the stock before you run short, so you always have some ready for use. Collect the used resin and recycle it when you collect a larger amount. This may come as a surprise, but these resins work better after recycling than those fresh from the bottle just obtained from a chemical supply house, even if delivered in the ionic form you need. Recycled resin is less likely to leach colored material, and as such, it is of better quality than new resin that has never been used.

CLEANING SINTERED GLASS FUNNELS

Keeping glassware clean is a prerequisite to avoiding cross-contamination and obtaining pure products. Only the last user of the glassware knows what it was used for, and therefore only that person can decide the proper method for cleaning it. That is why commercial dishwashers, which can be very helpful in laboratories dealing with only aqueous solutions, are virtually worthless in organic chemistry laboratories. Low wage workers, often employed as dishwashers, should be properly trained to avoid cross-contamination of products due to insolubility of many chemicals in more common solvents. Sintered glass funnels have to be washed by passing through them, in both directions, an adequate amount of proper solvent. To remove residues of organic (and most inorganic) material in hard-to-clean sintered glass funnels, the following procedure has been found adequate to avoid cross-contamination[*]:

1. Mount each sintered glass funnel on a suction flask by way of a rubber cone, and attach the hose connected to the water aspirator.
2. Without applying suction, transfer a layer (~0.5–1 cm high) of concentrated H_2SO_4 on the sintered glass, and add about the same volume of concentrated H_2O_2 on top of the acid.
3. With the sash as much down as practicable, manually sway the funnels to mix the acid with the peroxide. This generates heat, which helps the cleaning process.
4. Keep funnels containing the cleaning reagent in the hood overnight. The connection of the suction flask through the hose to the water aspirator with water *not* running helps the cleaning solution stay in the funnel for a longer time and do its job.

---

[*] All operations should be performed in a well-ventilated hood, and the worker should wear gloves, protective wear, and goggles.

5. Regardless of whether the cleaning solution did or did not run through the funnel overnight, apply suction and wash the funnel thoroughly with deionized water.

6. Fill the funnel with a solution of $NaHCO_3$ and let it slowly drip through, to ensure neutralization of residual acid that may be present, then wash the funnel successively with water and acetone, and pull enough air through to dry it.

PREPARATION OF FULLY ACTIVE, NEUTRAL SILICA GEL

The bulk silica gel that we buy to self-pack preparative columns contains variable amounts of water. This water is not an impurity, it serves a purpose. Without going into detail, suffice it to say that silica gel that contains a small amount of water is a much more powerful separation medium than dry silica gel. Commercially available silica gel is acidic in nature, and the presence of water increases its acidity to the point that, straight from the jar in which it was delivered to be used for chromatography, it can be used as a mild acidic reagent.[1,2] While this acidity is harmless and tolerable most of the time, there are situations when neutral silica gel is preferable. For example, neutral and dry silica gel is the proper medium to purify moisture-sensitive compounds (e.g., glycosyl halides or imidates). Many glycosyl chlorides and bromides, which would largely decompose when passed through ordinary silica gel columns, can be readily purified in this way. To convert commercial-grade silica gel to dry/neutral silica gel, we routinely use the protocol described below. The same protocol can be used to regenerate used silica gel.[*] Since the end-product is fully activated silica gel, it is recommended that, to prevent tailing, slightly more polar solvent should be used for elution, compared to what would be the optimum solvent for elution from commercial-grade silica gel.

1. Move silica gel to a suitably sized, corrosion-resistant vessel (e.g., 5-L Erlenmeyer flask or beaker) placed on a hot plate (safety precautions from above apply).

2. Add enough ~5% (v/v) HCl to easily stir the resulting suspension mechanically, and stir it at ~70–80°C for ~5 min.

3. Decant the supernatant containing superfine and colloidal material before it settles, and filter the suspension through a large, coarse porosity, sintered glass funnel.

4. Wash silica gel several times with hot deionized water, until the washings are almost neutral to litmus.

5. Move silica gel back to the original vessel (#1), and add enough deionized water to easily stir the resulting suspension mechanically. With stirring, add concentrated ammonia until the solution is strongly alkaline to litmus, stir the

---

[*] Some laboratories regenerate used silica gel. To avoid tailing on fully activated silica gel, 3–5% water is added after final drying, and the closed container is shaken until the lumps, which had formed upon addition of water, completely disintegrate. This indicates that the water added had been evenly distributed. Operations 2–4 are performed to regenerate used silica gel, and these can be omitted when dry/neutral silica gel is made from fresh, never used silica gel.

suspension at ~70–80°C for ~5 min, and repeat steps 3 and 4. pH ~8 is acceptable because residual ammonia will be removed during activation (read on).
6. Suck the silica gel dry, spread it in the hood on aluminum foil to form a
layer about 2–3 cm thick, and let it air-dry overnight. Move silica gel into a
clean, suitable, heat-resistant container and heat it in an oven at 140–160°C
overnight. Store in a tightly closed container until use.

PREPARATION OF AgClO₄

Some reagents work best when freshly prepared, such as some silver salts.
Simple, reliable protocols for making Ag_2O[3] and Ag_2CO_3[4] are readily available.
Silver perchlorate is used, for example, for activation of glycosyl fluorides[5] and
is often preferred to other reagents as a promoter in stereoselective syntheses
of 1,2-*cis*-glycosidic linkages.[5-10] Although AgClO₄ is available commercially,
the best results are obtained with freshly prepared reagent. Below is a simple
protocol for preparation of AgClO₄, which is a simplified version of previously
described procedures.[11,12]

Add perchloric acid (70%) with stirring at 100°C to a suspension of silver carbonate (22 g) in water (50 mL) until the effervescence ceases.* After filtration through a
medium-porosity, sintered glass funnel, to remove a small amount of insoluble material, concentrate the filtrate to dryness, and transfer the residue with the aid of a mini-
mal volume of benzene to a beaker (two layers may form). Add pentane with stirring,
collect white precipitate on a coarse, sintered glass funnel, and wash the solid with
pentane. Transfer the product quickly to a storage container, and dry it overnight at
105°/133 Pa. Break the lumps, if any, with a stainless-steel spatula or a heavy glass
rod. Final drying for 2 h at 160°C/133 Pa gives 30 g (90%) of white powder.

During these operations, protection from direct light is not necessary but long-
term storage in the dark is recommended. Although at least 15 different preparations
of AgClO₄ in our laboratory on this scale were uneventful, silver perchlorate has
been reported to detonate spontaneously.[12] Therefore, operations involving the dry
reagent should be carried out behind a safety shield.

THIN-LAYER CHROMATOGRAPHY (TLC) AND COLUMN CHROMATOGRAPHY

When considering the price, the time involved, and the amount of information an
organic chemist can obtain, no technique can compete with a *correctly performed*
thin-layer chromatography (TLC) experiment. Some useful hints listed below will
help you satisfy the italicized requirement:

1. The time spent in search for a better elution solvent system is time well spent.
2. The list of solvents you can try in various combinations is endless, but a few
proven combinations are more useful than others.

---

* Total volume should be only slightly more than the calculated amount. Allow HClO₄ to react completely before addition of the new portion to avoid having to remove a large excess of the acid.

3. Many inexperienced artisans use only CH$_2$Cl$_2$–MeOH or hexane–EtOAc as elution solvents for TLC and column chromatography, thereby depriving themselves of many benefits chromatography offers.

4. The best separation in a given solvent mixture is achieved when the compound in question shows Rf ~0.5.

5. Never use hot air to dry samples after they have been applied onto a silica gel plate *before* the plate has been developed.* When solutions in high-boiling solvents (DMSO, DMF, acetonitrile, water, or pyridine) are to be analyzed by TLC, it is helpful to move a drop or two of the solution into a vial and remove the solvent by a very strong stream of argon or nitrogen (air is often OK) *prior* to applying the sample on the plate. Only when solubility of the material is such that it has to be applied on the plate as a solution in such a high-boiling solvent, dry the *plate* with a stream of *cold* gas.

6. Mixtures of elution solvents for TLC should not be prepared in bulk quantities and stored for long periods of time. Reproducible results can be obtained only with freshly prepared solvent mixtures, allowing some time to saturate the small jar used to develop plates (1–2 minutes is plenty; simple shaking with no additional time allowed is an acceptable alternative). The practice of using filter paper to line the inner walls of TLC jars originated from the practice of paper chromatography. This sped up the development and maintained consistent atmosphere in TLC jars saturated with vaporized solvent. Chambers for paper chromatography are several gallons big, and the solvents used are often high boiling (e.g., pyridine or *n*-BuOH). That is almost never the case with the TLC technique. Thus, when the solvent mixture is moved into the small jar used for developing the plates, shaking the jar a little suffices, and lining the walls of the jar with filter paper is not necessary.

7. Use less-polar solvents for column chromatography, compared to those found to be best for resolution of the same mixture by TLC.

8. The use of a three- or three-+-component mixture for silica gel chromatography is almost never necessary. As a rule, two-component mixtures usually contain a larger proportion of the less-polar component, but exceptions exist.

9. Reactions should be monitored at reasonable intervals. Apply onto the same plate as many reference materials as you have available (starting material always, and also products or intermediates, if you have them).

10. A common myth about preparative column chromatography, virtually worshipped by less experienced chemists, is that the longer the column the better the separation. Very few exceptions aside, the opposite is the truth. Provided the column is well and uniformly packed, one has a better chance to successfully resolve a complex mixture using the same amount of silica gel in a *reasonably* fat, short column, onto which the sample can be applied as a thinner layer than would be the case using a longer, small-diameter column. Commercially available, machine-packed, disposable silica gel columns have

---

* Heat gun or hot plate should not be used if detection is to be done by ultraviolet (UV) light. Some compounds normally not detectable by UV light may decompose on exposure to heat forming UV light-detectable compounds, resulting in misinterpretation of the TLC results.

become more affordable and increasingly more popular. When self-made columns are used it is helpful, before the sample is applied, to place circular filter paper, cut tight but not extending the column diameter, on top of a well-packed column. This will prevent disturbing the top surface during addition of the sample, which has to stay flat during the separation. Plan ahead* so that you do not have to leave the column idle overnight before you collect all useful material. When material stays on the column for a long period of time, the diffusion forces will take over, and this impairs separation. Also, some compounds may decompose by staying on the column too long.

11. When the amount of material to be chromatographed is small (<200 mg), preparative TLC is an option worth trying, if detection of zones is feasible. The plate can be developed multiple times, if necessary, and it often gives better results than a small column.

CATALYTIC HYDROGENOLYSIS/HYDROGENATION

Irrespective of the luck you may have had in the past and have *not* started a fire when you have handled catalysts in a way *different* from the way described below, please follow the rules described here (a cautionary note in *Carbohydr. Res.*, 316, 1999, p. 65 is just one example of the extremely pyrophoric nature of Pd/C).

Common sense and chemical intuition dictates the following:

1. Never add dry catalyst to a solution of the compound, unless the solvent is water.
2. The catalyst should be added to a solution of other reactants in the form of a suspension in water or the same solvent in which the reaction is performed. Prepare the suspension in the following way:
   a. Transfer the appropriate amount (do not guess the amount, use a balance) of the catalyst into a small beaker. Move the beaker into a hood to perform operation b.
   b. When the solvent is other than water, add slowly (e.g., from a wash bottle), *under a gentle stream of an inert gas* (nitrogen or argon), enough solvent to cover all the catalyst. Swirl gently by hand to make a homogeneous suspension.
3. Transfer the suspension into the reaction vessel containing a solution of your material(s) in the requisite solvent.
4. Rinse the beaker with the same solvent into the reaction flask, and set up the hydrogenation depending on the equipment available.
5. When you intend to work up the reaction, have a jar labeled "Used Pd/C" (or other catalyst) ready to collect the used catalyst. After filtration of the reaction mixture, never suck much air through the layer of catalyst to let it air-dry. When the filtration is complete and the product has been collected, always wet the used catalyst with water, and move it into a jar designated

---

* This always helps.

for collection of used catalyst. Keep the jar containing suspension of the used catalyst in water closed, and dispose of its contents by way of chemical waste disposal.

CRYSTALLIZATION AND PREPARATION OF ANALYTICAL SAMPLES

Unlike in the more distant past, based on how results of synthetic organic chemistry have been often reported during the last 30 or so years, one might think that preparation of analytical samples is an anachronism and, therefore, something unimportant. This, sadly, is thought because correct results of combustion analysis, which is the *only* proof of purity of organic compounds, are only rarely reported in contemporary organic chemistry journals. More often than not, high-resolution mass spectrometry (HRMS) data are accepted in lieu of correct combustion analysis data, despite the fact that the former (which can be obtained even by the clumsiest chemist) provide no clue about the purity. However, there is more to synthetic organic chemistry than to pour solutions of reactants together, obtain the compound in dubious purity, rid it of solvents only to the point that one can report 99% yield, and claim by MS means that the desired compound is there. In order to bring such an *unfinished* experiment to completion, the reaction mixture must be properly processed, and the desired product(s) must be fished out and purified, identified, and characterized. It is not an absolute requirement but it is a good laboratory practice to attempt crystallization of stable, medium-size organic compounds. No single analytical technique can be relied upon to provide the total picture when one desires to fully characterize a sample. The present generation of chemists is fortunate because a number of spectroscopic tools are available which through combination can often be used to easily *identify* compounds (i.e., establish identity/structure), be these new compounds or compounds that have been previously synthesized. As far as *characterization*, the task includes establishment of purity, physical properties, elemental composition, and specific optical rotation, if the compound is chiral. All those must be determined with a pure substance. Quantitative elemental microanalysis is still regarded as the touchstone for purity of organic substances.[13] It complements other techniques such as NMR and other spectral techniques and, thus, cannot be replaced.[14] It enables one to determine the empirical formula,* which is the formula for a compound that contains the smallest set of integer ratios for the elements in the compound. An empirical formula determined by HRMS tells us the presence of a compound with that molecular formula but does not reveal any clue about the purity. Thus, if the expected ratio of elements is not confirmed by the analytical figures obtained by combustion analysis, the compound is impure, and any characteristics determined with such material may not be reliable. Evidently, regardless of the level of sophistication of the instrumentation used, spectroscopy cannot provide data that could be used as a criterion of purity in lieu of correct analytical figures. This does not mean that every new compound in a long sequence of reactions must be obtained

---

* This is the empirical formula of the *whole sample* analyzed, and not only of the compound that produced the HRMS peak of interest.

in analytical purity, but this should be attempted at least with key intermediates and final products, or simple, pure derivatives thereof (e.g., peracetate). Many large, unprotected, or some amorphous carbohydrates often do not produce correct analytical figures, but their fully acetylated derivatives do. To perform a multi-step carbohydrate synthesis on a few-milligram scale, and then not provide correct analytical figures and/or optical rotation data for any new compound, arguing that those extra few milligrams required are not available is absurd and should not be acceptable by any reputable scientific journal. Sadly, this is often not the case.

Reading recent synthetic literature makes one think that compounds went on strike and decided not to crystallize. Nothing could be further from the truth. The only reason for so few new crystalline compounds reported is that it is more convenient not to try to crystallize. Also, it is often evident within those rare cases when crystalline compounds are reported that there is a lot of confusion about crystallization and recrystallization. Let us therefore make sure we understand what is meant by "crystallization." Crystallization is *not* the most economical means of purification (in terms of waste of any precious material), but when properly executed it results in a material of the highest achievable purity; higher than can be achieved by any chromatography. "Crystallization" is a way of separating crude material into two portions: the *crystalline* material and a solution ("mother liquor") containing some amount of the same material plus soluble impurities that may have been present in the original crude sample. When you concentrate the eluate from a chromatography column and the residue after concentration crystallizes spontaneously, the process through which such crystalline material has been obtained is *not* crystallization. This is because there is no mother liquor, by way of which any impurities that may have been present would be separated from the purified, crystalline material. Even when you use chromatography to purify a compound, and dry it to the extent that NMR does not show the presence of any other material including residual, trace amount of solvents, crystallization of such material would still yield a purer substance. To obtain "recrystallized" material means to purify "twice" by crystallization.

Two myths about crystallization, very popular among less experienced chemists, are

1. The best way to perform crystallization, or to effect crystallization of a compound that resists crystallization, is to cool the solution to as low a temperature as possible.
2. The larger, better developed, and more beautiful the crystals are, the purer they are.

Concerning myth 1, it is common knowledge that some compounds crystallize easily, some have to be induced to crystallize, and some have never been crystallized. Useful, generally applicable hints concerning crystallization have been described elsewhere,[15] and there is only little to be added to that. Empirical evidence indicates that the most favorable temperature for the spontaneous formation of the first seed crystals is lower than the temperature at which the bulk of the material crystallizes in the highest yield. Therefore, when more common strategies fail, it can sometimes be useful to concentrate a solution of a substance to thick syrup, close the flask, and cool it in dry ice. After some time (~30 min), remove it from cooling, and allow the flask to slowly warm to room temperature.* Often, after a day or two, sometimes weeks or months, careful inspection of the syrup might reveal the presence of the first seeds.† When these are crushed with a spatula, more crystals are often formed much more quickly. Alternatively, with the aid of a spatula, the seeds with as little syrup as possible are moved into a vial, and the crystals should be mixed evenly within the residual syrup and divided into portions, which are then used as seed crystals when searching for a suitable solvent for crystallization.

Concerning myth 2, large crystals are usually formed when the material crystallizes slowly, undisturbed, without motion. Such crystals may occlude some of the mother liquor. Consequently, if the mother liquor contains impurities, these will be left behind when the sample is dried and all the solvent evaporates. Once crystals start forming, the suspension should be stirred. This aids crystallization, which often completes in a short time, forming sometimes powdered but still perfect crystalline material. When large crystals are formed, these should be pulverized before filtration by rubbing with a spatula, while still within the mother liquor.

When facing the task of preparing an analytical sample from a crystalline material to be submitted for combustion analysis, the following is a time-saving procedure that has been proven to yield good results, and which we use routinely.‡

It goes without saying that when an analytical sample is to be prepared it makes no sense to attempt crystallization from a solution that is cloudy, opaque, or contains solid impurities, such as fluffy residues, silica gel, or chipped glass from sintered glass funnels. Therefore:

1. Filter a *dilute* solution of the requisite material in a suitable solvent, preferably the solvent from which you plan to crystallize the material, into a small pear-shaped§ flask or any round-bottom flask that can be attached to a rotary evaporator. To filter¶ a small volume, it is convenient to use a very tight cotton plug in a disposable Pasteur pipette, or a syringe.
2. Concentrate the solution to a small volume, and induce crystallization either by seeding or, if seed crystals are not available, by rubbing the inside walls of the vessel with a spatula while applying cooling in an ice bath. Remove cooling and bring the solution to room temperature, if cooling results in cloudy solution or if the solute oils out.

---

* Do not try to open the flask before it and its content reach room temperature. Vacuum forms inside the flask as a result of cooling, and it would be difficult to open. Also, upon opening the cold container, the very cold content and inside wall of the flask would cause atmospheric moisture to condense inside the flask. That could be detrimental to anything you want to do next.
† It is a good practice to have a magnifier (~20×) handy in the laboratory.
‡ Because the method to obtain the analytical sample is broken down into several steps, it may look too complicated and time-consuming, but when you do it, repeat it a couple of times, and get used to it, you will agree that to get an analytical sample from the first crystals can take only a few minutes, when you have all utensils handy.
§ A pear-shaped flask is preferred to make it easier to scoop out the crystalline material with the aid of a bent spatula.
¶ The solution should be reasonably dilute to prevent the compound from crystallizing during filtration, and it will have to be concentrated after filtration.

If the solvent for crystallization has not been decided upon, here are a few simple empirical rules. Ethanol is by far the most popular and most effective solvent for crystallization of organic compounds,* and that should always be tried first. If the compound is too soluble in EtOH, the next solvent is difficult to suggest without knowing the polarity of the compound, but mixtures of solvents should be tried only after single solvents do not work. If mixtures of solvents have to be used, the compound should always be dissolved first in the minimum, reasonable amount of the solvent in which the compound is soluble, and the other solvent should be slowly added with manual swirling, *without cooling,*† *almost* to turbidity. If turbidity ensues, add a few drops of the first solvent until clear solution is obtained. Add a seed crystal, if available, and see if the seed dissolves. If so, you have to look for a different combination of solvents or add a drop or two of the solvent in which the compound is insoluble. If seed crystals are not available, occasionally rubbing the inside walls with the spatula and patience often help.

3. Once crystallization starts and the solvent is the proper one, it usually completes within a short time when helped by stirring. Collect the crystals by way of a clean, coarse sintered glass funnel and remove almost all the mother liquor by gentle suction, but *do not suck the crystals dry.* Wash the crystals with a small amount of suitable solvent‡ and now suck them dry, to obtain material **A**. (Materials obtained in this way are acceptable as analytical samples only when the material was pure according to TLC *before* crystallization, and so little material is available that recrystallization is impracticable. If this is the case, dry the material under vacuum at ~40°, measure melting point (m.p.), dry again at a temperature ~20°C lower than the m.p. just found, and re-measure m.p.) If the material was not pure before crystallization but **A** is (TLC), and much more material is available then proceed with the next step.

4. *Immediately* recrystallize a portion§ of **A** in the way described above, to end up with a reasonable amount of recrystallized substance, material **B**.

5. If the m.p. measured after exhaustive drying is satisfactory, divide material that has been crystallized twice (**B**) into two portions. Put ~3–5 mg (depending on how many elements the sample is going to be analyzed for) into a vial to be sent for combustion microanalysis. Store the rest of the recrystallized material in a separate vial. Dry both amounts at reduced pressure at ~40°C for 2–5 h. The time depends on the boiling temperature

---

* Contrary to the known rule that like dissolves like, many very polar and also many very non-polar compounds can be crystallized from EtOH.

† The addition of solvent in which the compound is insoluble decreases the solubility of the solute. The same would be achieved by cooling, and that was supposed to have already been tried and had not worked. To do both does not make much sense.

‡ Don't just pour some solvent onto the crystals but make a slurry by gentle stir. The temperature of the solvent with which you wash these crystals depends on the solubility of the material. Washing crystals with cold solvent and drying crystals by suction should be brief to avoid condensation of moisture.

§ The amount of recrystallized material should be sufficient to measure m.p., $[\alpha]_D$, subjecting a few mg for combustion analysis, measuring for records the final NMR, and keeping a few mg as a standard.

of the solvent of crystallization and the vacuum. Measure the melting points of **A** and **B**.

6. Keep both samples in the vacuum oven for ~2 h at 80°C or a temperature ~20°C lower than the measured melting temperature, whichever is lower. Measure m.p. of **A** and **B** again. If the m.p. of the recrystallized material (**B**) is more than 5°C higher than that of **A** (the material crystallized only once), another recrystallization should be attempted.

I am describing and recommending this protocol for crystallization of analytical samples, which includes recrystallization, not because I want to make the lives of carbohydrate chemists more difficult. The same protocol should be used for preparation of every new compound. It is important to note that only constant, reproducible melting "point"* information is worth publishing because m.p. is an important physical characteristic useful for identification of pure substances. That is the only purpose of measuring and reporting m.p., and such information can only be obtained in the manner described above. Needless to say, melting temperatures determined often with "crystalline residues obtained after evaporation of solvents" or with "amorphous powders" are worthless numbers.

Sadly, also in the so-called high-impact journals, we often find analytical data for *amorphous* compounds as hydrates, hemihydrates, etc., just because the analytical figures happen to fit that particular formula adjusted for solvation. This is scientifically unsound. In a crystal, *discrete* numbers of solvent molecules may partake in the lattice formation. In an *amorphous* substance, however, this is not so. Therefore, it is likely that it is possible to find an analysis that fits any compound using solvation. Therefore, an analysis of an *amorphous* compound is irrelevant if it needs solvation adjustment.

7. When the material produces correct analytical figures, measure $[\alpha]_D$ for the recrystallized material.

Amorphous compounds, often collected after purification by chromatography, that are meant to be sent for combustion analysis should be rid of any particular material by filtration through a very tight cotton plug in a disposable pipette or through a syringe filter. Use a filter of the lowest possible pore size (0.1 or 0.2 µm). Be sure to use the right type of filter. There are separate syringe filters meant for aqueous solutions and for solutions in organic solvents.

Aqueous solutions of compounds that could not be crystallized should be filtered and freeze-dried. Materials after freeze-drying can be dried more efficiently but may still contain some moisture. Therefore, after freeze-drying, the solid obtained should be further dried in a vacuum oven at elevated temperature, unless the solid is known to melt, decompose, or sublime under those conditions. Solutions in organic solvents of substances that resist crystallization should be concentrated, and attempts should

---

* We routinely refer to the temperature at which a substance melts as the "melting point." Technically, this is incorrect because regardless of how sharp the melting point is, it is always a temperature *range*.

be made to obtain solid foam. Repeated co-evaporation of a solution of many compounds with CCl_4 often results in the formation of foamy material. Also, quite often the freeze-drying of a solution in benzene results in obtaining a solid. The foam, amorphous solid or oil should be dried in a vacuum oven. The solid, amorphous material should not be confused with crystalline material. Solids that could not be *crystallized from a solvent* can be considered crystalline only when they show crystalline features under the microscope and also show a definite, reproducible melting point (range). The only objective criterion of crystallinity is, of course, x-ray analysis.

MEASURING $[\alpha]_D$ VALUES

There is not much one can do about improving the accuracy of the polarimeter, except to keep the instrument in good shape. The modern polarimeter is a highly computerized machine, and the reading is automatic and accurate, as long as the lamp is in good shape. One can, however, influence the *overall* accuracy of the measurement because the concentration of the solution is one of the variables in the calculation, and there are many stages in the sample preparation where the human factor is involved. It starts with drying the sample. Even though the final $[\alpha]_D$ value is not affected by a miniscule amount of moisture or residual solvents as much as combustion analysis, one has to make sure that the sample is very dry. This is why we normally measure optical rotation with the same material that produced correct analytical figures. It is then up to the chemist *not* to introduce additional, much larger errors. Carefully weighing the sample using the best balance available and measuring accurately the amount of the requisite solvent will make a difference. Values $[\alpha]_D$ of new compounds should be measured as soon as possible after purification, especially for compounds that are oily. Such compounds tend to (partially) decompose on storage, which may be left unnoticed and result in publication of unreliable data.

THE YIELDS: WHAT'S ACCEPTABLE, AND WHAT'S NOT

A yield of 10% is not something of which one has to be either ashamed or proud. Low yield may be justified and acceptable. Such yield does not necessarily mean that one is a clumsy or careless chemist. It may mean that one has not chosen the best method for synthesis or isolation, or that a better protocol has not yet been invented concerning synthesis of that particular substance.

Only those who do not work at the bench can be impressed by 99% yields; every good chemist should be suspicious about such figures because who knows how reliable other results are when the author is unrealistic about yields. The chemical transformation yield and the yield of the compound after isolation, purification, *and exhaustive drying* are two different things. Not even when the chemical transformation of the starting material is complete and a sole product is formed can the latter be isolated in 99% yield, as has been convincingly documented.[16] Thus, reporting such yields should not be acceptable by a respectable scientific journal. Similarly, reporting "quantitative" yield should not be acceptable, as there are always manipulative losses when anything other than evaporation of solvents is involved in the workup, no matter how careful the worker is. When the transformation of starting material is complete and the conversion is a one-product reaction, it is acceptable to say that the yield was "nearly theoretical." It is permissible and sometimes advisable to go from one synthetic step to another with *crude* product, but *yield of crude product* is a worthless number because one can target it before starting the experiment and then dry or not dry the *crude* product accordingly.

THINGS TO AVOID

First, here is some general advice: Do not assume that just because you see something in print, it is a priori true, proper, or correct, and by this I mean reading something in a respectable scientific journal. Do not mutilate trivial names of carbohydrates. For example, the term "lactose" is a trivial name reserved[17] for β-D-galactopyranosyl-(1 → 4)-α,β-D-glucopyranose, or 4-O-β-D-galactopyranosyl-α,β-D-glucopyranose. Thus, the terms "D-lactose," D-lactopyranose or D-lactopyranoside, which we sometimes encounter—also in carbohydrate chemistry–specialized journals—are notoriously incorrect, and their appearance in print is evidence of the cooperation of careless writers, referees, and editors. The "D" is redundant because of the definition, and "...pyran..." is obvious because of the (1→4)-linkage involved in the structure and is therefore unnecessary.

Do not confuse *concentration* and *evaporation*: Solvents are evaporated; solutions are concentrated. Solutions consist of solvents and solutes. Solvents are evaporated from the solutions to obtain solutes. The same can be accomplished by the concentration of solutions. Solutions are normally not evaporated, only when the solute also evaporates (e.g., during distillation or at a very high temperature, like when at ground zero of a nuclear explosion).

When writing a publication, do not confuse "Abstract" with "Introduction." The abstract should be a factual and concise description of what was done, found, and/or concluded. It should be free of general statements and citations and be fully understandable without the aid of graphics or artwork. Thus, stand-alone compound numbers should appear in the abstract only for substances that were previously named and had a number assigned to them or were described otherwise in the same abstract, so that there is no question what the numbers refer to. A "Conclusions" section usually only extends the paper unnecessarily by one paragraph. It is superfluous in a paper for which the abstract has been well written (c.f. the general advice above).

Without interpretation of at least structurally significant resonances, an NMR spectrum is often just a piece of paper with little value. Do not publish unassigned NMR data for small molecules. Such data are worthless in terms of providing proof or much support of the structure. Also, with wide availability of high-field NMR spectrometers, which normally come with user-friendly software, publishing unassigned data generated with such machines suggests to readers that authors do not know how to do the assignments.

Always remember the three basic axioms applicable to experimental sciences:

1. One hour spent in the library, which in these times may be only a few keystrokes away, can save you a month or longer in the laboratory.
2. What's easy has already been done.
3. Everything takes longer.

Have a sweet life with carbohydrates.

REFERENCES

1. Lehrfeld, J. *J. Org. Chem.,* 1967, *32,* 2544–2546.
2. Pathak, A. K.; Pathak, V.; Seitz, L. E.; Tiwari, K. N.; Akhtar, M. S.; Reynolds, R. C. *Tetrahedron Lett.* 2001, *42,* 7755–7757.
3. Hirst, E. L.; Percival, E. *Meth. Carbohydr. Chem.* 1963, *2,* 145–150.
4. Wolfrom, M. L.; Lineback, D. R. *Meth. Carbohydr. Chem.* 1963, *2,* 341–345.
5. Mukaiyama, T.; Murai, Y.; Shoda, S. *Chem. Lett.* 1981, 431–432.
6. Mukaiyama, T.; Hashimoto, Y.; Shoda, S. *Chem. Lett.* 1983, 935–938.
7. Kováč, P.; Lerner, L. *Carbohydr. Res.* 1988, *184,* 87–112.
8. Kováč, P.; Petráková, E.; Kočis, P. *Carbohydr. Res.* 1981, *93,* 144–147.
9. Igarashi, K.; Irisawa, J.; Honma, T. *Carbohydr. Res.* 1975, *39,* 213–225.
10. Kováč, P.; Palovčik, R. *Carbohydr. Res.* 1977, *54,* C11–C13.
11. Kováč, P.; Palovčík, R. *Chem. Zvesti* 1978, *32,* 501–513.
12. Wolfrom, M. L.; Thompson, A.; Lineback, D. R. *J. Am. Chem. Soc.* 1963, *28,* 860–861.
13. Harwood, L. M.; Moody, C. J.; Percy, J. M. Qualitative analysis of organic compounds; *Experimental Organic Chemistry: Standard and Microscale*; Harwood, L. M.; Moody, C. J.; Percy, J. M., Ed.; Blackwell, 2003, p. 716.
14. Etherington, K. J.; Rodger, A.; Hemming, P. *LabPlus International* 2001, 26–27.
15. Thompson, A.; Wolfrom, M. L. *Meth. Carbohydr. Chem.* 1962, *1,* 8–11.
16. Wernerova, M.; Hudlicky, T. *Synlet* 2010, 2701–2707.
17. McNaught, A. D. *Carbohydr. Res.* 1997, *297,* 1–92.

Pavol Kováč, Editor in Chief
Carbohydrate Chemistry: Proven Synthetic Methods

Series Editor

Pavol Kováč, Ph.D., Dr. h.c., with more than 40 years of experience in carbohydrate chemistry and more than 280 papers published in refereed scientific journals, several patents, and book chapters, is a strong promoter of good laboratory practices and a vocal critic of the publication of experimental chemistry lacking data that allow reproducibility. He obtained a M.Sc. in chemistry at the Slovak Technical University in Bratislava (Slovakia) and a Ph.D. in organic chemistry at the Institute of Chemistry, Slovak Academy of Sciences, Bratislava. After post-doctoral training at the Department of Biochemistry, Purdue University, Lafayette, Indiana (R.L. Whistler, advisor), he returned to the Institute of Chemistry and formed a group of synthetic carbohydrate chemists who had been active mainly in oligosaccharide  chemistry. After relocating to the United States (1981), he worked at Bachem, Inc., Torrance, California, where he established a laboratory for the production of oligonucleotides for automated synthesis of DNA. He joined the National Institutes of Health (NIH) (Bethesda, Maryland) in 1983, where he is currently one of the principal investigators and Chief of the Section on Carbohydrates (NIDDK, Laboratory of Bioorganic Chemistry), the world's oldest research group continuously working on chemistry, biochemistry, and immunology of carbohydrates, which was originally established by the American carbohydrate chemist Claude S. Hudson. Kováč's main interest is the development of conjugate vaccines for bacterial diseases from synthetic carbohydrate antigens.

Book Editors

Jeroen Codée obtained his Ph.D. from Leiden University, the Netherlands, under the guidance of Jacques van Boom and Stan van Boeckel in 2004. Codée then moved to the Eidgenössische Technische Hochschule (ETH) Zürich, Switzerland, for a two-year post-doctoral stay with Peter Seeberger. He subsequently returned to Leiden University, where he is now an associate professor in bio-organic chemistry. His research focuses on glycochemistry and glycobiology ranging from fundamental organic synthesis to vaccine/drug development and the development of glycobiology tools. He has authored or co-authored over 70 publications and seven book chapters.

Gijs van der Marel conducted his Ph.D. studies on DNA chemistry with Jacques van Boom and received his Ph.D. degree in 1981 from Leiden University. He has remained associated with this university, where he is now a full professor in synthetic organic chemistry. He has supervised over 70 Ph.D. students and authored or co-authored over 700 papers. His research focuses on all synthetic aspects of biopolymers, including nucleic acids, peptides, and carbohydrates, and hybrids and analogues thereof.

Contributors

Roger A. Ashmus
Department of Chemistry
Centennial Centre for Interdisciplinary
    Science (CCIS)
University of Alberta
Edmonton, Canada

Cecilia Attorresi
CIHIDECAR-CONICET
Departamento de Química Orgánica
Universidad de Buenos Aires
Buenos Aires, Argentina

Edward I. Balmond
School of Chemistry
University of Bristol
Bristol, United Kingdom

M. Teresa Barros
REQUIMTE, CQFB
Departamento de Química
Universidade Nova de Lisboa
Caparica, Portugal

Emiliano Bedini
Dipartimento di Scienze Chimiche
Università degli Studi di Napoli
    "Federico II"
Naples, Italy

Randy Benedict
Department of Chemistry
Iowa State University
Ames, Iowa

Markus Blaukopf
Department of Chemistry
University of Natural Resources and
    Life Sciences
Vienna, Austria

Luis Bohé
Centre de Recherche de Gif
Institut de Chimie des Substances
    Naturelles, CNRS
Gif-sur-Yvette, France

Thomas J. Boltje
Complex Carbohydrate Research
    Center
University of Georgia
Athens, Georgia

Geert-Jan Boons
Complex Carbohydrate Research
    Center
University of Georgia
Athens, Georgia

Siwarutt Boonyarattanakalin
School of Bio-Chemical Engineering
    and Technology
Sirindhorn International Institute of
    Technology
Thammasat University
Pathum Thani, Thailand

Anikó Borbás
Department of Pharmaceutical
    Chemistry
Medical and Health Science Center
University of Debrecen
Debrecen, Hungary

Isabelle Bossu
Département de Chimie Moléculaire
UMR-CNRS
Université Joseph Fourier
Grenoble, France

N.V. Bovin
Laboratory of Carbohydrates
Shemyakin and Ovchinnikov Institute
    of Bioorganic Chemistry of the
    Russian Academy of Sciences
Moscow, Russian Federation

Kevin Buffet
Département de Chimie, Laboratoire
    de Chimie Bio-Organique
Université de Namur (FUNDP)
Namur, Belgium

Audrey Caravano
Département de Chimie, Laboratoire
    de Chimie Bio-Organique
Université de Namur (FUNDP)
Namur, Belgium

Ivone Carvalho
Departamento de Ciências
    Farmacêuticas
Universidade de São Paulo
Ribeirão Preto-SP, Brazil
and
Dipartimento di Chimica Organica e
    Biochimica
Università degli Studi di Napoli
Naples, Italy

Riccardo Castelli
Leiden Institute of Chemistry
Leiden University
Leiden, The Netherlands

Yoann M. Chabre
Department of Chemistry
Université du Québec à Montréal
Québec, Canada

Lina Chan
Department of Chemistry
University of Toronto
Toronto, Ontario, Canada

Alexander A. Chinarev
Laboratory of Carbohydrates
Shemyakin and Ovchinnikov Institute
    of Bioorganic Chemistry of the
    Russian Academy of Sciences
Moscow, Russian Federation

Zoeisha Chinoy
University of Georgia
Athens, Georgia

Jeroen D.C. Codée
Leiden Institute of Chemistry
Leiden University
Leiden, The Netherlands

Paula Correia-da-Silva
REQUIMTE, CQFB
Departamento de Química
Universidade Nova de Lisboa
Caparica, Portugal
and
Instituto Superior de Ciências da Saúde
    Egas Moniz
Caparica, Portugal

Travis Coyle
Discipline of Chemistry
School of Biomedical, Biomolecular
    and Chemical Sciences
University of Western Australia
Crawley, Western Australia

David Crich
Centre de Recherche de Gif
Institut de Chimie des Substances
    Naturelles, CNRS
Gif-sur-Yvette, France

Antonella Dalla Cort
Dipartimento di Chimica
and
IMC-CNR Sezione Meccanismi di
    Reazione
Università La Sapienza
Roma, Italy

Emma M. Dangerfield
Malaghan Institute of Medical
    Research
Wellington, New Zealand
and
School of Chemical and Physical
    Sciences
Victoria University of Wellington
Wellington, New Zealand

Ana R. de Jong
Leiden Institute of Chemistry
Leiden University
Leiden, The Netherlands

Rosa M. de Lederkremer
CIHIDECAR-CONICET
Departamento de Química Orgánica
Universidad de Buenos Aires
Buenos Aires, Argentina

Alexei V. Demchenko
Department of Chemistry and
    Biochemistry
University of Missouri–St. Louis
St. Louis, Missouri

Cristina De Meo
Department of Chemistry
Southern Illinois University
    Edwardsville
Edwardsville, Illinois

Jérôme Désiré
Université de Poitiers
UMR CNRS 7285, IC2MP
Poitiers, France

Steven Dulaney
Department of Chemistry
Michigan State University
East Lansing, Michigan

Pascal Dumy
Département de Chimie Moléculaire
UMR-CNRS 5250 and ICMG FR 2607
Université Joseph Fourier
Grenoble, France

Maxime Durka
Département de Chimie, Laboratoire
    de Chimie Bio-Organique
Université de Namur (FUNDP)
Namur, Belgium

Guillaume Eppe
Département de Chimie, Laboratoire
    de Chimie Bio-Organique
Université de Namur (FUNDP)
Namur, Belgium

Ana-Paula Esteves
Centro de Química
Universidade do Minho
Braga, Portugal

Mark Farrell
School of Chemistry
National University of Ireland Galway
Galway, Ireland

Gianpiero Forte
Dipartimento di Chimica and IMC-
    CNR Sezione Meccanismi di
    Reazione
Università La Sapienza
Roma, Italy

Nicolas Galanos
Laboratoire de Chimie Organique 2
    - Glycochimie
UMR5246, CNRS
Institut de Chimie et Biochimie
    Moléculaires et Supramoléculaires
Université Lyon
Villeurbanne, France

Denis Giguère
Department of Chemistry
Université du Québec à Montréal
Québec, Canada

Chase Gobble
Chemistry Department
Southern Illinois University
    Edwardsville
Edwardsville, Illinois

Xavier Guinchard
Centre de Recherche de Gif
Institut de Chimie des Substances
  Naturelles, CNRS
Gif-sur-Yvette, France

Shivali A. Gulab
Carbohydrate Chemistry Group
Callaghan Innovation Research
  Limited
Lower Hutt, New Zealand

Bas Hagen
Leiden Institute of Chemistry
Leiden University
Leiden, The Netherlands

Scott J. Hasty
Department of Chemistry and
  Biochemistry
University of Missouri–St. Louis
St. Louis, Missouri

Mitchell Hattie
Discipline of Chemistry
School of Biomedical, Biomolecular
  and Chemical Sciences
University of Western Australia
Crawley, Western Australia

Mihály Herczeg
Research Group for Carbohydrates of
  the Hungarian Academy of Sciences
University of Debrecen
Debrecen, Hungary

Konrad Hohlfeld
School of Chemistry
University of Southampton
Southampton, United Kingdom

Min Huang
Centre de Recherche de Gif
Institut de Chimie des Substances
  Naturelles, CNRS
Gif-sur-Yvette, France

Xuefei Huang
Department of Chemistry
Michigan State University
East Lansing, Michigan

Yukishige Ito
RIKEN
Advanced Science Institute
Saitama, Japan
and
ERATO
Japan Science and Technology Agency
  (JST)
Saitama, Japan

Dilip V. Jarikote
School of Chemistry
National University of Ireland Galway
Galway, Ireland

Solen Josse
School of Chemistry
University of Southampton
Southampton, United Kingdom
and
Laboratoire des Glucides UMR6219
Université de Picardie Jules Verne
Amiens, France

Leonid O. Kononov
N.K. Kochetkov Laboratory of
  Carbohydrate Chemistry
N.D. Zelinsky Institute of Organic
  Chemistry of the Russian Academy
  of Sciences
Moscow, Russian Federation

Paul Kosma
Department of Chemistry
University of Natural Resources and
  Life Sciences
Vienna, Austria

Pavol Kováč
NIDDK, LBC
National Institutes of Health
Bethesda, Maryland

Viktors Kumpiņš
Material Science and Applied
    Chemistry
Riga Technical University
Riga, Latvia

László Lázár
Research Group for Carbohydrates of
    the Hungarian Academy of Sciences
University of Debrecen
Debrecen, Hungary

Doris Lee
Department of Chemistry
University of Toronto
Toronto, Ontario, Canada

Tianlei Li
Département de Chimie, Laboratoire
    de Chimie Bio-Organique
Université de Namur (FUNDP)
Namur, Belgium

Bruno Linclau
School of Chemistry
University of Southampton
Southampton, United Kingdom

Óscar López
Department of Organic Chemistry
University of Seville
Seville, Spain

Jevgeņija Mackeviča
Faculty of Material Science and
    Applied Chemistry
Riga Technical University
Riga, Latvia

N.N. Malysheva
N.D. Zelinsky Institute of Organic
    Chemistry of the Russian Academy
    of Sciences
Moscow, Russian Federation

Shino Manabe
RIKEN
Advanced Science Institute
Saitama, Japan

Carla Marino
CIHIDECAR-CONICET
Departamento de Química Orgánica
Universidad de Buenos Aires
Buenos Aires, Argentina

Hannes Mikula
Institute of Applied Synthetic
    Chemistry
Vienna University of Technology
Vienna, Austria

Silvia Miranda
Instituto de Química Orgánica General
    (IQOG-CSIC)
Madrid, Spain

Christopher J. Moore
Department of Chemistry
University of Guelph
Guelph, Ontario, Canada

Paul V. Murphy
School of Chemistry
National University of Ireland Galway
Galway, Ireland

V.V. Nasonov
Laboratory of Carbohydrates
Shemyakin and Ovchinnikov Institute
    of Bioorganic Chemistry of the
    Russian Academy of Sciences
Moscow, Russian Federation

Swati S. Nigudkar
Department of Chemistry and
    Biochemistry
University of Missouri–St. Louis
St. Louis, Missouri

Markus Ohlin
Organic Chemistry
Lund University
Lund, Sweden

Ciaran O'Reilly
School of Chemistry
National University of Ireland Galway
Galway, Ireland

Pāvels Ostrovskis
Material Science and Applied
  Chemistry
Riga Technical University
Riga, Latvia

Michelangelo Parrilli
Dipartimento di Scienze Chimiche
Università degli Studi di Napoli
  "Federico II"
Naples, Italy

G.V. Pazynina
Laboratory of Carbohydrates
Shemyakin and Ovchinnikov Institute
  of Bioorganic Chemistry of the
  Russian Academy of Sciences
Moscow, Russian Federation

Krasimira T. Petrova
REQUIMTE, CQFB
Departamento de Química
Universidade Nova de Lisboa
Caparica, Portugal

Sebastien Picard
Centre de Recherche de Gif
Institut de Chimie des Substances
  Naturelles, CNRS
Gif-sur-Yvette, France

Sylvain Picon
School of Chemistry
Cardiff University
Cardiff, United Kingdom

Mélanie Platon
Department of Chemistry
University of Pittsburgh
Pittsburgh, Pennsylvania

Papapida Pornsuriyasak
Department of Chemistry and
  Biochemistry
University of Missouri–St. Louis
St. Louis, Missouri

Jean-Pierre Praly
Institut de Chimie et Biochimie
  Moléculaires et Supramoléculaires
  (ICBMS)
Université de Lyon
Villeurbanne, France

Jani Rahkila
Laboratory of Organic Chemistry
Åbo Akademi University
Åbo, Finland

Olivier Renaudet
Département de Chimie Moléculaire
UMR-CNRS 5250 and ICMG FR 2607
Université Joseph Fourier
Grenoble, France

Mark B. Richardson
School of Chemistry
Bio21 Molecular Science and
  Biotechnology Institute
University of Melbourne
Parkville, Victoria

Jacques Rodrigue
Department of Chemistry
Université du Québec à Montréal
Québec, Canada

Myriam Roy
CNRS CINaM, UMR 7325
Université Aix-Marseille
Marseille, France

René Roy
Department of Chemistry
Université du Québec à Montréal
Québec, Canada

Somsak Ruchirawat
Laboratory of Medicinal Chemistry
Chulabhorn Research Institute
Center of Excellence on Environmental
    Health, Toxicology and
    Management of Chemicals
Bangkok, Thailand
and
Program in Chemical Biology
Chulabhorn Graduate Institute
Center of Excellence on Environmental
    Health, Toxicology and
    Management of Chemicals
Bangkok, Thailand

Kevin Sheerin
Science Centre
School of Chemistry and Chemical
    Biology
Belfield, Dublin

Tze Chieh Shiao
Department of Chemistry
Université du Québec à Montréal
Québec, Canada

Georg Sixta
Department of Chemistry
University of Natural Resources and
    Life Sciences
Vienna, Austria

Sameh E. Soliman
NIDDK, LBC
National Institutes of Health
Bethesda, Maryland

Christian Stanetty
Department of Chemistry
University of Natural Resources and
    Life Sciences
Vienna, Austria

Bridget L. Stocker
School of Chemical and Physical
    Sciences
Victoria University of Wellington
Wellington, New Zealand
and
Malaghan Institute of Medical
    Research
Wellington, New Zealand

Keith A. Stubbs
Discipline of Chemistry
School of Biomedical, Biomolecular
    and Chemical Sciences
University of Western Australia
Crawley, Western Australia

Mark S. Taylor
Department of Chemistry
University of Toronto
Toronto, Ontario, Canada

Baptiste Thomas
Département de Chimie Moléculaire
UMR-CNRS 5250 and ICMG FR 2607
Université Joseph Fourier
Grenoble, France

Mikael Thomas
Centre de Recherche de Gif
Institut de Chimie des Substances
    Naturelles
UPR 2301 CNRS
Gif-sur-Yvette, France

Abdellatif Tikad
Département de Chimie, Laboratoire
    de Chimie Bio-Organique
Université de Namur (FUNDP)
Namur, Belgium

Mattie S.M. Timmer
School of Chemical and Physical
    Sciences
Victoria University of Wellington
Wellington, New Zealand

Gopinath Tiruchinapally
Department of Chemistry
Michigan State University
East Lansing, Michigan

Huu-Anh Tran
Department of Chemistry
University of Alberta
Edmonton, Alberta, Canada

Māris Turks
Faculty of Material Science and
    Applied Chemistry
Riga Technical University
Riga, Latvia

A.B. Tuzikov
Laboratory of Carbohydrates
Shemyakin and Ovchinnikov Institute
    of Bioorganic Chemistry of the
    Russian Academy of Sciences
Moscow, Russian Federation

Gijsbert A. van der Marel
Leiden Institute of Chemistry
Leiden University
Leiden, The Netherlands

Monica Varese
Dipartimento di Scienze del Farmaco
Università degli Studi del Piemonte
    Orientale
Novara, Italy

Stéphane P. Vincent
Département de Chimie, Laboratoire
    de Chimie Bio-Organique
Université de Namur (FUNDP)
Namur, Belgium

Anne Geert Volbeda
Leiden Institute of Chemistry
Leiden University
Leiden, The Netherlands

Guitao Wang
Department of Chemistry
Michigan State University
East Lansing, Michigan
and
Department of Chemistry
Wayne State University
Detroit, Michigan

Chakree Wattanasiri
Laboratory of Medicinal Chemistry
Chulabhorn Research Institute
Center of Excellence on Environmental
    Health, Toxicology and
    Management of Chemicals
Bangkok, Thailand
and
Program in Chemical Biology
Chulabhorn Graduate Institute
Center of Excellence on Environmental
    Health, Toxicology and
    Management of Chemicals
Bangkok, Thailand

Patrick Wisse
Leiden Institute of Chemistry
Leiden University
Leiden, The Netherlands

Zhimeng Wu
Carbohydrate Chemistry Group
Callaghan Innovation Research
    Limited
Lower Hutt, New Zealand

Peng Xu
NIDDK, LBC
National Institutes of Health
Bethesda, Maryland

Bo Yang
Department of Chemistry
Michigan State University
East Lansing, Michigan

Jagodige P. Yasomanee
Department of Chemistry and
  Biochemistry
University of Missouri–St. Louis
St. Louis, Missouri

Zhaojun Yin
Department of Chemistry
Michigan State University
East Lansing, Michigan

Lei Zhang
College of Chemistry and Life
  Sciences
Zhejiang Normal University
Jinhua, P.R. China

Qingju Zhang
Shanghai Institute of Organic
  Chemistry
Shanghai, China

Xiangming Zhu
School of Chemistry and Chemical
  Biology
University College Dublin
Dublin, Ireland
and
College of Chemistry and Life
  Sciences
Zhejiang Normal University
Jinhua, P.R. China

Alexander I. Zinin
N. K. Kochetkov Laboratory of
  Carbohydrate Chemistry
N. D. Zelinsky Institute of Organic
  Chemistry of the Russian Academy
  of Sciences
Moscow, Russian Federation

Section I

Synthetic Methods

1 Highly Stereoselective 1,2-*cis*-Glycosylations Employing the C-2 (S)-(phenylthiomethyl) benzyl ether as a Chiral Auxiliary

Thomas J. Boltje
*Randy Benedict**
Geert-Jan Boons†

CONTENTS

* Checker under supervision of N. Pohl: npohl@iastate.edu.
† Corresponding author: gjboons@ccrc.uga.edu.

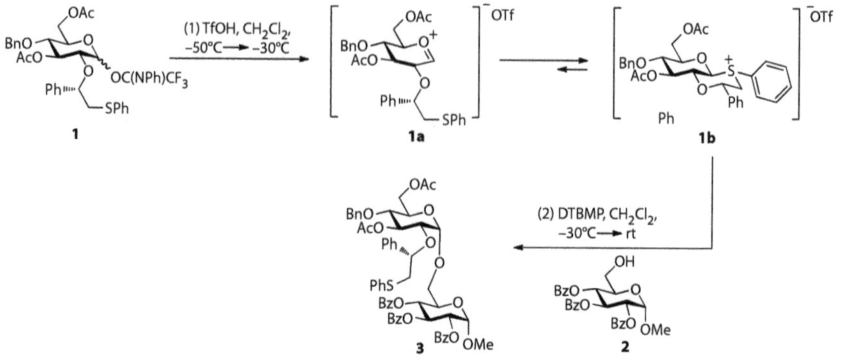

SCHEME 1.1 Activation of glycosyl donor **1** with a stoichiometric amount of triflic acid gives rise to the β-sulfonium ion intermediate **1b** which, upon addition of glycosyl acceptor **2**[1], reacts stereoselectively to form 1,2-*cis*-linked disaccharide **3**. (DTBMP, 2,6-di-*tert*-butyl-4-methyl-pyridine; TfOH, trifluoromethanesulfonic acid.)

The principal challenge in the chemical synthesis of oligosaccharides remains the stereoselective introduction of the glycosidic bond.[2–3] The stereoselective synthesis of 1,2-*trans* glycosidic linkages can be accomplished by employing neighboring group participation of a C-2 acyl functionality and is, therefore, relatively straightforward. The stereoselective synthesis of 1,2-*cis* linkages is much harder to achieve since no general strategy is available. Most commonly, a non-participating group at C-2 is needed in addition to a participating solvent system in order to achieve good 1,2-*cis* selectivity.[4] In general, though, this approach still leads to the formation of mixtures of anomers.[3]

A new approach to the synthesis of 1,2-*cis* glycosides is the use of C-2 chiral auxiliaries that feature two key elements (Scheme 1.1).[5] The first is a nucleophilic thiophenyl group that can attack the intermediate oxa-carbenium ion (**1a**) and form an intermediate sulfonium ion (**1b**).[6] The second is a chiral substituent that directs the thiophenyl head group to the equatorial (β) face and ensures that the *trans*-decalin system is formed as opposed to the *cis*-decalin. The equilibrium between the oxa-carbenium ion and β-sulfonium ion can be controlled by the selection of protecting groups and constitution pattern of the chiral auxiliary.[7] Subsequent addition of a glycosyl acceptor will result in attack from the α-face and formation of the 1,2-*cis*-linked disaccharide in high yield and with excellent stereoselectivity. The methodology is applicable to the solid-phase synthesis of oligosaccharides[8] and holds considerable promise to become a general methodology for the stereoselective synthesis of 1,2-*cis* glycosides.

EXPERIMENTAL

GENERAL METHODS

^{1}H and ^{13}C NMR spectra were recorded with a 300 MHz or 500 MHz spectrometer. Chemical shifts are reported in parts per million (ppm) relative to tetramethylsilane (TMS) as the internal standard. Nuclear magnetic resonance (NMR) data are

presented as follows: chemical shift, multiplicity (s = singlet, d = doublet, t = triplet, dd = doublet of doublet, m = multiplet and/or multiple resonances), coupling constant in Hertz (Hz). All NMR signals were assigned on the basis of ^{1}H NMR, ^{13}C NMR, COSY, and HSQC experiments. Mass spectra were recorded with a MALDI-TOF mass spectrometer. The matrix used was 2,5-dihydroxy-benzoic acid (DHB) and Ultamark 1621 as the internal standard. Column chromatography was performed on silica gel G60 (Silicycle, 60–200 μm, 60 Å). Thin-layer chromatography (TLC) analysis was conducted on Silicagel 60 F$_{254}$ (EMD Chemicals) with detection by UV light (254 nm) where applicable, and by charring with 20% sulfuric acid in ethanol or with a solution of $(NH_4)_6Mo_7O_{24} \cdot H_2O$ (25 g/L) in 10% sulfuric acid in ethanol. CH_2Cl_2 was freshly distilled from calcium hydride under nitrogen prior to use. All reactions were carried out under an argon atmosphere.

3,6-DI-*O*-ACETYL-4-*O*-BENZYL-2-*O*-[(*S*)-2-(PHENYLTHIOMETHYL)BENZYL]-α/β-D-GLUCOPYRANOSYL TRIFLUORO-[*N*-PHENYL]-ACETIMIDATE (1)

3,6-Di-*O*-acetyl-4-*O*-benzyl-2-*O*-[(*S*)-2-(phenylthiomethyl)benzyl]-α/β-D-glucopyranose[8] (0.61 g, 1.07 mmol) was dissolved in dry CH_2Cl_2 (10 mL) and the solution was cooled to 0°C. 2,2,2-Trifluoro-*N*-phenylacetimidoyl chloride (1.1 mL, 5.0 equiv.) and 1,8-diazabicyclo[5.4.0]undec-7-ene (0.32 mL, 2.0 equiv.) were added and the mixture was stirred for 10 min at 0°C. After concentration *in vacuo* and chromatography (0% → 10% – EtOAc in PE) 1 was obtained, 0.72 g (0.97 mmol, 91%) as a white foam. TLC: (EtOAc/PE, 1/2 v/v): R$_f$ = 0.7; [α]D +18.2 (*c* 0.7, CHCl$_3$); ^{1}H NMR (300 MHz, CDCl$_3$) δ 7.40–7.07 (m, 18H, Ar), 6.83–6.81 (m, 2H, Ar) 6.63 (bs, 1H, H-1), 5.57 (t, 1H, *J* = 9.6 Hz, H-3), 4.52–4.38 (m, 3H, H-7, CH$_2$ Bn), 4.32–4.1 (m, 2H, H-6a/b), 4.13–4.02 (m, 1H, H-5), 3.58–3.337 (m, 2H, H-2, H-4), 3.29 (dd, 1H, *J* = 8.4 Hz, *J* = 13.8 Hz, H-8a), 3.13 (dd, 1H, *J* = 4.8 Hz, *J* = 13.8 Hz, H-8b), 2.03 (s, 3H, COCH$_3$), 1.74 (s, 3H, COCH$_3$); ^{13}C NMR (CDCl$_3$) δ 170.4 (CO), 169.4 (CO), 143.4 (Cq, Ar), 139.9 (Cq, Ar), 136.9 (Cq, Ar), 136.2 (Cq, Ar), 129.0–119.2 (CH, Ar), 92.0 (C-1), 81.9 (C-7), 75.9 (C-2), 75.4 (C-4), 72.5 (C-3), 70.7 (C-5), 62.3 (C-6), 41.6 (C-8), 20.8 (CH$_3$, Ac), 20.7 (CH$_3$, Ac); HR MALDI-TOF MS: [M + Na]$^+$ calcd. for C$_{39}$H$_{38}$OF$_3$NO$_8$S, 760.2167; found, 760.2149.

Methyl 3,6-di-*O*-acetyl-4-*O*-benzyl-2-*O*-[(1*S*)-phenyl-2-(phenylsulfanyl) ethyl]-α-D-glucopyranosyl-(1 → 6)-2,3,4-tri-*O*-benzoyl-α-D-glucopyranoside (3): Glycosyl donor 1 (100 mg, 0.14 mmol) was dissolved in dry CH_2Cl_2 (3 mL) and activated 3Å molecular sieves* (~500 mg) were added. The mixture was stirred for 10 min at rt before cooling to –50°C. TfOH (12.0 μL, 1.0 equiv.) was added and the mixture was slowly warmed to –30°C over a period of 15 min. Thin-layer chromatography (TLC) analysis showed complete activation of glycosyl donor 1 (TLC: (EtOAc/PE, 1/2 v/v): R$_f$ = 0.7) and only hydrolyzed donor (TLC: (EtOAc/PE, 1/2 v/v): R$_f$ = 0.35) was observed. A solution of glycosyl acceptor 2[1] (82 mg, 1.2 equiv.) and 2,6-di-*tert*-butyl-4-methyl-pyridine (82 mg, 3.0 equiv.) in CH_2Cl_2 (2 mL), which was dried using activated 3Å molecular sieves* (~200 mg), was added dropwise. The resulting

---

* Molecular sieves (Aldrich [St. Louis, MO] AW-300, 1.6 mm pellets) were placed in a round-bottom flask fitted with a two-way adapter. The flask was heated under high vacuum with an electric heating mantle (30 min) or open flame (2 min). The molecular sieves were placed under argon and cooled to room temperature.

mixture was allowed to warm slowly to rt over a period of 3 h. The mixture was diluted with CH$_2$Cl$_2$ (10 mL) and sat. aq. NaHCO$_3$ (5 mL) was added. The organic layer was separated, dried (MgSO$_4$), and concentrated in vacuo. Silica gel chromatography (0%→10% – EtOAc in toluene) afforded 123 mg (0.12 mmol, 86%) of **3** as a white foam. TLC: (EtOAc/PE, 1/2 v/v): R$_f$ = 0.45; [α]D -14.6 (c 0.3, CHCl$_3$); ^{1}H NMR (500 MHz, CDCl$_3$) δ 7.99–7.85 (m, 6H, Ar), 7.52–7.09 (m, 24H, Ar), 6.17 (t, 1H, J = 10.0 Hz, H-3^I), 5.54 (t, 1H, J = 9.0 Hz, H-3^{II}), 5.42 (t, 1H, J = 10.0 Hz, H-4^I), 5.26 (dd, 1H, J = 4.0 Hz, J = 10.0 Hz, H-2^I), 5.19 (d, 1H, J = 4.0 Hz, H-1^I), 5.01 (d, 1H, J = 3.0 Hz, H-1^{II}), 4.48 (d, 1H, J = 11.0 Hz, CHHPh), 4.43 (d, 1H, J = 11.0 Hz, CHHPh), 4.38–4.36 (m, 2H, H-5^I, H-7^{II}), 4.27–4.10 (m, 3H, H-5^{II}, H-6a^{II}, H-6b^{II}), 3.89 (t, 1H, J = 10.0 Hz, H-6a^I), 3.75 (d, 1H, J = 10.0 Hz, H-6b^I), 3.51 (s, 3H, OCH$_3$), 3.38–3.34 (m, 2H, H-2^{II}, H-4^{II}), 3.26 (dd, 1H, J = 8.5 Hz, J = 14.0 Hz, H-8a^{II}), 3.03 (dd, 1H, J = 5.0 Hz, J = 14.0 Hz, H-8b^{II}), 2.04 (s, 3H, COCH$_3$), 1.70 (s, 3H, COCH$_3$); ^{13}C NMR (CDCl$_3$) δ 170.6 (CO), 169.6 (CO), 165.7 (CO), 165.7 (CO), 165.5 (CO), 140.6 (Cq, Ar), 137.5 (Cq, Ar), 136.4 (Cq, Ar), 133.4, 133.2, 133.0, 129.9–125.8 (CH, Ar), 96.5 (C-1^{II}), 96.1 (C-1^I), 81.6 (C-5^I), 77.4 (C-4^{II}), 76.1 (C-2^{II}), 73.7 (CH$_2$ Bn), 72.6 (C-3^{II}), 72.1 (C-2^I), 70.4 (C-3^I), 69.9 (C-4^I), 68.6 (C-7^{II}), 68.3 (C-5^{II}), 67.0 (C-6^I), 62.8 (C-6^{II}), 55.6 (CH$_3$, OMe), 41.3 (C-8^{II}), 20.9 (CH$_3$, Ac), 20.8 (CH$_3$, Ac); HR MALDI-TOF MS: [M + Na]$^+$ calcd. for C$_{59}$H$_{58}$O$_{16}$S, 1077.3342; found, 1077.3396.

ACKNOWLEDGMENTS

This research was supported by the National Institute of General Medical Sciences (NIGMS) of the National Institutes of Health (Grant No. 2R01GM065248).

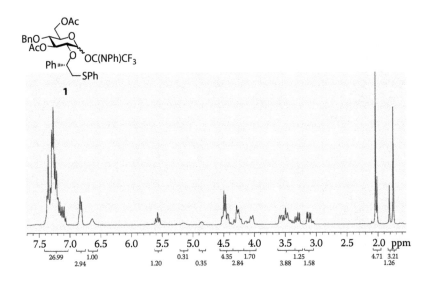

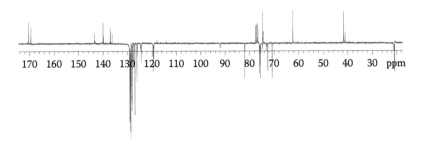

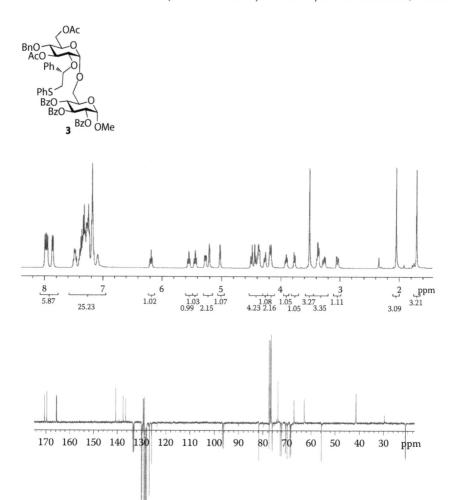

REFERENCES

1. Ziegler, T.; Kováč, P.; Glaudemans, C. P. J. *Carbohydr. Res.* 1989, *194*, 185–198.
2. Boltje, T. J.; Buskas, T.; Boons, G. J. *Nat. Chem.* 2009, *1*, (8), 611–622.
3. Zhu, X. M.; Schmidt, R. R. *Angew. Chem. Int. Ed.* 2009, *48*, (11), 1900–1934.
4. Demchenko, A. V. *Curr. Org. Chem.* 2003, *7*, (1), 35–79.
5. Kim, J. H.; Yang, H.; Khot, V.; Whitfield, D.; Boons, G. J. *Eur. J. Org. Chem.* 2006, (22), 5007–5028.
6. Kim, J. H.; Yang, H.; Park, J.; Boons, G. J. *J. Am. Chem. Soc.* 2005, *127*, 12090–12097.
7. Boltje, T. J.; Kim, J. H.; Park, J.; Boons, G. J. *Org. Lett.* 2011, *13*, (2), 284–287.
8. Boltje, T. J.; Kim, J. H.; Park, J.; Boons, G. J. *Nat. Chem.* 2010, *1*, (8), 611–622.

2 Regioselective Reductive Openings of 4,6-O-Benzylidene-Type Acetals using LiAlH$_4$-AlCl$_3$

Mihály Herczeg
László Lázár
*Markus Ohlin**
Anikó Borbás†

CONTENTS

* Checker under supervision of U. Ellervik: Ulf.ellervik@organic.lu.se.
† Corresponding author: borbas.aniko@unideb.hu.

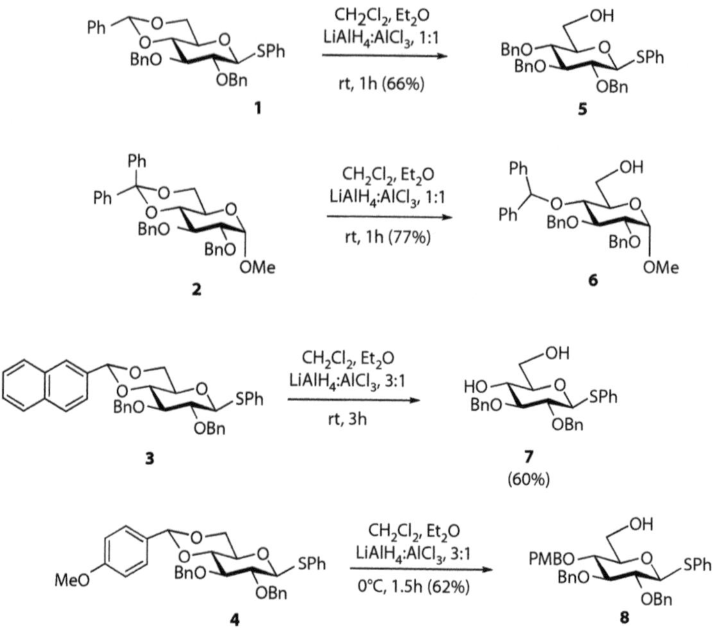

SCHEME 2.1

Regioselective reductive openings of the 4,6-O-benzylidene-type acetals of hexopyranosides to the corresponding benzyl-type ethers are important methods in carbohydrate chemistry. The transformation requires a hydride donor reagent in combination with a protic or a Lewis acid. Depending on the applied reagents and conditions, either the 4-OH or the 6-OH groups can be liberated. In the first synthetically applicable method, which was elaborated by Lipták et al.,[1] 4,6-O-benzylidene acetals of hexopyranosides were reacted with $LiAlH_4$ and $AlCl_3$ to give 4-O-benzyl ethers with a free 6-OH group. The $NaCNBH_3$-HCl reagent system introduced by Garegg et al.[2] resulted in opposite regioselectivity and thus became complementary to the $LiAlH_4$-$AlCl_3$ method. After publication of these two classical procedures, various reagents have been introduced for regioselective cleavage of 4,6-O-benzylidene-type acetals.[3–22] However, the regioselectivities and yields of the reactions are often influenced by the structure of the substrate and neighboring substituents. The $LiAlH_4$-$AlCl_3$ reagent in combination with CH_2Cl_2-Et_2O as solvent is often reliable in providing the 4-O-alkyl derivatives in high yield and regioselectivity. The limitation of this process is the incompatibility of ester, amide, and imide substituents with these reaction conditions.

It is important to note that to open different acetals, different ratios of $LiAlH_4$ to $AlCl_3$ have to be applied. An equimolar mixture of $LiAlH_4$ and $AlCl_3$ (forming AlH_2Cl) is used for the reductive openings of benzylidene[1,23] and diphenylmethylene[24] acetals, while 2-(naphthyl)methylene[25] p-methoxybenzylidene and anthracenylmethylene[26] acetals can be reduced with a mixture of $LiAlH_4$ to $AlCl_3$ in a 3:1 ratio (forming AlH_3) which does not affect the benzylidene and diphenylmethylene acetals.

Here, we present four examples for reductive openings of 4,6-*O*-acetals using the $LiAlH_4$-$AlCl_3$ reagent combination. The reactions were carried out in a CH_2Cl_2-Et_2O (3:2) solvent mixture instead of the original 1:1 mixture,[23–26] since we found that slight increase of the ratio of the dichloromethane significantly increases the reaction rate. Opening the benzylidene ring of **1** and diphenylmethylene ring of **2** with $LiAlH_4$-$AlCl_3$ (1:1) in CH_2Cl_2-Et_2O (3:2) went to completion at room temperature. Thus, the reflux temperature described in the literature[24,27] is unnecessary. In the case of **2**, formation of the 4,6-diol was also observed, but to a significantly lower extent than what is observed at higher temperature.[24]

The CH_2Cl_2-Et_2O (3:2) solvent system was convenient for ring-opening reactions of **3** and **4**, as well. Attempts to reductively cleave *p*-methoxybenzylidene acetal **4** in CH_2Cl_2-Et_2O (1:1) failed at 0°C, while at room temperature ring cleavage and degradation took place simultaneously. The opening of the 2-(naphthyl)methylene ring was rather slow; complete conversion of **3** required 3 hours even when working at three times higher concentration.

The use of different ratios of $LiAlH_4$ to $AlCl_3$ calls for different work-up procedures, which are discussed in the experimental section.

EXPERIMENTAL

GENERAL METHODS

Optical rotations were measured at room temperature with a Perkin-Elmer 241 automatic polarimeter. Melting points were determined on a Kofler hote-stage apparatus and are uncorrected. Thin-layer chromatography (TLC) was performed on DC-Alurolle Kieselgel 60 F_{254} (Merck, Whitehouse Station, NJ), and the spots were visualized under ultraviolet (UV) light and charring with 5% ethanolic sulfuric acid. Column chromatography was performed on Silica gel 60 (Merck, particle size 0.063–0.200 mm). Organic solutions were dried over $MgSO_4$ and concentrated in vacuum at 45°C (water bath). The ^1H (360 MHz) and ^{13}C NMR (90.54 MHz) spectra were recorded with Bruker DRX-360 spectrometer. Chemical shifts are referenced to Me_4Si (0.00 ppm for ^1H) or to the residual solvent signals (77.00 ppm for ^{13}C). Infrared (IR) spectra were recorded on a PerkinElmer (Waltham, MA) 16 PC Fourier transform infrared (FTIR) spectrometer. Elemental analyses (C, H, S) were performed using an Elementar (Germany) Vario MicroCube instrument.

GENERAL PROCEDURES

A: To a stirred solution of the 4,6-*O*-benzylidene-type derivative (1.0 mmol, **1** and **2**) in a mixture of dry CH_2Cl_2[*] (9 mL) and dry Et_2O[a] (3 mL) were added

---

[*] Dry CH_2Cl_2 from Sigma-Aldrich was used; Et_2O was dried over metallic Na.

successively* LiAlH$_4$[†] (171 mg, 4.5 mmol) and a solution of AlCl$_3$[‡] (600 mg, 4.5 mmol) in dry Et$_2$O (3 mL), under argon at room temperature. Proper monitoring of the progress of the reaction can be achieved by applying a micro-workup procedure of samples taken from the vigorously stirred reaction mixture.[§] When the TLC (95:5 CH$_2$Cl$_2$-EtOAc) indicated complete disappearance of the starting material (1 h), the reaction mixture was cooled in an ice bath, and the excess of reagent was decomposed by careful addition of EtOAc (20 mL) followed by H$_2$O (5 mL), and the stirring was continued for additional 5 min. The mixture obtained, consisting of a gray, non-filterable suspension and a clear organic phase, was poured into a separating funnel and diluted with EtOAc (50 mL). The layers were separated and the organic phase was washed with H$_2$O (3 × 30 mL), dried, and concentrated. The residue was chromatographed (100 g of silica gel/g of the crude product) using 95:5 CH$_2$Cl$_2$-EtOAc as eluent.

B: To a stirred solution of the 4,6-*O*-benzylidene-type derivative (1.0 mmol, **3** and **4**) in a mixture of dry CH$_2$Cl$_2$[a] (**3**: 3 mL, **4**: 9 mL) and dry Et$_2$O (**3**: 1 mL, **4**: 3 mL) were added under argon (for **3** at room temperature; for **4** at 0°C) successively[b] LiAlH$_4$[c] (171 mg, 4.5 mmol) and a solution of AlCl$_3$[d] (200 mg, 1.5 mmol) in dry Et$_2$O (**3**: 1 mL, **4**: 3 mL). When the TLC (95:5 CH$_2$Cl$_2$-EtOAc) indicated complete disappearance of the starting material (**3**: 3 h at room temperature; **4**: 0.5 h at 0°C),[e] the reaction mixture was cooled in an ice bath, and the excess of reagent was decomposed by careful addition, successively, of EtOAc (20 mL), and H$_2$O (5 mL), and the mixture was stirred for an additional 5 min. The obtained heterogeneous mixture containing finely precipitated solid was filtered through a pad of Celite, and the filter cake was washed with EtOAc (2 × 5 mL). The combined filtrates were transferred into a separating funnel and diluted with EtOAc (50 mL), the layers were separated and the organic phase was washed with H$_2$O (3 × 50 mL), dried, and concentrated. The residue was chromatographed (100 g of silica gel/g of the crude product) using CH$_2$Cl$_2$-EtOAc as eluent (98 : 2 for **7** and 95 : 5 for **8**).

PHENYL 2,3,4-TRI-*O*-BENZYL-1-THIO-β-D-GLUCOPYRANOSIDE (5)[28]

Prepared from **1**[27] (540 mg, 1.0 mmol) by method A. Yield, 358 mg (66%)[¶], mp 120–122°C (from EtOH)**; R$_f$ 0.62 (CH$_2$Cl$_2$-EtOAc 95:5); [α]$_D$ + 4.8 (*c* 0.2, CHCl$_3$)[g]; ^{1}H NMR (CDCl$_3$) δ 7.52–7.24 (m, 20H, arom), 4.92–4.83 (m, 4H, 2 × PhC*H*$_2$), 4.76–4.63

---

* The order of addition of the reagents is important; if the order is changed, the acidic character of AlCl$_3$ may affect the acid-sensitive protecting groups. On a large scale, LiAlH$_4$ has to be added in ~500 mg portions.

[†] LiAlH$_4$ from Sigma-Aldrich (St. Louis, MO) was used. (Caution: LiAlH$_4$ violently reacts with water, including atmospheric moisture, releasing gaseous hydrogen.)

[‡] Anhydrous AlCl$_3$ from Sigma-Aldrich was used. Preparation of ethereal solution of the AlCl$_3$: In a flask protected from atmospheric moisture with a calcium chloride tube, AlCl$_3$ was added with stirring to the cooled (0°C) dry Et$_2$O.

[§] To a ~50-100 μL sample taken from the reaction mixture, ethyl acetate (~200-400 μL) and water (1-2 drops) were added. After shaking of this mixture the organic phase is subjected to TLC analysis.

[¶] On larger scales higher yields were obtained for compounds **5** (84%), **7** (90%), and **8** (87%).

** Lit. 28, mp 121.5–122°C (from EtOH); [α]$_D$ +7.7 (*c* 3.4, CHCl$_3$).

(m, 3H, PhCH$_2$, H-1), 3.88–3.85 (m, 1H), 3.75–3.68 (m, 2H), 3.58 (t, 1H, J 9.3 Hz), 3.48 (t, 1H, J 9.3 Hz), 3.39–3.36 (m, 1H), 2.16 (s, 1H, OH); ^{13}C NMR (CDCl$_3$) δ 138.1, 137.8, 137.7, 133.4 (4 × C$_q$ arom.), 131.6–127.6 (arom.), 87.4 (C-1), 86.4, 80.9, 79.2, 77.4 (C-2, C-3, C-4, C-5), 75.7, 75.4, 75.0 (3 × PhCH$_2$), 61.9 (C-6); IR (KBr) v 3336, 3028, 2903, 2360, 1944, 1584, 1496, 1452, 1358, 1111, 1062, 1027, 911, 873, 833, 818, 745, 735, 695, 655, 558, 463 cm^{-1}. Anal. calcd. for C$_{33}$H$_{34}$O$_5$S: C, 73.04; H, 6.31; S, 5.91. Found: C, 73.27; H, 6.39; S, 5.74.

METHYL 2,3-DI-O-BENZYL-4-O-DIPHENYLMETHYL-α-D-GLUCOPYRANOSIDE (6)[24]

Prepared from 2[24] (538 mg, 1.0 mmol). Yield, 403 mg (75%), colorless syrup; R$_f$ 0.41 (CH$_2$Cl$_2$-EtOAc 95:5); [α]$_D$ +129.6 (c 0.3, CHCl$_3$)*; ^{1}H NMR (CDCl$_3$) δ 7.33–7.05 (m, 20H, arom.), 5.95 (s, 1H, Ph$_2$CH), 4.91, 4.77, 4.62, 4.55 (4d, 4H, 2 × PhCH$_2$), 4.53 (d, 1H, $J_{1,2}$ 3.6 Hz H-1), 4.07 (t, 1H, J 9.0 Hz, J 9.4 Hz), 3.65 (m, 1H), 3.56–3.45 (m, 3H), 3.53–3.27 (m, 4H), 0.94 (s, 1H, OH); ^{13}C NMR (CDCl$_3$) δ 142.7, 142.0, 138.4, 138.1 (4 × C$_q$ arom.), 128.6–126.0 (arom.), 98.0 (C-1), 83.7, 82.6, 80.1, 74.1, 70.5 (C-2, C-3, C-4, C-5, Ph$_2$CH), 75.8, 73.3 (2 × PhCH$_2$), 61.7 (C-6), 55.2 (OCH$_3$); IR (KBr) v 3584, 3493, 3087, 3062, 3029, 3007, 2922, 2844, 2071, 1953, 1884, 1811, 1755, 1602, 1586, 1546, 1495, 1454, 1358, 1329, 1191, 1157, 1066, 1031, 912, 868, 836, 741, 699, 665, 620, 602, 561, 505, 465 cm^{-1}. Anal. calcd. for C$_{34}$H$_{36}$O$_6$ (540.65): C, 75.53; H, 6.71. Found: C, 74.60; H, 6.60.

PHENYL 2,3-DI-O-BENZYL-4-O-(2-NAPHTHYL)METHYL-1-THIO-β-D-GLUCOPYRANOSIDE (7)

Prepared from 3 (590 mg, 1.0 mmol) by method B. Yield, 355 mg (60%)^f, mp 118–122°C (from EtOH); R$_f$ 0.52 (CH$_2$Cl$_2$-EtOAc 98:2); [α]$_D$ –8.6 (c 0.5, CHCl$_3$); ^{1}H NMR (CDCl$_3$) δ 7.79–7.24 (m, 22H, arom.), 4.99–4.71 (m, 7H, 2 × PhCH$_2$, NAPCH$_2$, H-1), 3.91–3.88 (m, 1H), 3.77–3.68 (m, 2H), 3.63 (t, 1H, J 9.3 Hz), 3.49 (t, 1H, J 9.5 Hz), 3.43-3.39 (m, 1H), 2.19 (s, 1H, OH); ^{13}C NMR (CDCl$_3$) δ 138.2, 137.8, 135.2, 133.4, 133.1, 132.9 (6 x C$_q$ arom.), 131.6-125.8 (arom.), 87.4 (C-1), 86.4, 80.9, 79.3, 77.4 (C-2, C-3, C-4, C-5), 75.7, 75.4, 75.0 (2 × PhCH$_2$, NAPCH$_2$), 61.9 (C-6); IR (KBr) v 3417, 3055, 3027, 2902, 2869, 2356, 1945, 1583, 1496, 1477, 1355, 1283, 1129, 1077, 1063, 1027, 986, 853, 815, 746, 734, 694, 621, 613, 557, 477 cm^{-1}. Anal. calcd. for C$_{37}$H$_{36}$O$_5$S: C, 74.97; H, 6.12; S, 5.41. Found: C, 75.47; H, 6.08; S, 5.19.

PHENYL 2,3-DI-O-BENZYL-4-O-(4-METHOXYBENZYL)-1-THIO-β-D-GLUCOPYRANOSIDE (8)

Prepared from 4[17] (570 mg, 1.0 mmol) by method B. Yield, 356 mg (62%)^f, mp 141–143°C (from MeOH); R$_f$ 0.38 (CH$_2$Cl$_2$-EtOAc 95:5); [α]$_D$ –0.8 (c 0.1, CHCl$_3$); ^{1}H NMR (CDCl$_3$) δ 7.52–6.81 (m, 19H, arom), 4.92–4.86 (m, 3H, PhCH$_2$), 4.78–4.56 (m, 4H, PhCH$_2$, H-1) 3.87–3.83 (m, 1H), 3.73 (s, 3H, OCH$_3$), 3.70–3.64 (m, 2H), 3.56 (t,

---

* Lit. 24, [α]$_D$ +129 (c 0.3, CHCl$_3$).

1H, J 9.3 Hz), 3.47 (t, 1H, J 9.3 Hz), 3.38–3.33 (m, 1H), 2.25 (t, 1H, J 6.5 Hz, OH); ^{13}C NMR (CDCl$_3$) δ 159.2, 138.2, 137.7, 133.4, 129.8 (5 × C$_q$ arom.), 131.5–113.7 (arom.), 87.4 (C-1), 86.4, 80.9, 79.2, 77.1 (C-2, C-3, C-4, C-5), 75.6, 75.4, 74.6 (3 × PhCH$_2$), 61.8 (C-6), 55.1 (OCH$_3$); IR (KBr) v 3322, 3058, 3029, 2905, 2345, 1944, 1879, 1612, 1585, 1515, 1401, 1358, 1253, 1130, 1062, 1040, 988, 824, 742, 695, 649, 621, 558, 536, 499 cm^{-1}. Anal. calcd. for C$_{34}$H$_{36}$O$_6$S: C, 71.30; H, 6.34; S, 5.60. Found: C, 71.61; H, 6.39; S, 5.53.

ACKNOWLEDGMENTS

The work is supported by the TÁMOP 4.2.1/B-09/1/KONV-2010-0007 project. The project is co-financed by the European Union and the European Social Fund.

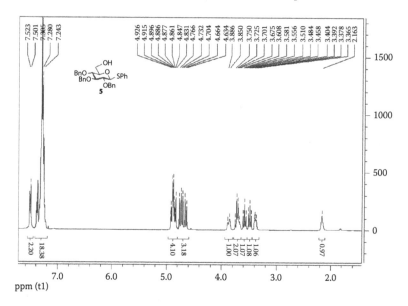

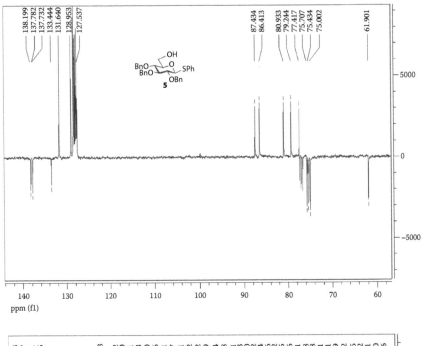

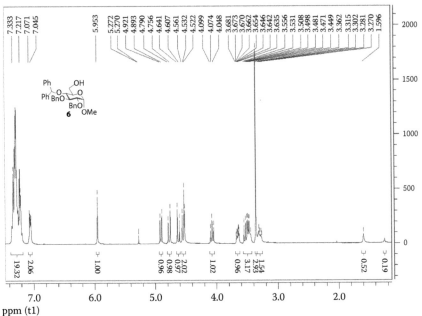

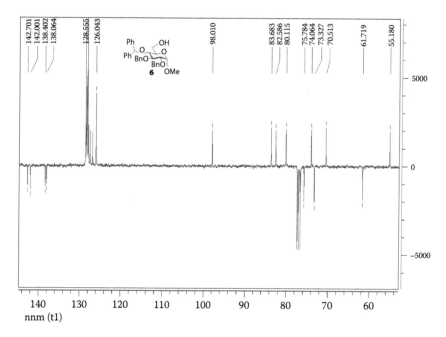

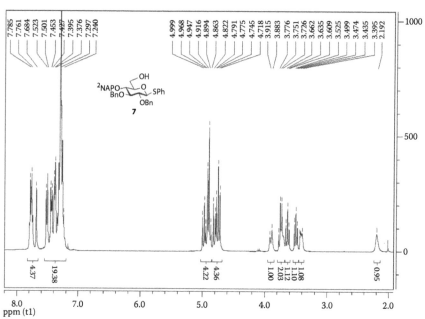

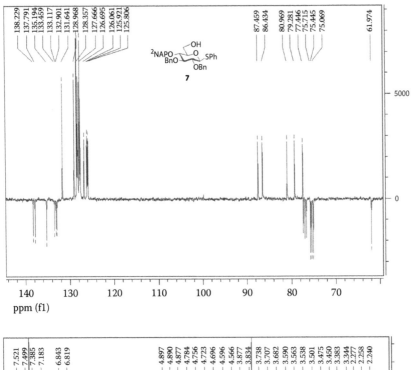

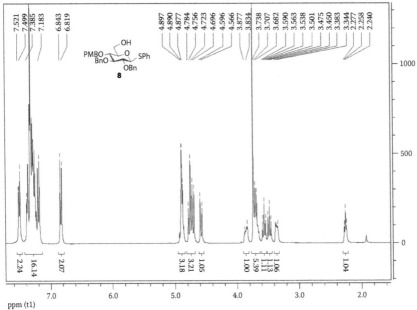

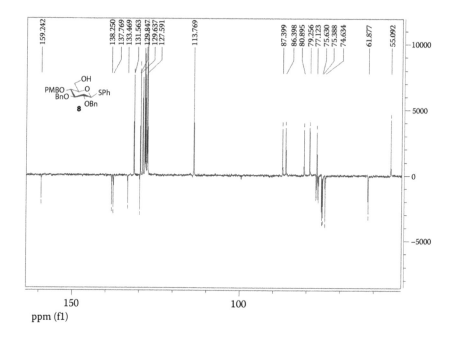

REFERENCES

1. Lipták, A.; Jodál, I.; Nánási, P. *Carbohydr. Res.* 1975, *44*, 1–11.
2. Garegg, P. J.; Hultberg, H. *Carbohydr. Res.* 1981, *93*, C10–C11.
3. Ek, M.; Garegg, P. J.; Hultberg, H.; Oscarson, S. *J. Carbohydr. Chem.* 1983, *2*, 305–311.
4. Mikami, T.; Asano, H.; Mitsunobu, O. *Chem. Lett.* 1987, 2033–2036.
5. Guindon, Y.; Girard, Y.; Berthiaume, S.; Gorys, V.; Lemieux, R.; Yoakim, C. *Can. J. Chem.* 1990, *68*, 897–902.
6. Tanaka, N.; Ogawa, I.; Yoshigase, S.; Nokami, J. *Carbohydr. Res.* 2008, *343*, 2675–2679.
7. Guindon, Y.; Girard, Y.; Berthiaume, S.; Gorys, V.; Lemieux, R.; Yoakim, C. *Can. J. Chem.* 1990, *68*, 897–902.
8. DeNinno, M. P.; Etienne, J. B.; Duplantier, K. C. *Tetrahedron Lett.* 1995, *36*, 669–672.
9. Oikawa, M.; Liu, W.-C.; Nakai, Y.; Koshida, S.; Fukase, K.; Kusumoto, S. *Synlett* 1996, 1179–1180.
10. Jiang, L.; Chan, T.-H. *Tetrahedron Lett.* 1998, *39*, 355–358.
11. Chandrasekhar, S.; Reddy, Y. R.; Reddy, C. R. *Chem. Lett.* 1998, 1273–1274.
12. Ghosh, M.; Dulina, R. G.; Kakarla, R.; Sofia, M. J. *J. Org. Chem.* 2000, *65*, 8387–8390.
13. Sakagami, M.; Hamana, H. *Tetrahedron Lett.* 2000, *41*, 5547–5551.
14. Debenham, S. D.; Toone, E. J. *Tetrahedron: Asymmetry* 2000, *11*, 385–387.
15. Wang, C.-C.; Luo, S.-Y.; Shie, C.-R.; Hung, S.-C. *Org. Lett.* 2002, *4*, 847–849.
16. Dilhas, A.; Bonnaffé, D. *Tetrahedron Lett.* 2004, *45*, 3643–3645.
17. Hernandez-Torres, J. M.; Achkar, J.; Wei, A. *J. Org. Chem.* 2004, *69*, 7206–7211.
18. Shie, C.-R.; Tzeng, Z.-H.; Kulkarni, S. S.; Uang, B.-J.; Hsu, C.-Y.; Hung, S.-C. *Angew. Chem. Int. Ed.* 2005, *44*, 1665–1668.
19. Tanaka, K.; Fukase, K. *Synlett* 2007, 164–166.
20. Tani, S.; Sawadi, S.; Kojima, M.; Akai, S.; Sato, K. *Tetrahedron Lett.* 2007, *48*, 3103–3104.

21. Zinin, A. I.; Malysheva, N. N.; Shpirt, A. M.; Torgov, V. I.; Kononov, L. O. *Carbohydr. Res.* 2007, *342*, 627–630.
22. Daragics, K.; Fügedi, P. *Tetrahedron Lett.* 2009, *50*, 2914–2916.
23. Lipták, A.; Imre, J.; Harangi, J.; Nánási, P. *Tetrahedron* 1979, *38*, 3721–3727.
24. Hajkó, J.; Szabovik, G.; Kerékgyártó, J.; Kajtár M.; Lipták, A. *Aust. J. Chem.* 1996, *49*, 357–363.
25. Borbás, A.; Szabó, Z. B.; Szilágyi, L.; Bényei, A.; Lipták, A. *Tetrahedron* 2002, *58*, 5723–5732.
26. Jakab, Zs.; Mándi, A.; Borbás, A.; Bényei, A.; Komáromi, I.; Lázár, L.; Antus, S.; Lipták, A. *Carbohydr. Res.* 2009, *344*, 2444–2453.
27. Lipták, A.; Jodál, I.; Harangi, J.; Nánási, P. *Acta. Chim. Hung.* 1983, *113*, 151–154.
28. Pfäffli, P. J.; Hixson, S. H.; Anderson, L. *Carbohydr. Res.* 1972, *23*, 195–206.

3 A Facile Method for Oxidation of Primary Alcohols to Carboxylic Acids in Carbohydrate Synthesis

Zhaojun Yin
Steven Dulaney
Gopinath Tiruchinapally
Bo Yang
*Guitao Wang**
Xuefei Huang†

CONTENTS

Uronic acids are common constituents of glycosaminoglycans (GAGs).[1,2] Because of their important roles in many biological processes,[3-5] there is high interest in synthesis of complex GAG oligosaccharides.[6-9] Due to the presence of the electron withdrawing carboxyl group, uronic acid building blocks often have low reactivities in glycosylation reactions. An alternative involves the use of pyranosides as building

---

* Checker under the supervision of Z. Guo: zwguo@chem.wayne.edu
† Corresponding author: xuefei@chemistry.msu.edu.

blocks, which can give higher glycosylation yields. However, a prerequisite for this strategy is the availability of a reliable method for the subsequent conversion of the primary alcohols to carboxylic acids.

Although oxidation of primary alcohols to carboxylic acids is a fundamental transformation in organic synthesis,[10] the presence of many protective groups and the acid labile glycosidic linkages in oligosaccharides severely limit available methodologies. In addition, the possible need to simultaneously convert multiple primary alcohols to carboxylic acids presents further challenges. During our studies on GAG synthesis, we had to oxidize the three primary hydroxyl groups to carboxylic acids in hexasaccharide **1**.[11,12] Using the established methods, such as pyridinium dichromate (PDC) mediated oxidation,[13–16] 2,2,6,6-tetramethylpiperidinyl-1-oxy (TEMPO)-catalyzed oxidations with co-oxidants such as NaOCl[2,17–25] and iodobenzene diacetate (BAIB),[26,27] NaClO$_2$ catalyzed by NaOCl/TEMPO[28] or a two-step process of Dess-Martin oxidation followed by NaClO$_2$ treatment,[29] multiple partially oxidized products were obtained with only a trace amount of the desired product. Finally, we discovered that a convenient two-step, one-pot protocol with TEMPO/NaOCl followed by treatment with NaClO$_2$[11] afforded the desired carboxylic acid in high yield and good purity. Herein, we report the selectivity and functional group compatibility of this oxidation protocol with four examples of representative monosaccharide building blocks **2–5**.

For our oxidation reaction, the substrate was dissolved in a mixture of dichloromethane and water, to which TEMPO and NaOCl were added together with phase transfer agents in a sodium bicarbonate solution. Upon complete consumption of the starting material (~1 h), the pH value of the reaction was brought to neutral by addition of aqueous 1N HCl. Sodium chlorite was then added with *tert*-butanol, 2-methyl 2-butene[30] and sodium dihydrogen phosphate. The desired carboxylic acid was obtained from the reaction mixture following workup and column chromatography. It has been reported that TEMPO/NaOCl oxidation of electron-rich aromatic rings could result in poor yield and purity due to significant chlorination of the aromatic groups.[28] With our protocol, the three PMB groups in the electron-rich glucoside **2** remained intact during oxidation, leading to glucuronic acid **6** in 81% yield (Table 3.1, entry 1). With the mild reaction condition, acid-sensitive isopropylidene groups in galactoside **3** were not affected (Table 3.1, entry 2). The allyl group, a popular linker for the bioconjugation of carbohydrates with proteins,[31] was shown to be incompatible with TEMPO oxidation due to its sensitivity to radicals.[13] However, under our biphasic reaction condition, the allyl mannosyl pyranoside **4** was converted to the desired mannosyl uronic acid **8** in excellent yield (Table 3.1, entry 3). Traditionally, oxidation of the primary hydroxyl group in a thioglycoside is prone to side reactions due to the ease of sulfoxide and sulfone formation.[13,32] We decided to test the electron-rich diol thioglycoside **5**, where the presence of a free secondary hydroxyl group would probe the selectivity of oxidation of the primary hydroxyl

TABLE 3.1

Oxidation of Monosaccharides

$$R-CH_2OH \xrightarrow[\text{then NaClO}_2,\ 2\text{-methyl-2-butene, } t\text{BuOH}]{\text{TEMPO, NaBr, } n\text{Bu}_4\text{NBr, NaOCl, NaHCO}_3} R-CO_2H$$

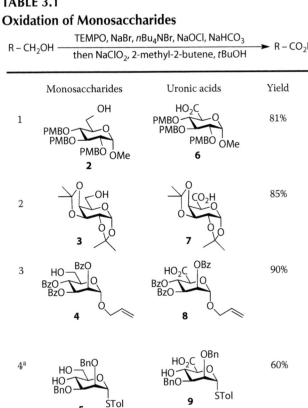

	Monosaccharides	Uronic acids	Yield
1	**2**	**6**	81%
2	**3**	**7**	85%
3	**4**	**8**	90%
4[a]	**5**	**9**	60%

[a]Ester **10** (~ 5%) was isolated from the reaction mixture.

group over secondary ones. The oxidation of compound **5** with our protocol provided the desired uronic acid product **9** in 60% yield (Table 3.1, entry 4). No sulfones or sulfoxides were obtained from the reaction mixture. Besides uronic acid **9**, ester **10** (~5%) was isolated as the major side product from the reaction mixture. In addition to monosaccharides, this oxidation protocol has been successfully applied by us as well as other groups to the oxidation of oligosaccharides and polysaccharides.[11,12,33–36]

The previously reported TEMPO/NaOCl procedure[17-19,23] failed to produce the desired carboxylic acids in our hands in several cases, while the two-step procedure of TEMPO/NaOCl followed by NaClO$_2$ was successful.[11,12] We propose that the failure of the TEMPO/NaOCl procedure could be attributed to the hydrophobici-ties of the substrates. It was proposed that aldehydes are intermediates in TEMPO-catalyzed NaOCl oxidation of primary alcohols to carboxylic acids (Scheme 3.1).[23] With a hydrophilic aldehyde, the carbonyl group can be hydrated to form a vicinal diol, which is subsequently oxidized by TEMPO/NaOCl to produce a carboxylic acid. However, in cases of protected hydrophobic carbohydrates, the hydration step may become slow, leading to low yields of carboxylic acid products. In contrast, the aldehyde could be quickly converted to the carboxylic acid by $NaClO_2$. A one-step procedure that involved the uses of TEMPO, NaOCl and $NaClO_2$ together was not successful, probably due to the instability of the $NaClO_2/NaOCl$ mixture.[28]

Scheme 3.1 Proposed intermediates for the oxidation.

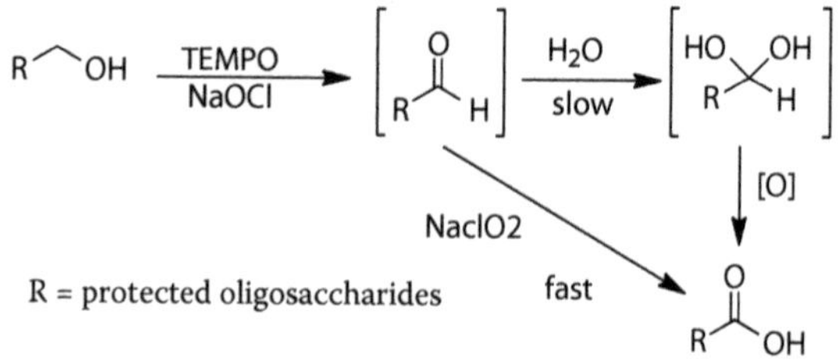

The biphasic condition (methylene chloride and water) for the oxidation is crucial for chemoselectivity. When oxidation of thioglycoside was carried out in a homoge-neous acetonitrile/water mixture, multiple side products *not* containing the thiotolyl, carboxylic acid, or the corresponding aldehyde group were formed. This suggests that the single-phase homogenous condition for TEMPO/NaOCl oxidation is less selective and can affect sensitive functional groups.

The formation of side product ester **10** in oxidation of diol **5** (Table 3.1, entry 4) can be explained with the intermediacy of aldehyde as well. Upon generation of aldehyde **11** from **5**, intermolecular nucleophilic attack of the carbonyl group by the free hydroxyl group produced hemiacetal **12**, oxidation of which led to carboxylic ester **10** (Scheme 3.2). Formation of similar by-products during oxidation of primary hydroxyl groups in carbohydrates has been observed previously.[37]

Scheme 3.2 Proposed mechanism for the formation of ester **10**.

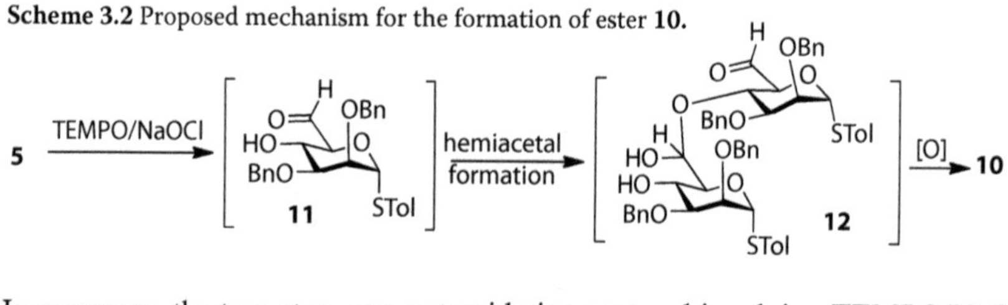

In summary, the two-step, one-pot oxidation protocol involving TEMPO/NaOCl followed by $NaClO_2$ can efficiently convert primary alcohols to carboxylic acids in good yields. The protocol is compatible with a wide range of functional groups (Table 3.1).

EXPERIMENTAL

GENERAL METHODS

The chemicals used were reagent grade, used as supplied, except where noted. Analytical thin-layer chromatography was performed using silica gel 60 F254 glass plates. Compounds were visualized by ultraviolet (UV) light (254 nm) and by staining [with a CAM solution containing $Ce(NH_4)_2(NO_3)_6$ (0.5 g) and $(NH_4)_6Mo_7O_{24} \cdot 4H_2O$ (24.0 g) in 6% H_2SO_4 (500 mL)]. Flash column chromatography was performed on silica gel 60 (230–400 Mesh). Nuclear magnetic resonance (NMR) spectra were referenced using Me_4Si (0 ppm), residual $CHCl_3$ (δ ^1H NMR 7.27 ppm, ^{13}C NMR 77.0 ppm). Peak and coupling constant assignments are based on ^1H-NMR, ^1H–^1H gCOSY, and (or) ^1H–^{13}C gHMQC experiments. Optical rotations were measured at 25°C. High-resolution mass spectra were recorded on a Micromass electrospray mass spectrometer equipped with an orthogonal electrospray source (Z-spray) operated in positive ion mode. Elemental analyses were measured by Columbia Analytical Services (Kelso, WA).

GENERAL PROCEDURE FOR OXIDATION

Aqueous solutions of NaBr (1 M, 25 μL), tetrabutylammonium bromide (1 M, 50 μL), TEMPO (2.2 mg, 0.014 mmol, 0.3 equiv per hydroxyl group), and saturated aqueous solution of $NaHCO_3$ (125 μL) were added at 0°C to a solution of alcohol (0.045 mmol) in CH_2Cl_2 (1 mL) and H_2O (170 μL). The mixture was treated with fresh aqueous NaOCl solution (150 μL, chlorine content was 4–5%). (The use of NaOCl solution having higher chlorine content or solution more than six months old led to inconsistent results.) The cooling was not removed until thin-layer chromatography (TLC) showed that the starting material was consumed (~1 hour).[*]

After adjusting the pH to near neutral (1 N HCl ~50 μL, pH 6–7) tBuOH (0.7 mL), 2-methyl 2-butene in THF (2 M, 1.4 mL) and a solution of $NaClO_2$ (50 mg, 0.44 mM) and NaH_2PO_4 (40 mg, 0.34 mM) in H_2O (0.2 mL) were added. The reaction mixture was kept at room temperature for 1–2 hours, diluted with saturated aqueous NaH_2PO_4 solution (5 mL), and extracted with EtOAc (3 × 10 mL). The combined organic layers were dried ($MgSO_4$) and concentrated, and the residue was chromatographed (1 : 1 hexanes–EtOAc containing 1% HOAc[38]).

METHYL 2,3,4-TRI-*O*-*P*-METHOXYBENZYL-α-D-GLUCOPYRANOSIDURONIC ACID (6)

Obtained as syrup (81%) from compound **2** (0.045 mmol) following the general oxidation procedure. $[\alpha]_D^{20}$ –21.4 (c = 1, $CHCl_3$); ^1H NMR (500 MHz, $CDCl_3$) δ 3.30 (s, 3 H, anomeric OMe), 3.49 (dd, 1 H, J = 9.0, 2.0 Hz, H-2), 3.62 (t, 1 H, J = 9.0 Hz, H-4),

---

[*] It was crucial to closely monitor the progress of the reaction especially for substrates with free secondary hydroxyl groups (e.g., diol **5**). Once the starting material is completely consumed, the reaction should be quenched by adjusting pH to neutral. In the case of diol **5**, prolonged reaction time resulted in formation of a large amount of ester **10** (>30%), which was presumably formed due to intermolecular attack of the newly formed aldehyde by the free secondary hydroxyl group, followed by further oxidation to ester.

3.70 (s, 3 H, PMB-O*Me*), 3.78 (s, 3 H, PMB-O*Me*), 3.79 (s, 3 H, PMB-O*Me*), 3.92 (t, 1 H, J = 9.0 Hz, H-3), 4.15 (d, 1 H, J = 10 Hz, H-5), 4.51–4.59 (m, 3 H, PMB-CH_2 and H-1), 4.71–4.65 (m, 3 H, PMB-CH_2), 4.82 (d, 1 H, J = 10.5 Hz, PMB-CH_2), 6.75–6.79 (m, 2 H, aromatic), 6.81–6.86 (m, 4 H, aromatic), 7.13–7.16 (m, 2 H, aromatic), 7.21–7.24 (m, 4 H, aromatic); ^{13}C NMR (125 MHz, CDCl$_3$) δ 55.15 (anomeric OCH$_3$), 55.23 (PMB-OCH$_3$), 55.69 (PMB-OCH$_3$), 69.83 (C-5), 73.13, 74.76 and 75.46 (3CH$_2$), 79.15 (C-2 and C-4), 81.06 (C-3), 98.51 (C-1), 113.76 (aromatic carbons), 113.79, 113.87, 128.20, 129.52, 129.64, 129.75, 130.02, 130.79, 159.18, 159.31, 159.44, 173.69 (COOH); HRMS C$_{31}$H$_{35}$O$_{10}$ [M-H]$^-$ calc. 568.2230 found 567.2228.

1,2:3,4-D$_I$-*O*-$_{ISOPROPYLIDENE}$-α-$_D$-$_{GALACTOPYRANOSIDURONIC ACID}$ (7)

Obtained from compound **3** (0.045 mmol) following the general oxidation procedure as syrup (85%), which was azotropically dried with toluene. m.p. 151–153°C (lit.[39] mp 152°C); [α]$_D^{20}$ –63 (*c* 0.1, CHCl$_3$), (lit.[40] mp 157°C, [α]D –84, *c* 1, CHCl$_3$), (lit.[41] mp 157–159°C); ^{1}H NMR (600 MHz, CDCl$_3$) δ 1.32 (s, 6 H, 2 × CH$_3$), 1.43 (s, 3 H, C*H*$_3$), 1.51 (s, 3 H, C*H*$_3$), 4.37 (dd, 1 H, J = 5.4, 3.0 Hz, H-2), 4.43 (d, 1 H, J = 2.4 Hz, H-5), 4.61 (dd, 1 H, J = 7.8, 2.4 Hz, H-4), 4.66 (dd, 1 H, J = 7.8, 3.0 Hz, H-3), 5.63 (d, 1 H, J = 5.4 Hz, H-1); ^{13}C NMR (150 MHz, CDCl$_3$) δ 24.37 (CH$_3$), 24.77 (CH$_3$), 24.98 (CH$_3$), 25.84 (CH$_3$), 68.25 (C-5), 70.51 (C-2), 70.54 (C-3), 71.63 (C-4), 96.35 (C-1), 109.39 (C(CH$_3$)$_2$), 110.08 (C(CH$_3$)$_2$), 171.11 (COOH); HRMS C$_{12}$H$_{17}$O$_7$ [M-H]$^-$ calcd. 273.0974 found 273.0964. The NMR data were consistent with literature.[11]

A$_{LLYL}$ 2,3,4-$_{TRI}$-*O*-$_{BENZOYL}$-α-D-$_{MANNOPYRANOSIDURONIC ACID}$ (8)

Obtained from compound **4** (0.045 mmol) following the general oxidation procedure as syrup, yield 90%. [α]$_D^{20}$ –84.2 (*c* = 1, CHCl$_3$); ^{1}H NMR (500 MHz, CDCl$_3$): δ 4.20 (dd, 1 H, J = 5.5, 13.0 Hz, -O-CH$_2$-CH = CH$_2$), 4.37 (dd, 1 H, J = 5.5, 13.0 Hz, -O-CH$_2$-CH = CH$_2$), 4.71 (d, 1 H, J = 9.5 Hz, H-5), 5.30 (d, 1 H, J = 2.0 Hz, H-1), 5.32 (ddd, 1 H, J = 1.2, 3.0, 10.5 Hz, -O-CH$_2$-CH = CH$_2$), 5.42 (ddd, 1 H, J = 1.5, 3.0, 17.2 Hz, -O-CH$_2$-CH = CH$_2$), 5.71 (dd, 1 H, J = 2.0, 3.0 Hz, H-2), 5.96 (dd, 1 H, J = 3.0, 9.5 Hz, H-3), 6.00 (m, 1H, -O-CH$_2$-C*H* = CH$_2$), 6.05 (t, 1 H, J = 9.5 Hz, H-4), 7.30–8.11 (m, 15 H, aromatic); ^{13}C NMR (125 MHz, CDCl$_3$): δ 67.51 (C-4), 69.43 (C-3), 69.52 (C-5), 69.56 (-O-CH$_2$-CH = CH$_2$), 70.04 (C-2), 96.94 (C-1), 118.87 (-O-CH$_2$-CH = CH$_2$), 128.37–129.97 (aromatic carbons), 165.29, 165.52, 165.57, 171.39; HRMS C$_{30}$H$_{25}$O$_{10}$ [M-H]$^-$ calcd. 545.1453 found 545.1447. The NMR data was consistent with literature.[11]

p-T$_{OLYL}$ 2,3-$_{DI}$-*O*-$_{BENZYL}$-1-$_{THIO}$-α-D-$_{MANNOPYRANOSYLURONIC ACID}$ (9)

Obtained from compound **5** (0.045 mmol) following the general oxidation procedure as syrup, yield 60%. [α]$_D^{20}$ –63.0 (*c* = 1, CHCl$_3$); ^{1}H NMR (600 MHz, CDCl$_3$) δ 2.34 (s, 3 H, C*H*$_3$PhS-), 3.76 (dd, 1 H, J = 9.2, 2.4 Hz, H-3), 3.96 (t, 1 H, J = 2.4 Hz, H-2), 4.35 (t, 1 H, J = 9.2 Hz, H-4), 4.59–4.71 (m, 5 H, H-5 and PhC*H*$_2$O-), 5.55 (d, 1 H, J = 2.4 Hz, H-1), 7.12–7.14 (m, 2 H, aromatic), 7.30–7.37 (m, 12 H, aromatic); ^{13}C NMR (150 MHz, CDCl$_3$) δ 21.07, 68.40 (C-4), 71.33 (C-5), 72.36 (PhCH$_2$O-), 72.58 (PhCH$_2$O-), 75.65 (C-2), 78.13 (C-3), 86.45 (C-1), 127.84, 127.87, 127.89, 127.98, 128.38, 128.45, 129.21, 130.00, 132.29, 137.44, 137.67, 138.25, 172.49; HRMS

$C_{27}H_{28}NaO_6S$ [M+Na]$^+$ calc. 503.1504 found 503.1507. Anal. calcd. for $C_{27}H_{28}O_6S$: C, 67.48; H, 5.87; Found: C 67.10; H, 6.25.

ACKNOWLEDGMENTS

We are grateful for financial support from the National Institutes of Health (R01-GM-72667).

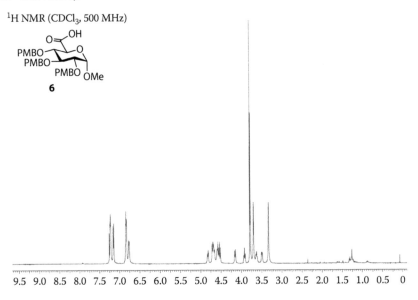

^{1}H NMR (CDCl$_3$, 500 MHz)

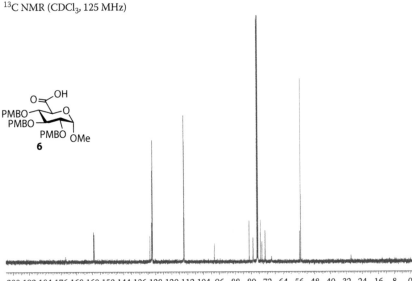

^{13}C NMR (CDCl$_3$, 125 MHz)

Chemical Shift (ppm)

^{1}H NMR (CDCl$_3$, 600 MHz)

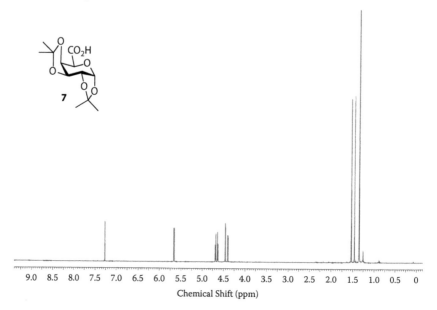

Chemical Shift (ppm)

^{13}C NMR (CDCl$_3$, 150 MHz)

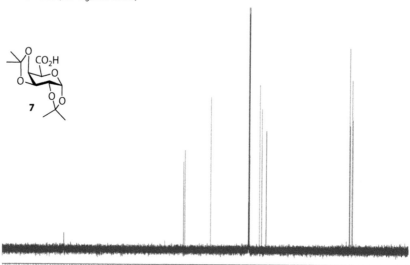

Chemical Shift (ppm)

¹H NMR (CDCl₃, 500 MHz)

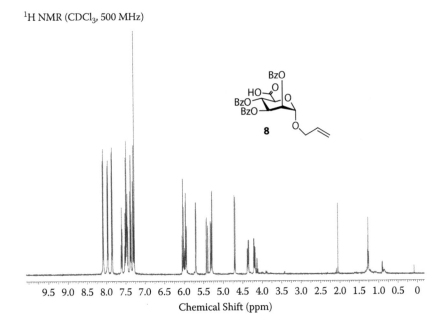

¹³C NMR (CDCl₃, 125 MHz)

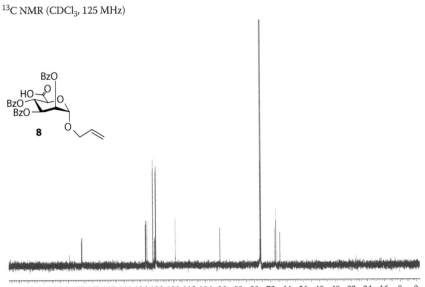

^{1}H NMR (CDCl$_3$, 600 MHz)

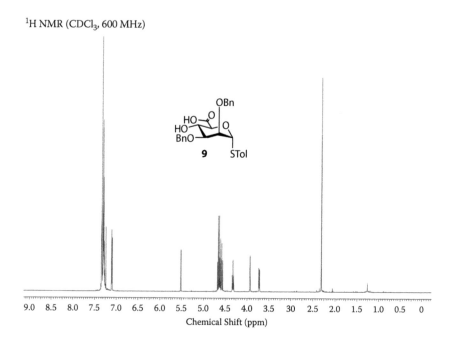

^{13}C NMR (CDCl$_3$, 150 MHz)

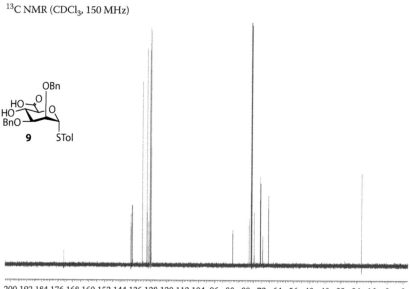

REFERENCES

1. van den Bos, L. J.; Codée, J. D. C.; Litjens, R. E. J. N.; Dinkelaar, J.; Overkleeft, H. S.; van der Marel, G. A. *Eur. J. Org. Chem.* 2007, 3963–3976.
2. Stachulski, A. V.; Jenkins, G. N. *Natural Product Rep.* 1998, *15*, 173–186.
3. Raman, R.; Sasisekharan, V.; Sasisekharan, R. *Chem. Biol.* 2005, *12*, 267–277.
4. Gama, C. I.; Hsieh-Wilson, L. C. *Curr. Opin. Chem. Biol.* 2005, *9*, 609–619.
5. Linhardt, R. J.; Toida, T. *Acc. Chem. Res.* 2004, *37*, 431–438.
6. Wang, Z.; Xu, Y.; Yang, B.; Tiruchinapally, G.; Sun, B.; Liu, R.; Dulaney, S.; Liu, J.; Huang, X. *Chem. Eur. J.* 2010, *10*, 8365–8375.
7. Baleux, F.; Loureiro-Morais, L.; Hersant, Y.; Clayette, P.; Arenzana-Seisdedos, F.; Bonnaffé, D.; Lortat-Jacob, H. *Nat. Chem. Biol.* 2009, *5*, 743–748.
8. Noti, C.; Seeberger, P. H. In *Chemistry and Biology of Heparin and Heparan Sulfate*; Garg, H. G., Linhardt, R. J., Hales, C. A., Eds.; Elsevier: Oxford, 2005, 79–142.
9. Petitou, M.; van Boeckel, C. A. A. *Angew. Chem. Int. Ed.* 2004, *43*, 3118–3133.
10. Larock, R. C. In *Comprehensive Organic Transformations*; Wiley-VCH: New York, 1999, 1646–1650.
11. Huang, L.; Teumelsan, N.; Huang, X. *Chem. Eur. J.* 2006, *12*, 5246–5252.
12. Huang, L.; Huang, X. *Chem. Eur. J.* 2007, *13*, 529–540.
13. Vermeer, H. J.; Halkes, K. M.; van Kuik, J. A.; Kamerling, J. P.; Vliegenthart, J. F. G. *J. Chem. Soc., Perkin Trans.* 2000, *1*, 2249–2263.
14. Kramer, S.; Nolting, B.; Ott, A.-J.; Vogel, C. *J. Carbohydr. Chem.* 2000, *19*, 891–921.
15. La Ferla, B.; Lay, L.; Guerrini, M.; Poletti, L.; Panza, L.; Russo, G. *Tetrahedron* 1999, *55*, 9867–9880.
16. Halkes, K. M.; Slaghek, T. M.; Hypponen, T. K.; Kruiskamp, P. H.; Ogawa, T.; Kamerling, J. P.; Vliegenthart, J. F. G. *Carbohydr. Res.* 1998, *309*, 161–174.
17. Litjens, R. E. J. N.; Heeten, R. D.; Timmer, M. S. M.; Overkleeft, H. S.; van der Marel, G. A. *Chem. Eur. J.* 2005, *11*, 1010–1016.
18. Chauvin, A.-L.; Nepogodiev, S. A.; Field, R. A. *J. Org. Chem.* 2005, *70*, 960–966.
19. Lee, J.-C.; Lu, X.-A.; Kulkarni, S. S.; Wen, Y.-S.; Hung, S.-C. *J. Am. Chem. Soc.* 2004, *126*, 476–477.
20. Haller, M.; Boons, G. J. *J. Chem. Soc., Perkin Trans.* 2001, *1*, 814–822.
21. Yeung, B. K. S.; Chong, P. Y. C.; Petillo, P. A. In *Glycochemistry. Principles, Synthesis, and Applications*; Wang, P. G., Bertozzi, C. R., Eds.; Marcel Dekker: New York, 2001, 425–492.
22. Baisch, G.; Ohrlein, R. *Carbohydr. Res.* 1998, *312*, 61–72.
23. de Nooy, A. E. J.; Besemer, A. C.; van Bekkum, H. *Synthesis* 1996, 1153–1174.
24. Li, K.; Helm, R. F. *Carbohydr. Res.* 1995, *273*, 249–253.
25. Davis, N. J.; Flitsch, S. L. *Tetrahedron Lett.* 1993, *34*, 1181–1184.
26. van den Bos, L. J.; Codee, J. D. C.; van der Toorn, J. C.; Boltje, T. J.; van Boom, J. H.; Overkleeft, H. S.; van der Marel, G. A. *Org. Lett.* 2004, *6*, 2165–2168.
27. Walvoort, M. T. C.; Sail, D.; van der Marel, G. A.; Codée, J. D. C. In *Carbohydrate Chemistry: Proven Synthetic Methods*, Paul Kovac, Ed., CRC Press: Boca Raton, FL, p. 67. 2011; Vol. 1.
28. Zhao, M.; Li, J.; Mano, E.; Song, Z.; Tschaen, D. M.; Grabowski, E. J. J.; Reider, P. J. *J. Org. Chem.* 1999, *64*, 2564–2566.
29. Clausen, M. H.; Madsen, R. *Chem. Eur. J.* 2003, *9*, 3821–3832.
30. 2-Methyl-2-butene was added as a chlorine scavenger to prevent potential side reactions and *t*-butanol presumably kept the reaction medium more homogeneous. See B. S. Bal, W. E. Childers and H. W. Pinnick, *Tetrahedron* 1981, *37*, 2091–2096.

31. Gilewski, T.; Ragupathi, G.; Bhuta, S.; Williams, L. J.; Musselli, C.; Zhang, X. F.; Bencsath, K. P.; Panageas, K. S.; Chin, J.; Hudis, C. A.; Norton, L.; Houghton, A. N.; Livingston, P. O.; Danishefsky, S. J. *Proc. Natl. Acad. Sci. U.S.A.* 2001, *98*, 3270–3275.
32. Allanson, N. M.; Liu, D.; Chi, F.; Jain, R. K.; Chen, A.; Ghosh, M.; Hong, L.; Sofia, M. J. *Tetrahedron Lett.* 1998, *39*, 1889–1892.
33. Rajput, V. K.; Mukhopadhyay, B. *J. Org. Chem.* 2008, *73*, 6924–6927.
34. Guchhait, G.; Misra, A. K. *Tetrahedron Asymmetry* 2009, *20*, 1791–1797.
35. Scanlan, E. M.; Mackeen, M. M.; Wormald, M. R.; Davis, B. G. *J. Am. Chem. Soc.* 2010, *132*, 7238–7239.
36. Wu, X.; Cui, L.; Lipinski, T.; Bundle, D. R. *Chem. Eur. J.* 2010, *16*, 3476–3488.
37. Lei, P. S.; Ogawa, Y.; Kováč, P. *J. Carbohydr. Chem.* 1996, *15*, 485–500.
38. HOAc is necessary for the purification. Otherwise, a significant amount of tetrabutylammonium salt of the product was obtained, which gave very broad NMR spectra.
39. Kloth, K.; Brünjes, M.; Kunst, E.; Jöge, T.; Gallier, F.; Adibekian, A.; Kirschning, A. *Adv. Syn. Catal.* 2005, *347*, 1423–1434.
40. Sell, H. M.; Link, K. P. *J. Am. Chem. Soc.* 1938, *60*, 1813–1814.
41. Kováč, P. *J. Carbohydr., Nucleosides, Nucleotides*, 1974, *1*, 183–186.

4 Ultrasonic Energy Promoted Allylation to Generate 1-C-(2,3,4,6-Tetra-O-Benzyl-α-D-Glucopyranosyl) prop-2-ene

Mark Farrell
Ciaran O'Reilly
Dilip V. Jarikote
*Paul V. Murphy**
Jani Rahkila[†]

CONTENTS

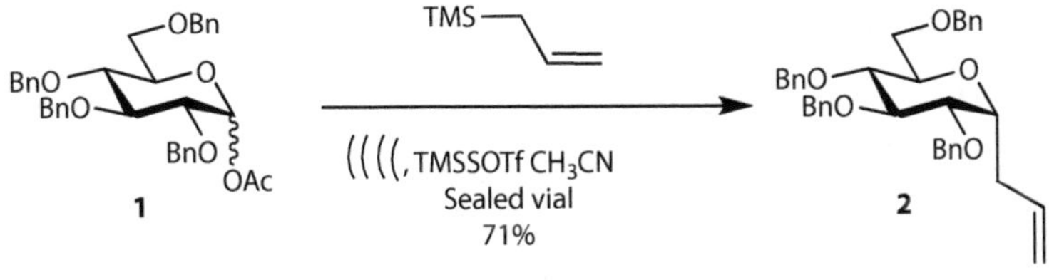

* Corresponding author: paul.v.murphy@nuigalway.ie.
[†] Checker under supervision of R. Leino: leino@abo.fi.

In recent years C-glycoside formation has been of interest.[1] It has been established that C-glycoside analogues of naturally occurring O-glycosides can often display interesting differences in their reactivity and biological activity.[2] The synthesis of many different C-glycosyl compounds commences from a glycosyl prop-2-ene.[3]

The use of ultrasonic energy[4] has led to improvements in yield of reactions which include protecting group manipulations,[6] copper-catalyzed azide-alkyne cycloaddition reactions,[7] acyl migrations,[6] glycosidation reactions,[5] and Suzuki/Heck–type reactions.[8] Reports of the C-allylation (or allenylation) of various protected glycosides are abundant in the literature. However many of the methods require prolonged reaction times and/or high temperatures. We recently investigated the effects of ultrasonic energy on the formation of glycosyl prop-2-enes with the purpose of shortening the reaction times and increasing the efficiency of preparation of these derivatives. As an example we provide detailed conditions for the allylation of 1-O-acetyl-2,3,4,6-tetra-O-benzyl-D-glucopyranoside **1**, which was carried out in the presence of TMSOTf and allyltrimethylsilane. The conversion of **1** to **2** was completed within 15 min when the reaction was enhanced by ultrasonic radiation in a sealed vial; the conversion of **1** to **2** was incomplete after 2 h using conventional heating. This showed that ultrasonic radiation significantly enhanced the rate of the reaction. The reaction has been successfully conducted in a sealed vial on both 100 mg and 1 g scales.

EXPERIMENTAL

GENERAL METHOD

The reaction vessel (sealed vial) was similar to those used in microwave-promoted reactions.[9] Chemical shifts are reported relative to internal Me_4Si in $CDCl_3$ (δ 0.0) for ^1H and (δ 77.0) for ^{13}C. The ^1H NMR (nuclear magnetic resonance) signals were assigned with the aid of Correlation Spectroscopy (COSY). Thin-layer chromatography (TLC) was performed on aluminum sheets precoated with silica gel, and spots were visualized by ultraviolet (UV) light and charring with H_2SO_4-EtOH (1:20), or cerium molybdate. Flash chromatography was carried out with silica gel 60 (0.040–0.630 mm). Acetonitrile, petroleum ether, and dichloromethane were used as purchased from Sigma-Aldrich.

PREPARATION OF 1-C-(2,3,4,6-TETRA-O-BENZYL-α-D-GLUCOPYRANOSYL)PROP-2-ENE (2)

Compound **1** (100 mg, 0.2 mmol) was placed in a Biotage microwave vial and sealed. The vial was then placed under a nitrogen atmosphere (using a water aspirator and a balloon of nitrogen). Dry acetonitrile (500 µL) and allyl trimethylsilane (113 µL, 0.71 mmol) were then added, followed by the dropwise addition of trimethylsilyl triflate (18 µL, 0.09 mmol) at 0°C. The reaction mixture was placed in an ultrasonic bath and sonicated for 15–20 min (monitoring by TLC), after which the reaction was quenched with saturated aqueous $NaHCO_3$, and the mixture was diluted with dichloromethane. The phases were separated and the aqueous layer was extracted with additional dichloromethane. The combined organic extracts were washed with brine,* dried (MgSO$_4$), and filtered and the solvent was removed under reduced pressure. Compound **2** (68 mg, 71% white solid) was obtained after chromatography of the residue (7:1 petroleum ether-EtOAc, R$_f$ 0.37; m.p. 59–61°C (lit[10] 59–60°C); ^{1}H NMR (600.13 MHz, CDCl$_3$, 25°C): δ 7.00–7.50 (m, 20 H, arom. H), 5.81 (dddd, 1 H, $J_{2',1'a}$ = 6.2 Hz, $J_{2',1'b}$ = 7.5 Hz, $J_{2',3'Trans}$ = 17.1 Hz, $J_{2',3'Cis}$ = 10.2 Hz, H-2'), 5.10 (dddd, 1 H, $J_{3'Trans,\,1'a}$ = –2.0 Hz, $J_{3'Trans,1'b}$ = 1.9 Hz, $J_{3'Trans,3'Cis}$ = 1.7 Hz, H-3'Trans), 5.06 (dddd, 1 H $J_{3'Cis,\,1'a}$ = 1.6 Hz, $J_{3'Cis,1'b}$ = 1.4, H-3'Cis), 4.93 and 4.81 (each d, each 1 H, J = 10.9 Hz, 3-CH_2Ph), 4.81 and 4.47 (each d, each 1 H, J = 10.7 Hz, 4-CH_2Ph), 4.68 and 4.62 (each d, each 1 H, J = –11.7 Hz, 2-CH_2Ph), 4.61 and 4.46 (each d, each 1 H, J = –12.1 Hz, 6-CH_2Ph), 4.13 (ddd, 1 H, $J_{1,2}$ = 5.8 Hz, $J_{1,1'a}$ = 3.7 Hz, $J_{1,1'b}$ = 11.5 Hz, H-1), 3.80 (dd, 1 H, $J_{3,2}$ = 9.5 Hz, $J_{3,4}$ = 8.8 Hz, H-3), 3.76 (dd, 1 H, H-2), 3.70 (dd, 1 H, $J_{6b,5}$ = 3.7 Hz, $J_{6b,6a}$ = 10.6 Hz, H-6b), 3.64 (dd, $J_{4,5}$ = 9.9 Hz, H-4), 3.62 (dd, 1 H, $J_{6a,5}$ = 2.1 Hz, H-6a), 3.61 (ddd, 1 H, H-5), 2.50 (ddddd, 1 H, $J_{1'b,1'a}$ = 15.2 Hz, H-1'b), 2.47 (ddddd, 1 H, H-1'a). ^{13}C NMR (150.9 MHz, CDCl$_3$): δ 138.9–138.2 and 127.7–128.5 (arom. C), 134.8 (C-2'), 117.0 (C-3'), 82.5 (C-3), 80.2 (C-2), 78.2 (C-4), 75.6 (3-CH_2Ph), 75.1 (4-CH_2Ph), 73.8 (C-1), 73.6 (6-CH_2Ph), 73.2 (2-CH_2Ph), 71.2 (C-5), 69.0 (C-6), 29.1 (C-1'). ESIMS calcd. for C$_{37}$H$_{41}$O$_5$ 565.7 (M + H)$^+$, found: 565.3; calcd. for C$_{37}$H$_{40}$O$_5$Na 587.7 (M + Na)$^+$, found: 587.3; IR ν$_{max}$/cm^{-1} : 3030, 2913, 2865, 1497, 1454, 1362, 1080, 697; [α]$_D^{20}$ = +38.9° (c 1, CH$_2$Cl$_2$). Calcd. for C$_{37}$H$_{40}$O$_5$: C, 78.69; H, 7.14; found: C, 78.75; H, 7.17.

ACKNOWLEDGMENTS

The authors thank Science Foundation Ireland (RFP/06/CHO32, PI/IN1/B966) and the European Commission (Marie Curie EIF Grant No. 220948 to DVJ) for funding.

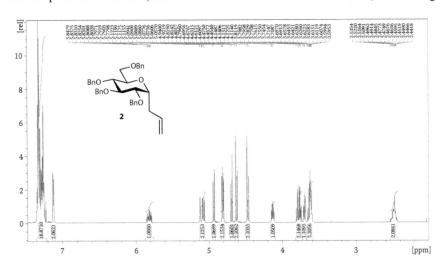

---

* *Caution:* Although we did not experience any problems with this procedure, sonication for longer than 10 min may lead to an increase in temperature inside the vial and thus an increase in pressure; hence, the reaction should be allowed to cool every 10 min.

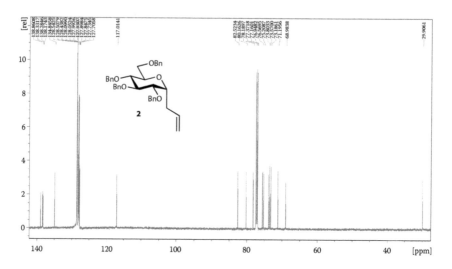

REFERENCES

1. (a) Terauchi, M.; Abe, H.; Matsuda, A.; Shuto, S. *Org. Lett.* 2004, *6*, 3751–3754; (b) Vieira, A. S.; Fiorante, P. F.; Hough, T. L. S.; Ferreira, F. P.; Lüdtke, D. S.; Stefani, H.A. *Org. Lett.* 2008, *10*, 5215–5218; (c) Cox, J. M.; Rainier, J. D. *Org. Lett.* 2001, *3*, 2919–2922; (d) Fletcher, S.; Jorgensen, M. R.; Miller, A. D. *Org. Lett.* 2004, *6*, 4245–4248.
2. (a) Kolympadi, M.; Fontanella, M.; Venturi, C.; André, S.; Gabius, H.-J.; Jiménez-Barbero, J.; Vogel, P. *Chem-Eur. J.* 2009, *15*, 2861–2873; (b) Chen, G.; Chien, M.; Tsuji, M.; Franck, R. W. *ChemBioChem* 2006, *7*, 1017–1022.
3. (a) Zou, W.; Shao, H.; Wu, S.-H. *Carbohydr. Res.* 2004, *339*, 2475–2485; (b) Xie, J. *Eur. J. Org. Chem.* 2002, *2002*, 3411–3418; (c) Girard, C.; Miramon, M.-L.; de Solminihac, T.; Herscovici, J. *Carbohydr. Res.* 2002, *337*, 1769–1774; (d) Kobertz, W. R.; Bertozzi, C. R.; Bednarski, M. D. *Tetrahedron Lett.* 1992, *33*, 737–740; (e) Brenna, E.; Fuganti, C.; Grasselli, P.; Serra, S.; Zambotti, S. *Chem-Eur. J.* 2002, *8*, 1872–1878.
4. (a) Lepore, S. D.; He, Y. *J. Org. Chem.* 2003, *68*, 8261–8263; (b) Jarikote, D. V.; Deshmukh, R. R.; Rajagopal, R.; Lahoti, R. J.; Daniel, T.; Srinivasan, K. V. *Ultrason. Sonochem.* 2003, *10*, 45–48; (c) Chen, M.-Y.; Lu, K.-C.; Lee, A. S.-Y.; Lin, C.-C. *Tetrahedron Lett.* 2002, *43*, 2777–2780; (d) Adewuyi, Y. G. *Ind. Eng. Chem. Res.* 2001, *40*, 4681–4715; (e) Kardos, N.; Luche, J. L. *Carbohydr. Res.* 2001, *332*, 115–131; (f) Gholap, A. R.; Venkatesan, K.; Daniel, T.; Lahoti, R. J.; Srinivasan, K. V. *Green Chem.* 2003, *5*, 693–696.
5. Deng, S.; Gangadharmath, U.; Chang, C.-W. T. *J. Org. Chem.* 2006, *71*, 5179–5185.
6. Zhang, J.; Chen, H.-N.; Chiang, F.-I.; Takemoto, J. Y.; Bensaci, M.; Chang, C.-W. T. *J. Comb. Chem.* 2007, *9*, 17–19.
7. (a) Rajagopal, R.; Jarikote, D. V.; Srinivasan, K. V. *Chem. Commun.* 2002, 616–617; (b) Deshmukh, R. R.; Rajagopal, R.; Srinivasan, K. V. *Chem. Commun.* 2001, 1544–1545.
8. (a) Jarikote, D. V.; Murphy, P. V. *Eur. J. Org. Chem.* (Microreview), 2010, 4929–4970; (b) Andre, S.; Velasco-Torrijos, T.; Leyden, R.; Gouin, S.; Tosin, M.; Murphy, P. V.; Gabius, H.-J. *Org. Biomol. Chem.* 2009, *7*, 4715–4725; (c) Leyden, R.; Velasco-Torrijos, T.; Andre, S.; Gouin, S.; Gabius, H.-J.; Murphy, P. V. *J. Org. Chem.* 2009, *74*,

9010–9026; (d) Pilgrim, W.; Murphy, P. V. *Org. Lett.* 2009, *11*, 939–942; (e) Murphy, P. V. *Eur. J. Org. Chem.* 2007, 4177–4187; (f) O'Brien, C.; Polakova, M.; Pitt, N.; Tosin, M.; Murphy, P. V. *Chem-Eur. J.* 2007, *13*, 902–909.

9. Jarikote, D. V.; O'Reilly, C.; Murphy, P. V. *Tetrahedron Lett.* 2010, *51*, 6776–6778.
10. McGarvey, G. J.; LeClair, C. A.; Schmidtmann, B. A. *Org. Lett.* 2008, *10,* 4728–4730.

5 Synthesis of a Multivalent Glycocyclopeptide using Oxime Ligation

Isabelle Bossu
Baptiste Thomas
René Roy*
Pascal Dumy
Olivier Renaudet†

CONTENTS

SCHEME 5.1 Synthesis of a tetravalent cluster of αGalNAc (**3**) by oxime condensation between N-acetylgalactopyranosylhydroxylamine (**2**) and aldehyde-containing cyclopeptide (**1**). (Ala, alanine; Gly, glycine; Lys, lysine; Pro, proline.)

The crucial roles of multivalent carbohydrate-protein interactions in biology have led to the development of a large panel of synthetic glycoclusters with the aim of exploring or inhibiting pathological events.[1] Among glycopolymers,[2] glycocalix-arenes,[3] glycocyclodextrins,[4] glycopeptide dendrimers,[5] poly(amidoamine) glycoden-drimers,[6] glycofullerenes,[7] or glyconanoparticles,[8] important biological properties of

* Checker: roy.rene@uqam.ca.
† Corresponding author: olivier.renaudet@ujf-grenoble.fr.

glycocyclopeptides have been discovered over the last decade, such as lectin binding or inhibition[9] or their utility as synthetic vaccines.[10] Such glycoclusters are prepared from conformationally restricted cyclopeptide scaffolds containing two β-turn inducers (Gly-Pro) and up to six orthogonally protected Lys residues, thus defining either one or two separate addressable domains.[11] The covalent conjugation of carbohydrates onto the upper domain of the scaffold is classically performed using an oxime ligation strategy.[12] After incorporating serines to the Lys side chains, treatment with sodium periodate provides aldehyde groups that react with high chemoselectivity with unprotected hydroxylamine-containing carbohydrates.[13] Here we demonstrate the efficiency of the oxime ligation through the formation of a tetravalent glycocluster displaying cancer-related Tn antigen analogues. The conjugation reaction between the glycosyl-hydroxylamine **2**[10a, 13a, 13b] and the cyclopeptide **1**[9a] is performed under mild aqueous acidic conditions. The corresponding glycocluster **3** is readily obtained in excellent yield (87%) after semi-preparative high-performance liquid chromatography (HPLC) and can be used for biological assays without further chemical transformation.

EXPERIMENTAL

GENERAL METHODS

All chemical reagents were purchased from Aldrich (Saint Quentin Fallavier, France) or Acros (Noisy-Le-Grand, France). The cyclic peptide (**3**) was prepared following the procedure described in reference 9a, using protected amino acids, Fmoc-Gly-Sasrin resin, and PyBOP from Advanced ChemTech Europe (Brussels, Belgium), Bachem Biochimie SARL (Voisins-Les-Bretonneux, France), and France Biochem S.A. (Meudon, France). The reaction progress was monitored by reverse-phase HPLC on Waters equipment at 1.3 mL/min (Macherey-Nagel (Hoerdt, France) Nucleosil 300 Å 5μm C_{18} particles, 125 × 3 mm) with ultraviolet (UV) monitoring at 214 nm and 250 nm using a linear A-B gradient (solvent A: water containing 0.09% trifluoroactic acid (TFA); solvent B: acetonitrile containing 0.09% TFA and 9.91% H_2O). Preparative separation was operated at 22 mL/min (Delta-Pak™ 100 Å 15 μm C_{18} particles, 200 × 25 mm) with UV monitoring at 214 nm and 250 nm using a linear A-B gradient. R_t indicates the peak retention time. Mass spectra were acquired using electrospray ionization in positive mode on an Esquire 3000+ Bruker Daltonics. ^{1}H and ^{13}C NMR (nuclear magnetic resonance) spectra were recorded in D_2O at 400 MHz with a Bruker Avance 400. Chemical shifts (δ) were reported in parts per million (ppm). Spectra were referenced to the residual proton solvent peaks relative to the signal of D_2O (δ 4.79 ppm for ^{1}H NMR). Proton assignment was done using gradient COrrelation SpectroscopY (gCOSY) experiment.

C[LYS(HC = N-O-αGALNAC)-ALA-LYS(HC = N-O-αGALNAC)-PRO-GLY-LYS(HC = N-O-αGALNAC)-ALA-LYS(HC = N-O-αGALNAC)-PRO-GLY] (3)

Hydrated aldehyde-containing cyclopeptide[9a] (**1**, 5.0 mg, 3.97 μmol)* and glycopyranosylhydroxylamine[10a, 13a, 13b] (**2**, 7.5 mg, 31.8 μmol) were dissolved in 0.1% TFA in

---

* ^{1}H NMR indicated that the aldehyde functions in **1** were hydrated; therefore, one water molecule per
  aldehyde functionality was included in the molecular mass of aldehyde-containing cyclopeptide **1**.

H_2O (400 μL, 10 mM). After stirring for 1 h at room temperature, analytical HPLC indicated complete conversion of **1** into **3** (R_t = 8.66 min, linear A-B gradient from 5 to 40% of B over 20 min). Without any workup, the crude mixture was directly purified by preparative HPLC (R_t = 10.7 min, linear A-B gradient from 5 to 40% of B over 30 min) to give, after freeze-drying **3** (7.1 mg, 3.45 μmol, 87%) as a flocculent powder. ^1H NMR (D_2O) δ 7.73 (s, 2 H, H_{ox}), 7.71 (s, 2 H, H_{ox}), 5.56–5.52 (m, 4 H, 4 × H-1), 4.66–4.59 (m, 2 H, 2 × $H_{\alpha Lys}$), 4.40–4.24 (m, 10 H, 2 × $H_{\alpha Ala}$, 2 × $H_{\alpha Lys}$, 2 × $H_{\alpha Pro}$, 4 × H-2), 4.11–3.88 (m, 14 H, 4 × $H_{\alpha Gly}$, 1 × $CH_{2\delta Pro}$, 4 × H-4, 4 × H-3), 3.88–3.43 (m, 14 H, 1 × $CH_{2\delta Pro}$, 4 × H-6, 8 × H-6), 3.32–3.12 (m, 8 H, 4 × $CH_{2\epsilon Lys}$), 2.33–2.21 (m, 2 H, 2 × $H_{\beta Pro}$), 2.10–1.22 (m, 48 H, 4 × CH_{3NHAc}, 4 × $CH_{2\beta Lys}$, 4 × $CH_{2\delta Lys}$, 2 × $H_{\beta Pro}$, 2 × $CH_{2\gamma Pro}$, 4 × $CH_{2\gamma Lys}$, 2 × CH_{3Ala}); ^{13}C NMR (D_2O) δ 174.9, 174.6, 174.6, 174.1, 173.2, 171.6, 171.46, 162.6, 146.5, 117.8, 114.9, 100.4, 72.0, 68.4, 67.5, 61.1, 52.9, 51.0, 49.7, 48.9, 48.1, 42.7, 39.2, 39.1, 30.9, 29.2, 28.1, 28.1, 27.7, 24.8, 22.6, 22.2, 22.0, 15.9; ESI-MS: $[M + H]^+$ calcd. for $C_{84}H_{135}N_{22}O_{38}$, 2061.1; found, 2061.1.

ACKNOWLEDGMENTS

This work was supported by the Université Joseph Fourier and the Centre National de la Recherche Scientifique (CNRS). We are grateful to the NanoBio program for access to the facilities of the Synthesis platform. The checker is thankful to M. Tze Chieh Shiao for the HPLC, NMR, and mass spectrometry (MS) analyses.

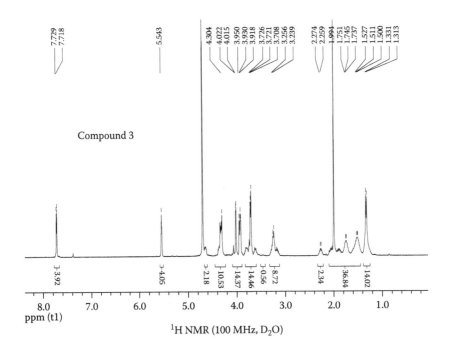

Compound 3

^1H NMR (100 MHz, D_2O)

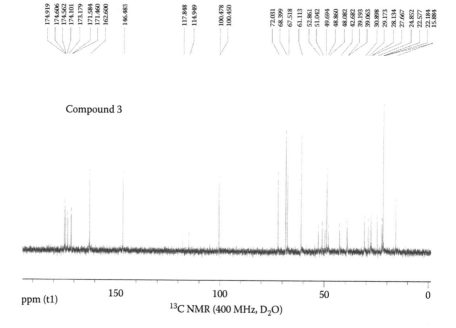

Compound 3

ppm (t1)

^{13}C NMR (400 MHz, D$_2$O)

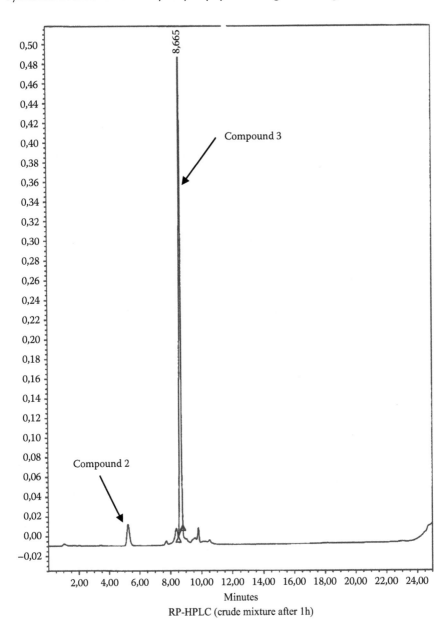

RP-HPLC (crude mixture after 1h)

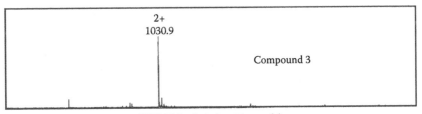

ESI-MS Analysis (positive mode)

REFERENCES

1. (a) Chabre, Y. M.; Roy, R. *Adv. Carbohydr. Chem. Biochem.* 2010, *63*, 165–393. (b) Doores, K. J.; Gamblin, D. O.; Davis, B. G. *Chem. Eur. J.*, 2006, *12*, 656–665. (c) Renaudet, O.; Spinelli, N. (Eds), *Synthesis and Biological Applications of Glycoconjugates*, Bentham Science, Oak Park, IL, 2011.
2. For recent reviews on glycopolymers, see: (a) Ladmiral, V.; Melia, E.; Haddleton, D. M. *Eur. Polym. J.* 2004, *40*, 431–449. (b) Voit, B.; Appelhans, D. *Macromol. Chem. Phys.* 2010, *211*, 727–735.
3. For recent reviews on glycocalixarenes, see: (a) Baldini, L.; Casnati, A.; Sansone, F.; Ungaro, R. *Chem. Soc. Rev.* 2007, *36*, 254–266. (b) Dondoni, A.; Marra, A. *Chem. Rev.* 2010, *110*, 4949–4977.
4. For a review on glycocyclodextrines, see: Fulton, D. A.; Stoddart, J. F. *Bioconjugate Chem.* 2001, *12*, 655–672.
5. For recent reviews on glycopeptide dendrimers, see: (a) Niederhafner, P.; Šebestík, J.; Ježek, J. *J. Pept. Sci.* 2008, *14*, 2–43. (b) Niederhafner, P.; Šebestík, J.; Ježek, J. *J. Pept. Sci.* 2008, *14*, 44–65.
6. For reviews on PAMAM glycodendrimers, see: (a) Roy, R. *Trends Glycosci. Glycotechnol.* 2003, *15*, 291–310. (b) Cloninger, M. J. *Curr. Opin. Chem. Biol.* 2002, *6*, 742–748.
7. For selected references on glycofullerenes, see: (a) Nierengarten, J.-F.; Iehl, J.; Oerthel, V.; Holler, M.; Illescas, B. M.; Munoz, A.; Martin, N.; Rojo, J.; Sanchez-Navarro, M.; Cecioni, S.; Vidal, S.; Buffet, K.; Durka, M.; Vincent, S. P. *Chem. Commun.* 2010, *46*, 3860–3862. (b) Kato, H.; Yashiro, A.; Mizuno, A.; Nishida, Y.; Kobayashi, K.; Shinohara, H. *Bioorg. Med. Chem. Lett.* 2001, *11*, 2935–2939. (c) Kato, H.; Kaneta, N.; Nii, S.; Kobayashi, K.; Fukui, N.; Shinohara, H.; Nishida, Y. *Chem. Biodiversity* 2005, *2*, 1232–1241.
8. For a selected review on glyconanoparticles, see: Marradi, M.; Martín-Lomas, M.; Penadés, S. *Adv. Carbohydr. Chem. Biochem.* 2010, *64*, 211–290.
9. (a) Renaudet, O.; Dumy, P. *Org. Lett.* 2003, *5*, 243–245. (b) Dubois, M.-P.; Gondran, C.; Renaudet, O.; Dumy, P; Driguez, H.; Fort, S.; Cosnier, S. *Chem. Commun.* 2005, *34*, 4318–4320. (c) Duléry, V.; Renaudet, O.; Wilczewski, M.; Van der Heyden, A.; Labbé, P. Dumy, P. *J. Comb. Chem.* 2008, *10*, 368–371. (d) Dendane, N.; Hoang, A.; Renaudet, O.; Vinet, F.; Dumy, P.; Defrancq, E. *Lab Chip* 2008, *8*, 2161–2163. (e) Wilczewski, M.; Van der Heyden, A.; Renaudet, O.; Dumy, P.; Coche-Guérente, L.; Labbé, P. *Org. Biomol. Chem.* 2008, *6*, 1114–1122. (f) Pujol, A.; Cuillel, M.; Renaudet, O.; Lebrun, C.; Charbonnier, P.; Cassio, D.; Gateau, C.; Dumy, P.; Mintz, E.; Delangle, P. *J. Am. Chem. Soc.* 2011, *133*, 286–296. (g) Bossu, I.; Šulc, M.; Křenek, K.; Dufour, E.; Garcia, J.; Berthet, N.; Dumy, P.; Křen, V.; Renaudet, O. *Org. Biomol. Chem.*, 2011, *9*, 1948–1959. (h) André, S.; Renaudet, O.; Bossu, I.; Dumy, P.; Gabius, H.-J. *J. Pept. Sci.* 2011, *17*, 427–437.
10. (a) Grigalevicius, S.; Chierici, S.; Renaudet, O.; Lo-Man, R.; Dériaud, E.; Leclerc, C.; Dumy, P. *Bioconjugate Chem.* 2005, *5*, 1149–1159. (b) Renaudet, O.; Křenek, K.; Bossu, I.; Dumy, P.; Kádek, A.; Adámek, D.; Vaněk, O.; Kavan, D.; Gažák, R.; Šulc, M.; Bezouška, K.; Křen, V. *J. Am. Chem. Soc.* 2010, *132*, 6800–6808. (c) Renaudet, O.; BenMohamed, L.; Dasgupta, G.; Bettahi, I.; Dumy, P. *ChemMedChem* 2008, *3*, 737–741. (d) Bettahi, I.; Dasgupta, G.; Renaudet, O.; Chentoufi, A. A.; Zhang, X.; Carpenter, D.; Yoon, S.; Dumy, P.; BenMohamed, L. *Cancer Immunol. Immunother.* 2009, *58*, 187–200. (e) Renaudet, O.; Dasgupta, G.; Bettahi, I.; Shi, A.; Nesburn, A. B.; Dumy, P.; BenMohamed, L. *PLoS One* 2010, *5*, e11216.

11. (a) Renaudet, O. *Mini-Rev. Org. Chem.* 2008, *5*, 274–286. (b) Boturyn, D.; Defrancq, E.; Dolphin, G. T.; Garcia, J.; Labbé, P.; Renaudet, O.; Dumy, P. *J. Pept. Sci.* 2008, *14*, 224–240. (c) Renaudet, O.; Garcia, J.; Boturyn, D.; Spinelli, N.; Defrancq, E.; Labbé, P. Dumy, P. *Int. J. Nanotechnol.* 2010, *7*, 738–752.

12. (a) Hang, H.; Bertozzi, C. R. *Acc. Chem. Res.* 2001, *34*, 727–736. (b) Peri, F.; Nicotra, F. *Chem. Comm.* 2004, 623–627. (c) Langenhan, J. M.; Thorson, J. S. *Curr. Org. Synth.* 2005, *2*, 59–81.

13. (a) Shiao, T. C.; Papadopoulos, A.; Renaudet, O.; Roy, R. *Carbohydrate Chemistry: Proven Synthetic Methods*, Vol. 1. P. Kovac, Ed., CRC Press/Taylor & Francis Group, 2012, Chap. 34, pp. 289–293. (b) Bossu, I.; Richichi, B.; Dumy, P.; Renaudet, O. *Carbohydrate Chemistry: Proven Synthetic Methods*, Vol. 1. P. Kovac, Ed., CRC Press/ Taylor & Francis Group, 2012,Chap. 42, pp. 377–386.

6 Reductive Amination Methodology for Synthesis of Primary Amines from Unprotected Synthons

Emma M. Dangerfield
*Zhimeng Wu**
Mattie S. M. Timmer[†]
Bridget L. Stocker[†]

CONTENTS

SCHEME 6.1 One-pot Vasella-reductive amination.

Amines are an important class of organic compounds and are found in many natural products, pharmaceuticals, and other valuable materials including agrochemicals and dyes.[1,2] There are many synthetic transformations that can be used to prepare amines,[1] but methodology for the synthesis of primary amines is more limited. Notable methodologies include reaction of molecular hydrogen with imines in the presence of a suitable catalyst,[3,4] and reductive aminations whereby a metal hydride

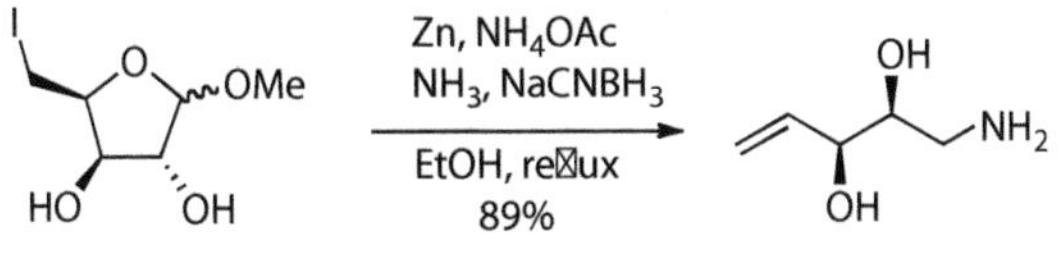

---

* Checker under supervision of P. Tyler: peter.tyler@callaghaninnovation.govt.nz
† Corresponding authors: Mattie.Timmer@vuw.ac.nz; Bridget.Stocker@vuw.ac.nz

reducing agent is employed.[5] These methodologies are useful, but some functional groups may be incompatible with the use of hydrogen gas. The use of metal hydrides suffers from lack of selectivity when the primary amine is the ultimate target, unless a nitrogen protecting group is used and then removed at a later stage of the synthesis.[6] The reason for lack of selectivity is the enhanced reactivity of the initially generated primary amine toward the starting aldehyde (or ketone), thus resulting in the formation of a secondary amine product. Attempts have been made to circumvent this over-alkylation via the use of ammonia and titanium(IV) isopropoxide-$NaHB_4$ complex. However, this methodology failed when aldehydes were used as starting materials.[7]

Given the importance of primary amines as synthetic products and intermediates, we became interested in developing new methodology for the preparation of amines from aldehydes without the need for protecting groups. Such a protecting-group-free approach would not only reduce the number of synthetic steps in a sequence but would also result in a decreased amount of waste. Key to our methodology is the use of excess NH_4OAc, at pH ~12, which can be provided by 20 mL of sat. NH_4OAc in ethanol per mmol of aldehyde.[8] Using these conditions, selective preparation of a primary amine can be achieved by reductive amination, both when performed in one step starting from an aldehyde or through two-step one-pot reductive amination according to Vasella using methyl iodo-glycosides as the starting material.[8–11]

Herein, we highlight the potential of our easy-to-perform reductive amination methodology by presenting a representative Vasella-reductive amination procedure for the conversion of methyl iodo-glycoside **1**, readily prepared in two steps from d-ribose,[12] into the corresponding alkenylamine **2**, a valuable synthetic intermediate with illustrated application in the synthesis of azasugars.[9,13] Primary alkenylamines have wide application as synthetic intermediates[14–17] and as substrates for ring-closing metathesis.[18–20] Briefly, a suspension of methyl iodo-glycoside **1**, Zn, $NaCNBH_3$, NH_4OAc (excess), and NH_3 in EtOH is refluxed overnight. The salts are removed by Dowex-H^+ ion exchange chromatography to afford the desired material. Should further purification be required, this can be achieved by way of silica gel chromatography (DCM→DCM/EtOH/MeOH/30% aq NH_3, 5/2/2/1). Using both purification protocols, alkenylamine **2** can be obtained in >90% yield. Perhaps the most challenging aspect of the synthesis is working with a "deprotected" substrate. Overloading the Dowex-H^+ column or underloading the silica gel column often results in low yields.

EXPERIMENTAL

GENERAL METHODS

This reaction was performed under atmospheric air. H_2O, and MeOH (Pure Science, Mana, Porirura) were distilled prior to use. EtOH (absolute, Pure Science, Mana, Porirura), DCM (LabServ, Auckland, New Zealand), 30% aqueous NH_3 (J. T. Baker Chemical Co, Deventer, The Netherlands), $NaCNBH_3$ (Sigma-Aldrich, Milwaukee, WI), and NH_4OAc (Global Science, Auckland, New Zealand) were used as received. Zn dust was activated by careful* addition of conc. H_2SO_4, followed by washing the product with EtOH (3×) and hexanes (3×), and storing it under dry hexanes. Solvents

---

* The addition of H_2SO_4 to Zn dust is exothermic. Adequate ice cooling should be applied, and the reaction should be carried out in a well-ventilated hood.

were removed by evaporation under reduced pressure. The reaction was monitored by thin-layer chromatography (TLC) on Macherey-Nagel silica gel-coated aluminum sheets (0.20 mm, silica gel 60) with detection by ultraviolet (UV) light (254 nm), by charring with 20% H_2SO_4 in EtOH, by dipping in I_2 in silica, or by spraying with a solution of ninhydrin in EtOH followed by heating at ~150°C. Column chromatography was performed on silica gel (Pure Science, 40–63 micron). Dowex® W50-X8 H^+ resin (Sigma-Aldrich, Milwaukee, WI) was used for ion exchange chromatography. High-resolution mass spectra were recorded on a Waters Q-TOF Premier™ Tandem Mass Spectrometer using positive electrospray ionization. Specific optical rotations were recorded using a Perkin-Elmer 241 polarimeter. Infrared spectra were recorded as thin films using a Bruker Tensor 27 Fourier transform infrared (FTIR) spectrometer, equipped with an attenuated total reflectance (ATR) sampling accessory, and are reported in wave numbers (cm^{-1}). Nuclear magnetic resonance spectra were recorded at 20°C in D_2O using a Varian Unity-INOVA operating at 300 and 75 MHz for ^1H and ^{13}C, respectively. Chemical shifts are given in ppm (δ) relative to tetramethylsilane. Nuclear magnetic resonance (NMR) peak assignments were made using COSY, HSQC, and HMBC experiments.

(2S,3S)-1-Amino-pent-4-ene-2,3-diol hydrochloride (2)

Activated Zn (120 mg, 1.83 mmol), NaCNBH$_3$ (69 mg, 1.10 mmol) and 30% aqueous NH$_3$ (3 mL) were added to a solution of iodo pyranoside **1** (100 mg, 0.365 mmol) in EtOH (7.5 mL) containing 3.75 g NH$_4$OAc. The mixture was stirred at reflux for 18 h, cooled to room temperature, filtered, and the filtrate was concentrated under reduced pressure. The residue was dissolved in H_2O (~3 × 2 mL), loaded onto a column of Dowex H^+ ion exchange resin (~25 g dry weight, freshly recycled),* and eluted. The eluent was reloaded onto the same column (× 2),† before washing the column with 100 mL H_2O to remove excess salt. The amine product was then eluted with 15% (50 mL) and 30% (100 mL) aqueous NH$_3$. The eluate was concentrated under reduced pressure and further purified using gradient flash chromatography (DCM/EtOH/MeOH/30% aqueous NH$_3$, 25/2/2/1→5/2/2/1, v/v/v/v). For characterization purposes, the isolated product was converted to the HCl salt.‡ 1m HCl (5 mL) was added to the alkenylamine product and the mixture was concentrated in vacuo (excess HCl is removed in the evaporation process) to give hydrochloride **2** as an amorphous white solid. Yield: 50 mg (89%). R_f = 0.37 (DCM/EtOH/MeOH/30% aqueous NH$_3$, 5/2/2/1, v/v/v/v); $[\alpha]_D^{20}$ = −40.0 (c = 0.07, EtOH); IR (film) 3405, 3212, 2888, 1555, 1434, 1016 cm^{-1}. ^1H NMR (300 MHz, D$_2$O) δ 5.71 (ddd, $J_{3,4}$ = 6.4 Hz, $J_{4,5\text{-}cis}$ = 10.5 Hz, $J_{4,5\text{-}trans}$ = 17.1 Hz, 1H, H-4), 5.34 (d, $J_{4,5\text{-}trans}$ = 17.1 Hz, 1H, H-5-*trans*), 5.28 (d, $J_{4,5\text{-}cis}$ = 10.5 Hz, 1H, H-5-*cis*), 4.10 (dd, $J_{2,3}$ = 5.0 Hz, $J_{3,4}$ = 6.4 Hz, 1H, H-3), 4.05 (s, 1H, NH), 3.81 (ddd, $J_{1a,2}$ = 3.3 Hz, $J_{2,3}$ = 5.0 Hz, $J_{1b,2}$ = 9.7 Hz, 1H, H-2), 3.15 (dd, $J_{1a,2}$ = 3.3 Hz, $J_{1a,1b}$ = 13.2 Hz, 1H, H-1a), 2.98

---

* If this reaction is scaled up to ~1.5 mmol, the amount of Dowex used should also be increased (~80 g dry weight).

† The eluent is loaded onto the same column to ensure complete retention of the product on the column material.

‡ The hydrochloride salt and free amine products differ significantly in their ^1H and ^{13}C NMR characteristics; by converting the amine into the hydrochloride, consistent data can be obtained.

(dd, $J_{1b,2}$ = 9.7 Hz, $J_{1a,1b}$ = 13.2 Hz, 1H, H-1b); ^{13}C NMR (75 MHz, D$_2$O) δ 135.8 (C4), 117.9 (C5), 73.9 (C3), 72.0 (C2), 42.0 (C1); HRMS(ESI) *m/z* calcd. for [C$_5$H$_{11}$O$_2$N+H]$^+$: 118.0863, obsd.: 118.0871.

ACKNOWLEDGMENTS

The authors thank the Wellington Medical Research Fund (BLS, MSMT), New Zealand Lotteries Health Research Project Grant (BLS, MSMT), and the Tertiary Education Commission, NZ (EMD, Top Achiever Doctoral Scholarship).

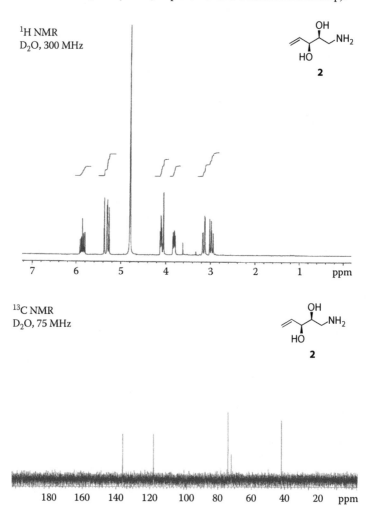

REFERENCES

1. Lawrence, S. A. *Amines: Synthesis, Properties and Applications*; Cambridge University Press, Cambridge, 2004.
2. Salvatore, R. N.; Yoon, C. H.; Jung, K. W. *Tetrahedron* 2001, *57*, 7785–7811.
3. Bódis, J.; Lefferts, L.; Müller, T. E.; Pestman, R.; Lercher, J. A. *Catal. Lett.* 2005, *104*, 23–28.
4. (a) Gomez, S.; Peters, J. A.; van der Waal, J. C.; van der Brink, P. J.; Maschmeyer, T. *Appl. Catal., A* 2004, *261*, 119–125; (b) Gomez, S.; Peters, J. A.; Maschmeyer, T. *Adv. Synth. Catal.* 2002, *344*, 1037–1057.
5. Baxter, E.; Reitz, A. *Organic Reactions* 2002, Vol. 59.
6. See for example: Underbakke, E. S.; Zhu, Y.; Kiessling, L. L. *Angew. Chem. Int. Ed.* 2008, *47*, 9677–9680.
7. Miriyala, B.; Bhattacharyya, S.; Williamson, J. S. *Tetrahedron* 2004, *60*, 1463–1471.
8. Dangerfield, E. M.; Plunkett, C. H.; Win-Mason, A. L.; Stocker, B. L.; Timmer, M. S. M. *J. Org. Chem.* 2010, *75*, 5470–5477.
9. Dangerfield, E. M.; Timmer, M. S. M.; Stocker, B. L. *Org. Lett.* 2009, *11*, 535–538.
10. Dangerfield, E. M.; Plunkett, C. H.; Stocker, B. L.; Timmer, M. S. M. *Molecules* 2009, *14*, 5298–5307.
11. Dangerfield, E. M.; Gulab, S. A.; Plunkett, C. H.; Timmer, M. S. M.; Stocker, B. L. *Carbohydrate Res.* 2010, *345*, 1360–1365.
12. Hansen, F. G.; Bundgaard, E.; Madsen, R. *J. Org. Chem.* 2005, *70*, 10139–10142.
13. Stocker, B. L.; Dangerfield, E. M.; Win-Mason, A.-L.; Haslett, G. W.; Timmer, M. S. M. *Eur. J. Org. Chem.* 2010, 1615–1637.
14. Kurasaki, H.; Okamoto, I.; Morita, N.; Tamura, O. *Org. Lett.* 2009, *11*, 1179–1181.
15. Paderes, M. C.; Chemler, S. R. *Org. Lett.* 2009, *11*, 1915–1981.
16. Zeng, W.; Chemler, S. R. *J. Org. Chem.* 2008, *73*, 6045–6047.
17. Fukumoto, H.; Takahashi, K.; Ishihara, J.; Hatakeyama, S. *Agnew. Chem. Int. Ed.* 2006, *45*, 2731–2734.
18. Pavlyuk, O.; Teller, H.; McMills, M. C. *Tetrahedron Lett.* 2009, *50*, 2716–2718.
19. Wang, H.; Matsuhashi, H.; Doan, B. D.; Goodman, S. N.; Ouyang, X.; Clark, W. M. *Tetrahedron* 2009, *65*, 6291–6303.
20. Verhelst, S. H. L.; Martinez, B. P.; Timmer, M. S. M.; Lodder, G.; Van der Marel, G. A.; Overkleeft, H. S.; Van Boom, J. H. *J. Org. Chem.* 2003, *68*, 9598–9603.

7 General Preparation of Imidazole-1-Sulfonate Esters

Travis Coyle
*Mark B. Richardson**
Mitchell Hattie
Keith A. Stubbs†

CONTENTS

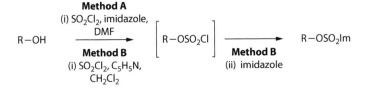

SCHEME 7.1   General method for preparation of imidazole-1-sulfonate (imidazylate) esters.

Nucleophilic substitution reactions are an important part of synthetic carbohydrate chemistry as they allow for the installation at a specifically activated position a variety of different functional groups, usually with inversion of stereochemistry. The

* Checker under the supervision of S. Williams: sjwill@unimelb.edu.au.
† Corresponding author: keith.stubbs@uwa.edu.au.

activation of alcohols in carbohydrate chemistry usually involves one of two different methods. The first involves a direct conversion from the alcohol through a transient species, usually an alkoxyphosphonium salt (e.g., Mitsunobu reaction). The second, a more general method, is through the initial formation of a sulfonate, which is an excellent leaving group. The best-known examples of these leaving groups are the methanesulfonates, p-toluenesulfonates, and trifluoromethanesulfonates. A related functional group that has only seen limited use in the carbohydrate literature is the imidazole-1-sulfonate (imidazylate), despite the group allowing preparation of a variety of carbohydrate-based molecules.[1–10] The imidazylate group has advantages over common sulfonates as they are more stable and amenable to chromatographic purification and they can be prepared from readily available and inexpensive reagents. Also, upon substitution they result in the formation of sulfate and imidazole as byproducts, which are not classified as genotoxic, in contrast to methanesulfonate, p-toluenesulfonate, and trifluoromethanesulfonate.[11]

Imidazylates are prepared through two different methods,[3,7] the first involving treatment of a partially-protected carbohydrate with sodium hydride in the presence of N,N'-sulfuryldiimidazole. Unfortunately, some common carbohydrate protecting groups are not compatible with the strong base required for this approach. The second, far milder, method involves treatment of a partially-protected carbohydrate with sulfuryl chloride in the presence of imidazole and an inert organic solvent (e.g., CH_2Cl_2, DMF) at low temperature ($-40°C$).[3] A chlorosulfate ester is formed initially[12,13] but as the solution is warmed to room temperature the chlorine atom is substituted by imidazole. A slight modification of this method initially uses a different base (e.g., pyridine), which is not imidazole, to form the intermediate chlorosulfate, and the mixture is subsequently treated with an excess of imidazole.

This protocol has been successfully applied to various protected carbohydrates (Table 7.1), demonstrating its utility.

TABLE 7.1
Examples of Various Imidazole-1-Sulfonate Esters

Ref. 1	Ref. 2	Ref. 3	Ref. 3	Ref. 4

Ref. 5	Ref. 8	Ref. 9	Ref. 10

EXPERIMENTAL

GENERAL METHODS

¹H and ¹³C nuclear magnetic resonance (NMR) spectra were obtained with a Bruker (Billerica, MA) AV600 spectrometer (600 MHz for ¹H and 150.8 MHz for ¹³C). For solutions in deuteriochloroform, CHCl₃ (¹H, δ 7.26) or CDCl₃ (¹³C, δ 77.0) were used as internal standards. Optical rotations were measured at room temperature, with a Perkin-Elmer (Waltham, MA) 141 polarimeter. Mass spectra were recorded with a VG-Autospec spectrometer using the fast atom bombardment (f.a.b.) technique, with 3-nitrobenzyl alcohol as a matrix, located at The University of Western Australia. Microanalyses were performed by the Microanalytical Unit, Australian National University, Canberra, A.C.T. Flash chromatography was performed on BDH silica gel with the specified solvents. Thin-layer chromatography (TLC) was performed on Merck silica gel 60 F_{254} aluminium-backed plates, and products were visualized with 5% sulfuric acid in EtOH. All solvents except DMF were distilled prior to use and dried according to the methods of Burfield.[14–17]

GENERAL REACTION CONDITIONS

Method A
Sulfuryl chloride (1.5 mmol) was added dropwise at −40°C to the alcohol (1 mmol) and imidazole (6 mmol) in DMF (5 mL), and the mixture was stirred for 1 h at −40°C and for 1 h at room temperature. The mixture was diluted with 1:1 CH_2Cl_2:H_2O and the organic extract washed with water, dried over $MgSO_4$, filtered, and concentrated. The residue was chromatographed to give the imidazylate.

Method B
Sulfuryl chloride (1.4 mmol) in CH_2Cl_2 (1 mL) was added dropwise at −35°C to the alcohol (1.0 mmol) and pyridine (2.5 mmol) in CH_2Cl_2 (10 mL), and the mixture was stirred for 1 h. Imidazole (5.0 mmol) was added, the mixture was allowed to warm to room temperature and stirred for 16 h. The solution was diluted (CH_2Cl_2), washed with water, dried over $MgSO_4$, filtered, and concentrated. The residue was chromatographed to give the imidazylate.

3-O-(Imidazole-1-sulfonyl)-1,2:5,6-di-O-isopropylidene-α-D-glucofuranose
Using Method A and 1,2:5,6-di-O-isopropylidene-α-D-glucofuranose with flash chromatography (EtOAc/petrol, 3:7) gave the title imidazylate as colorless needles in 81–91% yield. mp 96–98°C (Et₂O/petrol), lit.[7] mp 98–99°C (Et₂O/petrol). $[\alpha]_D$ −75.1° (CH_2Cl_2), lit.[7] $[\alpha]_D$ −76.3° (CH_2Cl_2). ¹H NMR (CDCl₃, 600 MHz) 1.20, 1.25, 1.31, 1.48 (12H, 4s, CH₃), 3.92–3.94 (1H, m, H6), 4.02–4.05 (3H, m, H4,5,6), 4.72 (d, $J_{1,2}$ = 3.6 Hz, H2), 4.86 (d, $J_{3,4}$ = 2.4 Hz, H3), 5.94 (d, H1), 7.13–7.14 (1H, m, Im), 7.35–7.31 (1H, m, Im), 7.97–7.98 (1H, m, Im). ¹³C NMR (150.9 MHz, CDCl₃) δ 24.6, 26.1, 26.5, 26.7 (4C, CH₃), 67.2 (C6), 71.6, 79.5, 82.6, 86.2 (C2,3,4,5), 105.1 (C1), 109.7, 112.9 [$C(CH_3)_2$], 118.4, 130.9, 137.1 (3C, Im). m/z 391.1163; [M + H]⁺ requires 391.1175. Anal. calcd. for $C_{15}H_{22}N_2O_8S$: C, 46.15; H, 5.68. Found: C, 46.13; H, 5.63%.

Benzyl 4-*O*-(imidazole-1-sulfonyl)-2,3-*O*-isopropylidene-α-L-xylopyranoside

Using Method B and benzyl 2,3-*O*-isopropylidene-α-L-xylopyranoside[9] with flash chromatography (EtOAc/petrol/Et₃N, 15:84:1) gave the title imidazylate as colorless needles in 70–84% yield. mp 81–82°C (Et₂O/petrol), lit.[9] mp 80–81°C (Et₂O/petrol). $[\alpha]_D$ −114° (MeCN), lit.[9] $[\alpha]_D$ −115° (MeCN). ^{1}H NMR (CDCl₃, 600 MHz) δ 1.33, 1.34 (6H, 2s, CH₃), 3.40 (dd, $J_{1,2}$ = 3.0 Hz, $J_{2,3}$ = 9.6 Hz, H2), 3.52 (dd, $J_{4,5}$ ≈ $J_{5,5}$ = 10.8 Hz, H5), 3.83 (dd, $J_{4,5}$ = 5.4 Hz, H5), 4.03 (dd, $J_{3,4}$ = 9.6 Hz, H3), 4.62, 4.74 (ABq, J = 12.0 Hz, CH_2Ph), 4.73 (ddd, H4), 5.18 (d, H1), 7.13–7.14 (1H, m, Im), 7.30–7.38 (6H, m, Im, Ph), 7.96–7.97 (1H, m, Im). ^{13}C NMR (150.9 MHz, CDCl₃) δ 26.1, 26.5 (2C, CH₃), 59.2 (C5), 70.2 (CH_2Ph), 73.1, 75.7, 81.1 (C2,3,4), 95.9 (C1), 111.5 [C(CH₃)₂], 118.0, 127.6, 128.0, 128.5, 131.0, 136.7, 137.3 (Im, Ph). *m/z* 411.1229; [M + H]⁺ requires 411.1226. Anal. calcd. for C₁₈H₂₂N₂O₇S: C, 52.67; H, 5.40. Found: C, 52.60; H, 5.42%.

ACKNOWLEDGMENTS

KAS and SJW thank the Australian Research Council for funding.

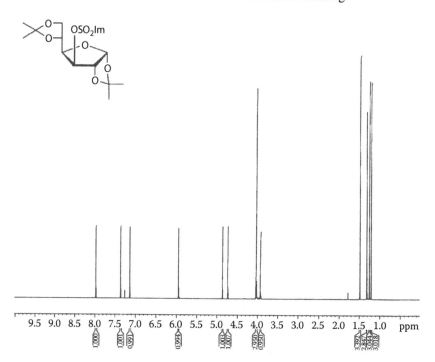

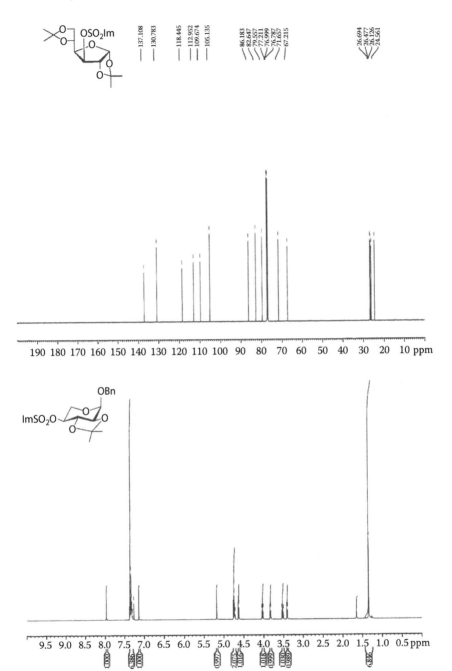

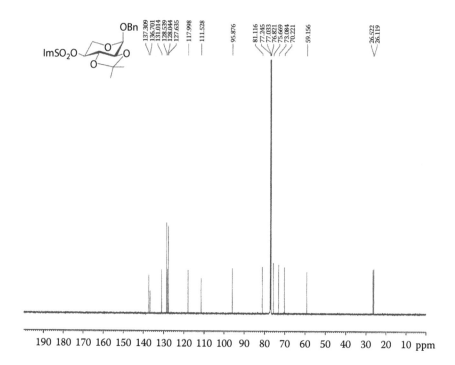

REFERENCES

1. David, S.; Malleron, A.; Cavaye, B. *New. J. Chem.* 1992, *16*, 751.
2. Baker, D. C.; Horton, D.; Tindall Jr., C. G. In *Methods in Carbohydrate Chemistry*; Whistler, R. L., BeMiller, J. N., Eds.; Academic Press: Orlando, FL, 1976; Vol. 7, p. 3.
3. Hanessian, S.; Vatèle, J. M. *Tetrahedron Lett.* 1981, *22*, 3579.
4. Tann, C. H.; Brodfuehrer, P. R.; Brundidge, S. P.; Sapino Jr., C.; Howell, H. G. *J. Org. Chem.* 1985, *50*, 3644.
5. Hashimoto, H.; Araki, K.; Saito, Y.; Kawa, M.; Yoshimura, J. *Bull. Chem. Soc. Jpn.* 1986, *59*, 3131.
6. Ahmed, F. M.; David, S.; Vatèle, J.M. *Carbohydr. Res.* 1986, *155*, 19.
7. Vatèle, J.M.; Hanessian, S. *Tetrahedron* 1996, *52*, 10557.
8. Chandrasekhar, S.; Tiwari, T.; Prakash, S. J. *Arkivoc* 2006, *11*, 155.
9. Goddard-Borger, E. D.; Stick, R. V. *Aust. J. Chem.* 2007, *60*, 211.
10. Sivets, G. G.; Kalinichenko, E. N.; Mikhailopulo, I. A.; Detorio, M. A.; McBrayer, T. R.; Whitaker, T.; Schinazi, R. F. *Nucleosides, Nucleotides & Nucleic Acids* 2009, *28*, 519.
11. Albaneze-Walker, J.; Raju, R.; Vance, J. A.; Goodman, A. J.; Reeder, M. R.; Liao, J.; Maust, M. T.; Irish, P. A.; Espino, P.; Andrews, D. R. *Org. Lett.* 2009, *11*, 1463.
12. Helferich, B. *Chem. Ber.* 1921, *56*, 1082.
13. Jennings, H. J.; Jones, J. K. N. *Can. J. Chem.* 1965, *43*, 3018.
14. Burfield, D. R.; Lee, K.-H.; Smithers, R. H. *J. Org. Chem.* 1977, *42*, 3060.
15. Burfield, D. R.; Gan, G.-H.; Smithers, R. H. *J. Appl. Chem. Biotechnol.* 1978, *28*, 23.
16. Burfield, D. R.; Smithers, R. H. *J. Org. Chem.* 1978, *43*, 3966.
17. Burfield, D. R.; Smithers, R. H.; Tan, A. S. C. *J. Org. Chem.* 1981, *46*, 629.

8 Regioselective Monoacylation and Monoalkylation of Carbohydrates Catalyzed by a Diarylborinic Ester

Doris Lee
*Christopher J. Moore**
Lina Chan
Mark S. Taylor†

CONTENTS

Selectively protected carbohydrates are key intermediates in oligosaccharide synthesis, and efficient methods for their preparation have been pursued intensively.[1] Two distinct approaches for selective monoprotection of carbohydrates involve (1) exploiting intrinsic differences in steric[2] or electronic[3] properties of hydroxyl (OH) groups and (2) the use of activating agents to enhance the nucleophilicity of specific positions. Organotin(IV) derivatives of carbohydrates have played an important role in the latter class of reactions: the reactivity of trialkylstannyl ethers and stannylene acetals toward electrophiles such as acyl and alkyl halides forms the basis of broadly applied protocols for regioselective protection.[4] In light of the lipophilicity and potential toxicity of organotin reagents, alternative activating reagents—including organoboron derivatives[5] and transition metal salts[6]—have been explored. The prospect of catalyst-controlled,

---

* Checker, under the guidance of F.-I. Auzanneau: fauzanne@uoguelph.ca.
† Corresponding author: mtaylor@chem.utoronto.ca.

regioselective manipulation of carbohydrate OH groups is of particular interest, and examples employing Lewis acids,[7,8] Lewis bases,[9] and enzymes[10] have been reported. However, catalytic methods that provide the combination of high levels of selectivity and broad substrate scope represent an important, unmet need in carbohydrate chemistry.

Diphenylborinic acid derivative **1** has recently been identified as an efficient precatalyst for selective acylation,[11] alkylation,[12] sulfonylation,[13] and glycosylation[14] of carbohydrate-derived substrates. These operationally simple methods employ a commercially available organoboron precatalyst and are characterized by high levels of selectivity for functionalization of the equatorial hydroxy groups of *cis*-1,2-diol moieties. The key intermediates in the proposed catalytic cycle are tetracoordinate borinate complexes generated by reversible covalent interactions between the organoboron catalyst and the carbohydrate substrate.[13] Representative monoprotections of galacto- and manno-configured substrates—namely, selective benzoylation of methyl β-D-arabinopyranoside (Scheme 8.1) and benzylation of methyl α-L-rhamnopyranoside (Scheme 8.2)—are presented herein. The range of pyranoside substrates and acyl or alkyl halide electrophiles that are tolerated by these catalytic protocols is illustrated in Table 8.1.[11,12]

As depicted in Scheme 8.1, treatment of commercially available methyl β-D-arabinopyranoside with borinate **1** (10 mol%), benzoyl chloride (PhCOCl, 1.5 equivalents) and Hünig's base (*i*-Pr$_2$NEt, 1.5 equivalents) in acetonitrile at ambient temperature gave monobenzoate **2** in 86% yield, after purification by flash chromatography. As judged by thin-layer chromatography (TLC) analysis, the reaction requires roughly 4 hours to reach completion at this catalyst loading. There is no evidence for the formation of regioisomeric or multiple benzoylated byproducts under these conditions: 2-benzamidoethyl benzoate (derived from the ethanolamine ligand of the precatalyst) is observed in the crude reaction mixture, but its separation from **2** by chromatography is straightforward. In the absence of catalyst, under conditions optimized for selective monoacylation of secondary hydroxy groups in carbohydrate substrates (1.2 equivalents of PhCOCl, pyridine solvent, –30°C), **2** is obtained in less than 3% yield. Acylation by the tributylstannyl ether method has been used to prepare **2** in 51% yield;[15] more recently, MoO$_2$(acac) has been used as a catalyst to prepare **2** in 78% yield.[8]

Regioselective, catalyst-controlled benzylation (Scheme 8.2) is accomplished by treatment of methyl α-L-rhamnopyranoside with borinate ester **1** (5 mol%), benzyl bromide (BnBr, 1.5 equivalents), silver(I) oxide (1.1 equivalents) as a halide abstracting agent and base, in acetonitrile solvent at 40°C. After 48 hours, conventional workup and purification by flash chromatography gave **3** in 89% yield. The 2-OBn regioisomer is the major byproduct of this reaction and was isolated in 11% yield

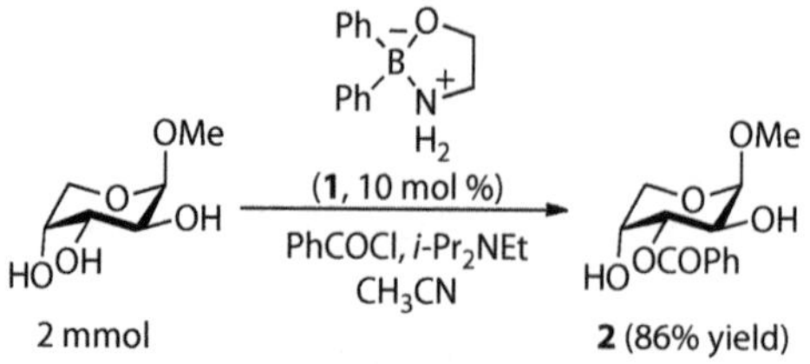

SCHEME 8.1   Catalyst-controlled monobenzoylation of methyl β-D-arabinopyranoside.

TABLE 8.1

Selective Acylations and Alkylations of Carbohydrate Derivatives Catalyzed by Diphenylborinate Ester 1

	R	% yield		R	% yield
(structure: OTBS, OH, HO, RO, OMe)	R = Bz	86	(structure: HO, OTBS, RO, HO, Si-Pr)	R = Bz	80
	R = Fmoc	69			
	R = Bn	91			
	R = 4-BrBn	99			
	R = Nap	89			
	R = BOM	84			
(structure: OH, OR, OH)	R = Bn	73	(structure: OH, RO, OH)	R = Bn	76
	R = 4-BrBn	73		R = 4-BrBn	72
	R = Nap	71		R = Nap	74
	R = BOM	63		R = BOM	66
(structure: OBn, OH, BnO, BnO, OR)	R = Bn	72	(structure: OMe, Me, HO, OR)	R = Bz	92
				R = Bn	94
				R = 4-BrBn	89
				R = Nap	86
				R = BOM	98
(structure: OMe, Me, HO, RO, OH)	R = Bn	94	(structure: HO, RO, HO, OMe)	R = Bz	91
	R = Bn	83		R = CO (CH$_2$)$_2$CH$_3$	69
	R = 4-BrBn	73		R = COCH$_2$CH(CH$_3$)$_2$	88
	R = Nap	99		R = COCH(CH$_3$)$_2$	75
	R = BOM	83		R = Cbz	69
(structure: HO, OTBS, RO, HO, OMe)	R = Bz	95		R = Fmoc	73
	R = Bn	77		R = Bn	74
	R = 4-BrBn	95		R = 4-BrBn	84
	R = Nap	82		R = Nap	71
	R = BOM	90		R = BOM	82
(structure: HO, OTBS, RO, HO, OMe)	R = Bn	74	(structure: HO, OR, RO)	R = 4-BrBn	77
	R = 4-BrBn	91			
	R = Nap	99			
	R = BOM	88			

Yields of isolated product on 0.2–1.0 mmol scale.
Abbreviations: Bz = benzoyl, Fmoc = [(9-fluorenylmethyl)oxy]carbonyl, Bn = benzyl, 4-BrBn = 4-bromobenzyl, Nap = 2-naphthylmethyl, BOM = (benzyloxy)methyl, Cbz = carbobenzyloxy.

after chromatography. The stannylene acetal method has previously been used to prepare **3** in 71%[16a] and 83% (with addition of CsF)[16b] yield.

EXPERIMENTAL

GENERAL METHODS

Acetonitrile (high-performance liquid chromatography [HPLC] grade, EMD chemicals) was purified by passage through two columns of activated alumina under argon in a solvent purification system (Innovative Technology, Inc.). Methyl β-D-arabinopyranoside was purchased from Carbosynth; methyl α-L-rhamnopyranoside was purchased from Alfa Aesar; benzoyl chloride (>99%,

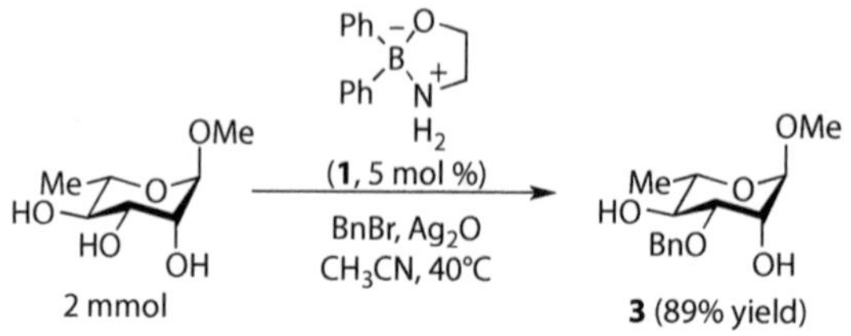

SCHEME 8.2 Catalyst-controlled monobenzylation of methyl α-L-rhamnopyranoside.

reagent grade), 2-aminoethyl diphenylborinate, diisopropylethylamine (99.5%, purified by redistillation), benzyl bromide (98%, reagent grade), and silver(I) oxide (99%, reagent grade) were purchased from Sigma-Aldrich. TLC analysis was conducted using silica gel–coated aluminum sheets (Merck, silica gel 60, F245). Compounds were visualized by ultraviolet (UV) light (245 nm) and by dipping into basic, aqueous potassium permanganate solution (6 g $KMnO_4$, 40 g K_2CO_3, and 10 mL 5% aq. NaOH in 600 mL water), followed by heating at 130°C. Flash chromatography was carried out on silica gel (P60, 40–63 μM, 60 Å, Silicycle). ^{1}H and ^{13}C NMR spectra were recorded for solutions in CD_3OD or $CDCl_3$ using a Bruker Avance III 400 MHz or Varian Mercury 400 MHz spectrometer, referenced to residual solvent peak. Spectral features are tabulated in the following order: chemical shift (δ, ppm); multiplicity (s-singlet, br s-broad singlet, d-doublet, t-triplet, q-quartet, m-complex multiplet); coupling constants (J, Hz); number of protons; assignment. Assignments are based on analysis of coupling constants, correlation spectroscopy (COSY) and heteronuclear single quantum coherence spectroscopy (HSQC). In cases of uncertain assignments, structural confirmation was secured through nuclear Overhauser effect spectroscopy (NOESY) experiments. High-resolution mass spectra (HRMS) were conducted on an AB/Sciex QStar spectrometer (ESI). Infrared (IR) spectra were obtained on a Perkin-Elmer Spectrum 100 instrument equipped with a single-reflection diamond/ZnSe ATR accessory, either in the solid state or as neat liquids, as indicated. Spectral features are tabulated as follows: wavenumber (cm^{-1}); intensity (s-strong, m-medium, w-weak, br-broad).

METHYL 3-*O*-BENZOYL-β-D-ARABINOPYRANOSIDE (2)

2-Aminoethyl diphenylborinate 1 (45 mg, 0.2 mmol) and methyl β-D-arabinopyranoside (328 mg, 2.0 mmol) were transferred to a 50 mL round-bottom flask containing a magnetic stir bar. The flask was then sealed with a septum and purged with argon. Anhydrous acetonitrile (10 mL) was added, followed by N,N-diisopropylethylamine (0.52 mL, 3.0 mmol) and benzoyl chloride (0.35 mL, 3.0 mmol). The resulting mixture was stirred at room temperature for 4 h. The mixture was transferred to a separatory funnel containing water (30 mL), brine (5 mL), and ethyl acetate (30 mL). The organic layer was separated, and the aqueous layer was extracted with ethyl acetate (2 × 30 mL). The combined organic layers were washed with brine (30 mL), dried over $MgSO_4$, filtered, and concentrated in vacuo. Chromatography (pentane/ethyl acetate, 40:60) afforded the title compound[8,15] as a white solid. Yield, 461 mg (86%), R$_f$ = 0.25 (pentane/ethyl acetate, 40:60); [α]$_D$ –181° (c 1.13, CHCl$_3$), (lit.[15] [α]$_D$ –179°). IR

(Powder, cm^{-1}): 3453 (br), 2930 (w), 1692 (m), 1452 (w), 1264 (m), 1134 (s), 1057 (s), 993 (s), 835 (m), 760 (m), 713 (s); ^{1}H NMR (400 MHz, CD$_3$OD): δ 8.12 (dd, J = 7.4 Hz and 1.2 Hz, 2H, PhH), 7.61 (td, J = 7.4 Hz and 1.2 Hz, 1H, PhH), 7.48 (dd, J = 7.4 Hz and 7.4 Hz, 2H, PhH), 5.24 (dd, J = 10.3 Hz and 3.3 Hz, 1H, H-3), 4.79 (d, J = 3.7 Hz, 1H, H-1), 4.20–4.16 (m, 2H, H-2 and H-4), 3.93 (d, J = 12.3 Hz, 1H, H-5a), 3.63 (dd, J = 12.3 Hz and 2.2 Hz, 1H, H-5b), 3.45 (s, 3H, OCH_3); ^{13}C NMR (100 MHz, CD$_3$OD): δ 167.9 (C = O), 134.2 (CH_{arom}), 131.6 (C_q Bz), 130.9 (CH_{arom}), 129.4 (CH_{arom}), 102.2 (C-1), 74.8 (C-3), 68.7 (C-2), 67.8 (C-4), 64.0 (C-5), 55.9 (OCH_3); HRMS (ESI): Calculated for C$_{13}$H$_{16}$O$_6$Na [M + Na]$^+$: 291.0845; Found: 291.0839.

Methyl 3-*O*-benzyl-α-L-rhamnopyranoside (3)

2-Aminoethyl diphenylborinate **1** (22.5 mg, 0.1 mmol), methyl α-L-rhamnopyranoside (356 mg, 2.0 mmol) and Ag$_2$O (510 mg, 2.2 mmol) were transferred to a 50-mL round-bottomed flask containing a magnetic stir bar. The flask was then sealed with a septum and purged with a stream of argon for 10 min. Anhydrous acetonitrile (20 mL) was then added to the flask, followed by benzyl bromide (0.36 mL, 3.0 mmol). The reaction mixture was heated to 40°C and stirred vigorously (500 rpm) for 48 h and then allowed to cool to room temperature. The solution was diluted with dichloromethane (20 mL), filtered through Celite® washing with dichloromethane (3 × 15 mL), and concentrated to dryness. Flash chromatographic purification (180 g silica gel; pentane/diethyl ether, 50:50 [300 mL] to 20:80 [2.5 L]) afforded the title compound[17,18] as a white solid. Yield, 478 mg (89%), R$_f$ = 0.35 (pentane/diethyl ether, 20:80); mp 79–81°C (cyclohexane), lit.[18] mp 80°C; [α]$_D$ –23.2° (*c* 0.745, CHCl$_3$), (lit.[18] [α]$_D$ –26.1°). IR (Neat, cm^{-1}): 3430 (br), 2909 (m), 1497 (w), 1453 (m), 1197 (w), 1104 (s), 1053 (s), 985 (s), 905 (m), 740 (m); ^{1}H NMR (400 MHz, CDCl$_3$): δ 7.39–7.29 (m, 5H, ArH), 4.71 (d, J = 1.6 Hz, 1H, H-1), 4.70 (d, J = 11.6 Hz, 1H, ArCH_2), 4.55 (d, J = 11.6 Hz, 1H, ArCH_2), 4.02–4.00 (m, 1H, H-2), 3.68–3.59 (m, 2H, H-3 and H-5), 3.54 (apparent td, J = 9.2, 2.4 Hz, 1H, H-4), 3.36 (s, 3H, OCH_3), 2.40 (d, J = 2.4 Hz, 1H, C$_2$-OH), 2.22 (d, J = 2.8 Hz, 1H, C$_4$-OH), 1.31 (d, J = 6.4 Hz, 3H, CHCH_3); ^{13}C NMR (100 MHz, CDCl$_3$): δ 137.7 (C_q Bn), 128.7 (CH_{arom}), 128.2 (CH_{arom}), 127.9 (CH_{arom}), 100.4 (C-1), 79.8 (C-3), 71.6 (CH_2Ar), 71.5 (C-4), 67.7 (C-2), 67.5 (C-5), 54.8 (OCH_3), 17.6 (C-6); HRMS (ESI): Calculated for C$_{14}$H$_{24}$NO$_5$ [M+NH$_4$]$^+$: 286.1654. Found: 286.1650.

Methyl 2-*O*-benzyl-α-L-rhamnopyranoside[18,19] was isolated as the major side product. Yield, 57 mg (11%), R$_f$ = 0.25 (pentane/diethyl ether, 20:80); ^{1}H NMR (400 MHz, CDCl$_3$): δ 7.38–7.28 (m, 5H, ArH), 4.74–4.70 (m, 2H, ArCH_2, H-1), 4.50 (d, J = 11.7 Hz, 1H, ArCH_2), 3.70–3.57 (m, 3H, H-2, H-3, H-5), 3.39 (t, J = 9.2 Hz, 1H, H-4), 3.34 (s, 3H, OCH_3), 1.30 (d, J = 6.1 Hz, 3H, CHCH_3); ^{13}C NMR (100 MHz, CDCl$_3$): δ 137.5 (C_q Bn), 128.6 (CH_{arom}), 128.1 (CH_{arom}), 127.9 (CH_{arom}), 97.9 (C-1), 79.8 (C-2), 74.0 (C-4), 73.0 (CH_2Ar), 71.5 (C-3), 67.5 (C-5), 54.8 (OCH_3), 17.8 (C-6).

ACKNOWLEDGMENTS

This work was supported by NSERC (Discovery Grants Program, graduate scholar-ships to D.L. and L.C.), Merck Research Laboratories (CADP Grant), the CFI, and the Province of Ontario.

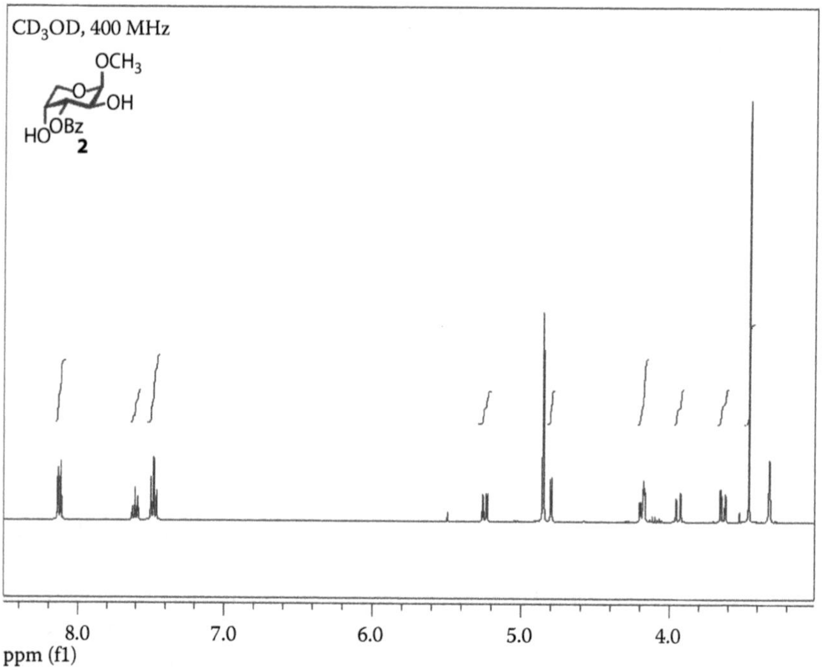

CD3OD, 400 MHz
OCH3
O
OH
HO
OBz
2
8.0
7.0
6.0
5.0
4.0
ppm (f1)

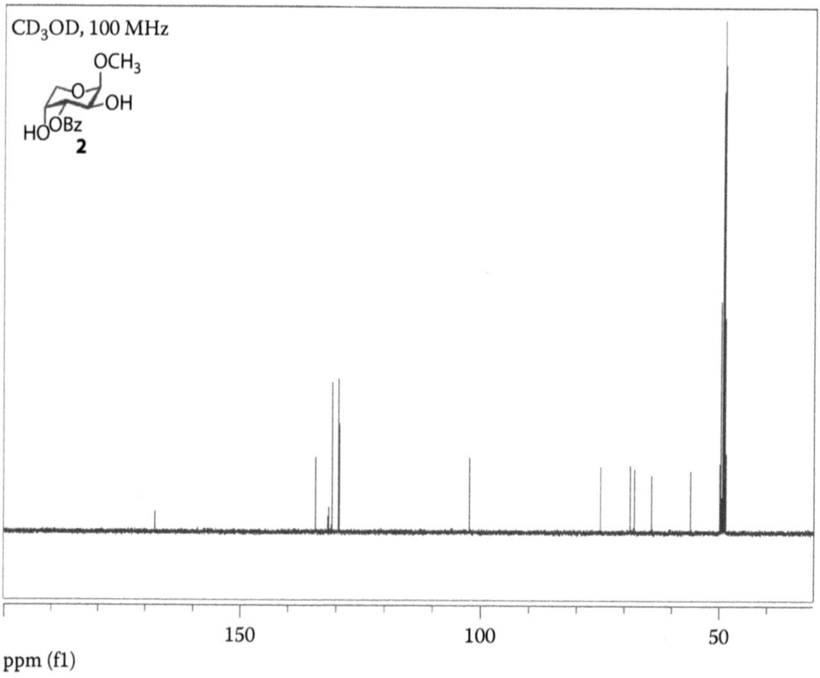

CD3OD, 100 MHz
OCH3
O
OH
HO
OBz
2
150
100
50
ppm (f1)

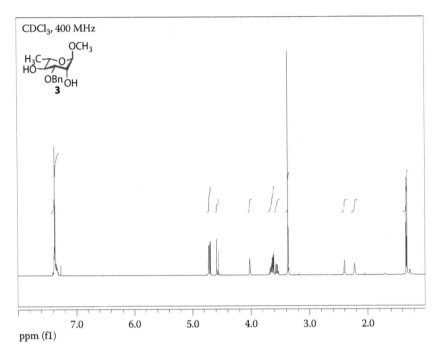

CDCl$_3$, 400 MHz

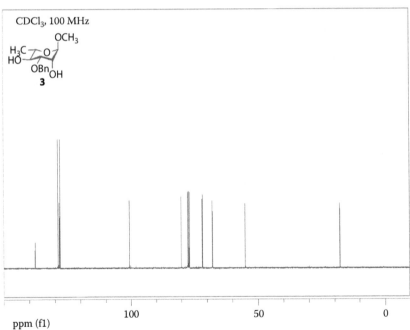

CDCl$_3$, 100 MHz

REFERENCES

1. Gómez, A. M. In *Glycoscience*; Fraser-Reid, B., Tatsuka, K., Thiem, J., Eds.; Springer-Verlag: Berlin, 2008; pp. 103–177.
2. Jiang, L.; Chan, T.-H. *J. Org. Chem.* 1998, *63*, 6035–6038.
3. Garegg, P. J. *Pure Appl. Chem.* 1984, *56*, 845–858.
4. David, S.; Hanessian, S. *Tetrahedron* 1985, *41*, 643–663; Grindley, T. B. *Adv. Carbohydr. Chem. Biochem.* 1998, *53*, 17–142.
5. Oshima, K.; Kitazono, E.-i.; Aoyama, Y. *Tetrahedron Lett.* 1997, *38*, 5001–5004.
6. Eby, R.; Webster, K. T.; Schuerch, C. *Carbohydr. Res.* 1984, *129*, 111–120; Osborn, H. M. I.; Brome, V. A.; Harwood, L. M.; Suthers, W. G. *Carbohydr. Res.* 2001, *332*, 157–166; Gangadharmath, U. B.; Demchenko, A. V. *Synlett* 2004, *12*, 2191–2193.
7. Wang, C.-C.; Lee, J.-C.; Luo, S.-Y.; Kulkami, S. S.; Huang, Y.-W.; Lee, C.-C.; Chang, K.-L.; Hung, S.-C. *Nature* 2007, *446*, 896–899; Français, A.; Urban, D.; Beau, J.-M. *Angew. Chem. Int. Ed.* 2007, *46*, 8662–8665; Demizu, Y.; Kubo, Y.; Miyoshi, H.; Maki, T.; Matsumura, Y.; Moriyama, N.; Onomura, O. *Org. Lett.* 2008, *10*, 5075–5077; Dhiman, R. S.; Kluger, R. *Org. Biomol. Chem.* 2010, *8*, 2006–2008.
8. Evtushenko, E. V. *J. Carbohydr. Chem.* 2010, *29*, 369–378.
9. Griswold, K. S.; Miller, S. J. *Tetrahedron* 2003, *59*, 8869–8875; Hu, G.; Vasella, A. *Helv. Chim. Acta* 2003, *86*, 4369–4390; Kawabata, T.; Muramatsu, W.; Nishio, T.; Shibata, T.; Schedel, H. *J. Am. Chem. Soc.* 2007, *129*, 12890–12895.
10. Therisod, M.; Klibanov, A. M. *J. Am. Chem. Soc.* 1987, *109*, 3977–3981; Wang, Y.-F.; Lalonde, J. J.; Momongan, M.; Bergbreiter, D. E.; Wong, C.-H. *J. Am. Chem. Soc.* 1988, *110*, 7200–7205; Lewis, J. C.; Bastian, S.; Bennett, C. S.; Fu, Y.; Mitsuda, Y.; Chen, M. M.; Greenberg, W. A.; Wong, C.-H.; Arnold, F. H. *Proc. Natl. Acad. Sci. U.S.A.* 2009, *106*, 16550–16555.
11. Lee, D.; Taylor, M. S. *J. Am. Chem. Soc.* 2011, *133*, 3724–3727.
12. Chan, L.; Taylor, M. S. *Org. Lett.* 2011, *13*, 3090–3093.
13. Lee, D.; Williamson, C. L.; Chan, L.; Taylor, M. S. *J. Am. Chem. Soc.* 2012, *134*, 8260–8267.
14. Gouliaras, C.; Lee, D.; Chan, L.; Taylor, M. S. *J. Am. Chem. Soc.* 2011, *133*, 13926–13929.
15. Tsuda, Y.; Haque, M. E.; Yoshimoto, K. *Chem. Pharm. Bull.* 1983, *31*, 1612–1624; Holzapfel C. W.; Koekomoer, J. M.; Marais, C. F. *S. Afr. J. Chem.* 1984, *37*, 9–26.
16. (a) Yang, G.; Kong, F.; Zhou, S. *Carbohydr. Res.* 1991, *211*, 179–182; (b) Kováč, P.; Edgar, K. J.; *J. Org. Chem.*, 1992, *57*, 2455–2467.
17. Rana, S. S.; Barlow, J. J.; Matta, K. L. *Carbohydr. Res.* 1980, *85*, 313–317; Lipták, A.; Nánási, P.; Neszélyi, A.; Wagner, H. *Tetrahedron* 1980, *36*, 1261–1268.
18. Toman, R.; Rosík, J., Zikmund, M. *Carbohydr. Res.* 1982, *103*, 165–169.
19. Crich, D.; Vinogradova, O. *J. Org. Chem.* 2007, *72*, 3581–3584.

9 The Synthesis of Carbamates from Alkenylamines

Emma M. Dangerfield
*Shivali A. Gulab**
Bridget L. Stocker[†]
Mattie S. M. Timmer[†]

CONTENTS

SCHEME 9.1 Carbamate annulation.

Carbamates are an important class of compounds and find wide utility as pharmaceuticals[1] and agrochemicals,[2] as intermediates in organic synthesis,[3] as protecting groups during peptide chemistry[4] and as linkers in combinatorial chemistry.[5] Such a versatile range of applications has resulted in many strategies being developed for the synthesis of carbamates. These include the use of phosgene and derivatives thereof,[6] reactions involving the use of carbon dioxide (including gaseous, electrochemical, and supercritical carbon dioxide),[7] reactions commencing with organic carbonates,[8] rearrangement reactions (including modifications to the Hoffmann, Curtius, and Lossen rearrangement),[9] and reactions using metal/non-metal carbonates/bicarbonates.[6]

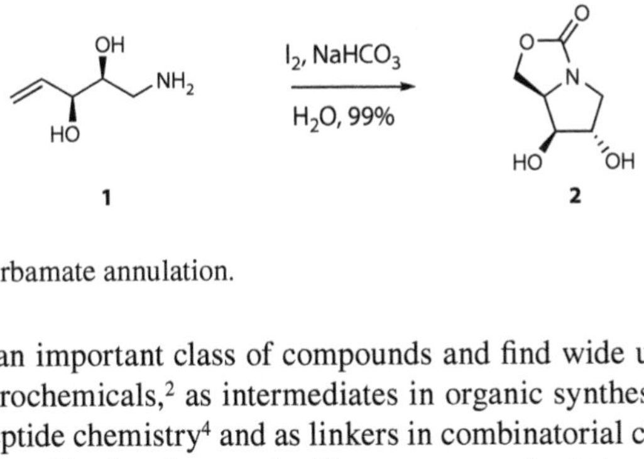

* Checker: shivali.gulab@callaghaninnovation.govt.nz
† Corresponding author: bridget.stocker@vuw.ac.nz, mattie.timmer@vuw.ac.nz.

With an interest in the development of efficient routes for the preparation of aza-sugars,[10] we investigated methodology for the synthesis of cyclic carbamates from alkenylamine precursors, such as **1** (Scheme 9.1). It has previously been reported that cyclic carbamates can be prepared, in modest yield, from the reaction of acyclic amines equipped with a halogen-leaving group and either tetraethylammonium bicarbonate[11] or sodium carbonate.[12] Accordingly, we subjected a series of alkenyl-amines (themselves readily prepared in three steps from a variety of natural sugars)[13] to I_2 and $NaHCO_3$ in H_2O and reported on the smooth transformation to the cor-responding carbamate in excellent yield and with >20:1 diastereoselectivity in favor of the 4,5-*cis* diastereomer (in the case of the five-membered pyrrolidine skeleton).[14] For the example given in Scheme 9.1, alkenylamine **1** was converted to carbamate **2** in 93% yield after stirring at rt for 18 h. We were also able to demonstrate that this annulation methodology was applicable for protected and substituted alkenyl-amine precursors,[15] and for the cyclization of alkenylamine homologues to give the piperidine carbon framework in certain examples,[16] though the diastereoselectivity of the latter is reduced (*ca.* 3:1). The carbamate annulation reaction itself is easy to perform and the resultant carbamate can be purified by silica gel chromatography, if necessary.

EXPERIMENTAL

GENERAL METHODS

This reaction was performed under atmospheric air. H_2O, MeOH (Pure Science, Mana, Porirura), and EtOAc (Pure Science, Mana, Porirura) were distilled prior to use. I_2 (Pure Science, Mana, Porirura), and $NaHCO_3$ (Pure Science, Mana, Porirura) were used as received. All solvents were removed by evaporation under reduced pressure. Reactions were monitored by TLC on Macherey-Nagel silica gel–coated aluminum sheets (0.20 mm, silica gel 60) with detection by UV-absorption (254 nm), by charring with 20% H_2SO_4 in EtOH, by dipping in I_2 in silica, or by spraying with a solution of ninhydrin in EtOH followed by heat-ing at ~150°C. Column chromatography was performed on Pure Science silica gel (40–63 micron). High-resolution mass spectra were recorded on a Waters Q-TOF Premier™ Tandem Mass Spectrometer using positive electrospray ion-ization. Specific optical rotations were recorded using a Perkin-Elmer 241 polar-imeter. Infrared spectra were recorded as thin films using a Bruker Tensor 27 Fourier transform infrared (FTIR) spectrometer, equipped with an attenuated total reflectance (ATR) sampling accessory, and are reported in wave numbers (cm^{-1}). Nuclear magnetic resonance (NMR) spectra were recorded at 20°C in D_2O using a Varian Unity-INOVA instrument operating at 300 and 75 MHz for ^1H and ^{13}C, respectively. Chemical shifts are given in ppm (δ) relative to tet-ramethylsilane. NMR peak assignments were made using COSY, HSQC, and HMBC experiments.

(6S,7S,7AR)-6,7-Dihydroxy-tetrahydro-pyrrolo[1,2-c]oxazol-3-one (2)

To a solution of the alkenylamine hydrochloride **1** (259 mg, 1.12 mmol) in water (5 mL) was added $NaHCO_3$ (141 mg, 1.68 mmol) and I_2 (313 mg, 1.23 mmol). The solution was stirred 18 h at room temperature and then filtered and concentrated under reduced pressure. The product was purified by gradient flash chromatography* (EtOAc→90:10 EtOAc/MeOH, v/v) to give carbamate **2** as an amorphous white powder. Yield, 167 mg (99%). $[\alpha]_D^{19}$ = −26.1 (c = 0.4, EtOH); IR (film) 3425, 3243, 3125, 1680, 1435, 1194, 1130, 844, 799, 717 cm⁻¹. ¹H NMR (300 MHz, D_2O) δ 4.64 (t, $J_{4,5a} = J_{5a,5b}$ = 9.3 Hz, 1H, H-5a), 4.51 (dd, $J_{4,5b}$ = 3.7 Hz, $J_{5a,5b}$ = 9.3 Hz, 1H, H-5b), 4.40 (d, $J_{1a,2}$ = 5.3 Hz, 1H, H-2), 4.31 (dt, $J_{4,5a}$ = 8.5 Hz, $J_{4,5b} = J_{3,4}$ = 3.7 Hz, 1H, H-4), 4.05 (d, $J_{3,4}$ = 3.7 Hz, 1H, H-3), 3.80 (dd, $J_{1a,2}$ = 5.3 Hz, $J_{1a,1b}$ = 12.6 Hz, 1H, H-1a), 3.13 (d, $J_{1a,1b}$ = 12.6 Hz, 1H, H-1b); ¹³C NMR (75 MHz, D_2O) δ 164.4 (C6), 76.5 (C2), 74.3 (C3), 64.0 (C5), 62.0 (C4), 52.2 (C1); HRMS(ESI) m/z calcd. for $[C_6H_9O_4N+H]^+$: 160.0604, obsd.: 160.0605.

ACKNOWLEDGMENTS

The authors would like to thank the Wellington Medical Research Fund (BLS, MSMT), New Zealand Lotteries Health Research Project Grant (BLS, MSMT), and the Tertiary Education Commission, NZ (EMD, Top Achiever Doctoral Scholarship).

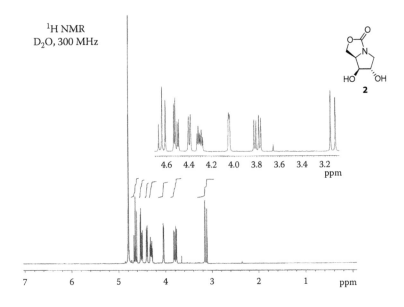

¹H NMR
D_2O, 300 MHz

---

* The carbamate product does not stain by TLC in H_2SO_4, I_2, permanganate or ninhydrin. However I⁻ does stain bright red in ninhydrin and this is often misleading. Purification of the carbamate product from the salts is best achieved using a short column (7 cm long silica on a 2 to 3 cm wide column for the described scale), beginning in EtOAc and increasing to 99/1 (EtOAc/MeOH), then 98/2 then 95/5 and a final flush with 90/10. The majority of the product should elute in 98/2.

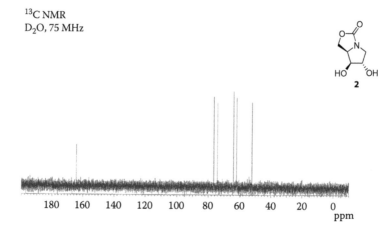

^{13}C NMR
D$_2$O, 75 MHz

REFERENCES

1. (a) Hutchinson, D. K. *Curr. Top Med. Chem.* 2003, *3*, 1021; (b) Alaxander, J. P.; Cravatt, B. F. *Chem. Biol.* 2005, *12*, 1179; (c) Tully, D. C.; Liu, H.; Chatterjee, A. K.; Alper, P. B.; Williams, J. A.; Roberts, M. J.; Mutnick, D.; Woodmansee, D. H.; Hollenbeck, T.; Gordon, P.; Chang, J.; Tutland, T.; Tumanut, C.; Li, J.; Harris, J. L.; Karenwasky, D. S. *Bioorg. Med. Chem. Lett.* 2006, *16*, 5107.

2. Ma, J.; Lu, N.; Qin, W.; Xu, R.; Wang, Y.; Chen, X. *Ecotoxicol. Environ. Safety* 2006, *63*, 268.

3. Smith, A. B.; Freez, B. S.; LaMarche, M. J.; Hirose, T.; Brouard, I.; Rucker, R. V.; Xian, M.; Sundermann, K. F.; Shaw, S. J.; Burlingame, M. A.; Horwitz, S. B.; Myles, D. C. *Org. Lett.* 2005, *7*, 311.

4. (a) Greene, T. W.; Wuts, P. G. M. *Protective Group in Organic Synthesis*, 4th Ed.; John Wiley, New York, 2007; (b) Kociensiki, P. J. *Protective Groups*, 3rd Ed., Thieme Verlag, Stuttgart, 2003.

5. (a) Buchstaller, H. P. *Tetrahedron* 1998, *54*, 3465; (b) Holte, P. T.; Thijs, L.; Zwanenburg, B. *Tetrahedron Lett.* 1998, *39*, 7404.

6. (a) Chaturvedi, D.; Mishra, N.; Mishra V. *Curr. Org. Syn.* 2007, *4*, 308; (b) Chaturvedi, D.; Ray, S. *Monatshefte für Chemie* 2006, *137*, 127.

7. (a) Chaturvedi, D.; Ray, S. *Curr. Org. Chem.* 2007, *11*, 987; (b) Yoshida, M.; Hara, N.; Okuyama, S. *Chem. Commun.* 2000, 151.

8. Aresta, M.; Berloco, C.; Quaranta, E. *Tetrahedron*, 1995, *51*, 8073.

9. (a) Moriarty, R. M.; Chany II, C. J.; Vaid, R. K.; Prakash, O.; Tuladhar, S. M. *J. Org. Chem.* 1993, *58*, 2478; (b) Richter, L. S.; Andersen, S. *Tetrahedron Lett.* 1998, *39*, 8747; (c) Vasantha, B.; Hematha, H. P.; Sureshbabu, V. V. *Synthesis* 2010, *17*, 2990.

10. Stocker, B. L.; Dangerfield, E. M.; Win-Mason, A.-L.; Haslett, G. W.; Timmer, M. S. M. *Eur. J. Org. Chem.* 2010, 1615–1637.

11. Inesi, A.; Mucciante, V.; Rossi, L. *J. Org. Chem.* 1998, *63*, 1337.

12. Hassner, A.; Burke, S. S. *Tetrahedron* 1974, *30*, 2613.

13. Dangerfield, E. M.; Plunkett, C. H.; Win-Mason, A. L.; Stocker, B. L.; Timmer, M. S. M. *J. Org. Chem.* 2010, *75*, 5470.

14. (a) Dangerfield, E. M.; Timmer, M. S. M.; Stocker, B. L. *Org. Lett.* 2009, *11*, 535; (b) Dangerfield, E. M.; Gulab, S. A.; Plunkett, C. H.; Timmer, M. S. M.; Stocker, B. L. *Carbohydrate Res.* 2010, *345*, 1360; (c) Dangerfield, E. M.; Plunkett, C. H.; Stocker, B. L.; Timmer, M. S. M. *Molecules* 2009, *14*, 5298.

10 Hydrolysis of Thioglycosides using Anhydrous NIS and TFA

Ana R. de Jong
Bas Hagen
Kevin Sheerin[*]
Jeroen D.C. Codée[†]
Gijsbert A. van der Marel[†]

CONTENTS

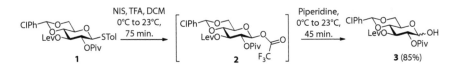

SCHEME 10.1 Hydrolysis of the anomeric thio-functionality.

[*] Checker under supervision of S. Oscarson: stefan.oscarson@ucd.ie.
[†] Corresponding authors: jcodee@chem.leidenuniv.nl; marel_g@chem.leidenuniv.

Thioglycosides are important building blocks in the synthesis of oligosaccharides and glycoconjugates.[1] Besides being used as glycosyl donors, they can be exploited as precursors for other donor types. To this end, the anomeric thio-functionality acts as a temporary masking group for the anomeric hydroxyl group, to be used in glycosylation reactions featuring dehydrative coupling conditions[2] or as an intermediate for trichloroacetimidate[1a,b] and trifluoro-(N-phenyl)acetimidate[3] donor synthesis. Many conditions for the hydrolysis of anomeric thio-functionalities have been reported. Examples of anomeric thio-hydrolyzing systems include N-iodosaccharin in wet acetonitrile,[4] N-bromosuccinimide (NBS) in wet acetone,[5] and N-iodosuccinimide (NIS) and trifluoroacetic acid (NIS/TFA) in wet dichloromethane (DCM).[6]

In our study of the automated synthesis of (1→3)-β-glucans, we required a large amount of hemiacetal **3**, which we accessed from the corresponding thioglycoside **1** (Scheme 10.1). For the hydrolysis of the anomeric thiofunction in **1**, we first explored the use of NBS in wet acetone. Although the anomeric thio-functionality was cleaved, migration of the pivaloyl group[7] to the anomeric hydroxyl prevented the isolation of the desired compound. Next we tried to prepare **3** using NIS/TFA in wet DCM, but this led to a mixture of at least three compounds. The main product in this mixture proved to be the thiocresol glucoside from which the chlorobenzylidene acetal had been cleaved.

To prevent the above described side reactions, we have explored a new method for the hydrolysis of the anomeric thio-functionality in donor **1**. The presence of the acid-labile chlorobenzylidene protective group in donor **1** directed us to a method that excludes any water. To replace the anomeric thio-functionality, we used the combination of NIS and TFA, as we described previously,[6] but the reaction was performed under anhydrous conditions to provide the intermediate anomeric trifluoroacetate **2**. Performing the reaction in deuterated methylene chloride allowed monitoring of the reaction by nuclear magnetic resonance (NMR) spectroscopy which revealed that the intermediate β-trifluoroacetyl glucose **2** was formed as a single diastereomer.[*] The cleavage of the anomeric trifluoroacetyl moiety was best accomplished by addition of three equivalents of piperidine to the reaction mixture.[†] Following this protocol, target lactol **1** was obtained in 78% yield.

EXPERIMENTAL

GENERAL METHODS

Dichloromethane was refluxed over P_2O_5, distilled, and dried over 3Å molecular sieves before use. Petroleum ether (boiling range 40–60°C) and ethyl acetate (EtOAc) were distilled prior to use. Piperidine was stored over KOH pellets. All other chemicals were used as received. Traces of water in the starting material were removed by

---

[*] NMR spectroscopy also revealed the formation of 4,6-O-(p-chlorobenzylidene)-3-O-levulinoyl-1-pivaloyl-α-D-glucopyranose, resulting from the migration of the pivaloyl function of the C-2 hydroxyl to the C-1 hydroxyl. The pivaloyl group migrates back to the C-2 OH upon treatment with piperidine in the next step of the reaction sequence.

[†] Quenching the reaction with a mixture of saturated $NaHCO_3$ and $Na_2S_2O_3$ led to substantial C-2-OPiv → C-1-OPiv migration.

coevaporation with toluene. Thin-layer chromatography (TLC) analyses were performed on silica gel 60 F_{254} plates (Merck), with irradiation with ultraviolet (UV) light (λ: 254 nm) and subsequent charring with a solution of $(NH_4)_6Mo_7O_{24} \cdot 4H_2O$ 25 g/L, $(NH)_4Ce(SO_4)_4 \cdot 2H_2O$ 10 g/L 10% H_2SO_4 in H_2O. One- and two-dimensional ^1H and ^{13}C NMR spectra were recorded for solutions in $CDCl_3$ on a Bruker AV-400 (400 MHz and 100 MHz, respectively). Chemical shifts (δ) are reported in ppm relative to TMS (δ: 0) for ^1H NMR or $CDCl_3$ (δ: 77.2 ppm) for ^{13}C NMR.

4,6-O-(P-CHLOROBENZYLIDENE)-3-O-LEVULINOYL-2-O-PIVALOYL-α,β-D-GLUCOPYRANOSE (3)

A 100 mL round-bottom flask was charged with p-tolyl 4,6-O-(p-chlorobenzylidene)-3-O-levulinyl-2-O-pivaloyl-1-thio-β-D-glucopyranoside (1, 1 mmol, 591 mg) in CH_2Cl_2 (10 mL, 0.1 M). The solution was magnetically stirred under argon, cooled to 0°C, and N-iodosuccinimide (1.1 mmol, 247 mg, 1.1 eq.) and trifluoroacetic acid (1.1 mmol, 0.08 mL, 1.1 eq.) were added. The ice bath was removed and the mixture was stirred for 75 minutes, when TLC analysis (Toluene/EtOAc, 4:1 v/v) indicated complete conversion of the starting material. The deep-red mixture was cooled in an ice-bath, and piperidine (3 mmol, 0.3 mL, 3 eq.) was added. The ice bath was removed and the mixture was warmed to room temperature and stirred for 45 minutes (TLC: PE/EtOAc 4:1 v/v). The reaction was quenched by addition of sat. aq. $Na_2S_2O_3$ (5 mL) and stirred vigorously until the mixture became colorless. The mixture was transferred to a 100 mL separatory funnel, the organic phase was diluted with 10 mL of CH_2Cl_2, the layers were separated and the organic layer was washed successively with 1M HCl (10 mL), H_2O (10 mL) and brine (10 mL). The organic phase was dried over $MgSO_4$, filtered, concentrated in vacuo and purified by column chromatography (PE/EtOAc 1:0 $\rightarrow$ 7:3) gave the title compound as a white solid (378 mg, 0.78 mmol, 78%). ^1H NMR (α anomer, 400 MHz): δ 7.39 (d, 2H, J_{vic} = 8.4 Hz, CH arom), 7.32 (d, 2H, J_{vic} = 8.4 Hz, CH arom), 5.65 (t, 1H, $J_{3,2}$ = $J_{3,4}$ = 10.0 Hz, H-3), 5.48 (s, 1H, CH benzylidene), 5.44 (t, 1H, J_{1-OH} = $J_{1,2}$ = 3.2 Hz, H-1), 5.84 (dd, 1H, $J_{2,1}$ = 3.8 Hz, $J_{2,3}$ = 9.8 Hz, H-2), 4.27 (dd, $J_{6a,6b}$ = 10.2 Hz, $J_{6a,5}$ = 4.8 Hz, H-6a), 4.14 (dt, 1H, $J_{5,4}$ = $J_{5,6b}$ = 9.8 Hz, J_{5-6a} = 4.8 Hz, H-5), 3.73 (t, 1H, J_{6b-5} = J_{6b-6a} = 10.0 Hz, H-6b), 3.65 (t, J_{4-3} = J_{4-5} = 9.8 Hz, H-4), 3.24 (d, J_{OH-1} = 3.2 Hz, OH), 2.71–2.57 (m, 4H, CH_2CH_2 Lev), 2.15 (s, 3H, CH_3), 1.20 (s, 9H, CH_3 Piv). ^{13}C NMR (100 MHz) δ 206.3 (C = O Lev ketone), 178.0 (C = O Piv), 171.7 (C = O Lev), 135.4–135.0 (C_q arom) 128.5 (CH arom), 127.8 (CH arom), 100.8 (CH benzylidene), 91.1 (C-1), 79.3 (C-4), 71.7 (C-2), 69.0 (C-3), 68.9 (C-6), 62.5 (C-5), 38.9 (C_q Piv), 37.9 (CH_2 Lev), 30.0 (CH_3 Lev), 28.0 (CH_2 Lev), 27.0 (CH_3 Piv). IR (neat): 3300, 2979, 2870, 1746, 1722, 1703, 1601, 1479, 1368, 1323, 1306, 1281, 1219, 1184, 1152, 1090, 1053, 1013, 986, 968, 943, 837, 820, 773 cm^{-1}. HRMS: calcd. $[C_{23}H_{29}ClO_9 + Na]^+$: 507.13978, found 507.13923.

ACKNOWLEDGMENTS

We thank the Netherlands Organization for Scientific Research (NWO) and Top Institute Pharma (TIP) for financial support.

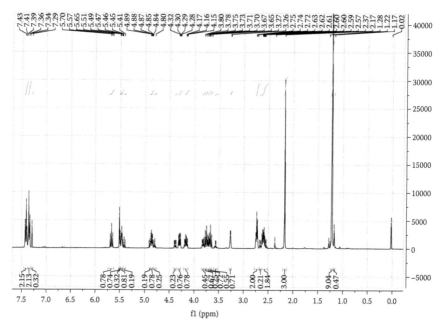

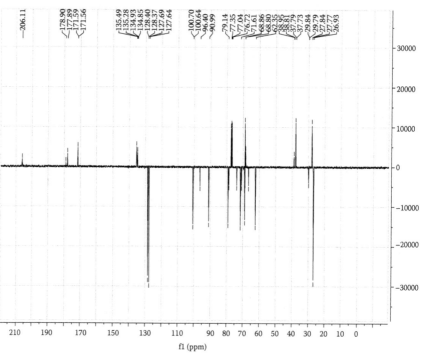

REFERENCES

1. (a) Levy, D. E., Fügedi, P., Eds., *The Organic Chemistry of Sugars*; CRC Press, Boca Raton, FL, 2006. (b) Van der Marel, G. A., Van den Bos, L. J., Overkleeft, H. S., Litjens, R. E. J. N., Codée, J. D. C. *Frontiers in Modern Carbohydrate Chemistry*, 2007, **Chapter 12.** (c) Demchenko, A. V., Ed., *Handbook of Chemical Glycosylation*; Wiley-VCH, Weinheim, 2008. (d) Garegg, P. J. *Adv. Carbohydr. Chem. Biochem.* 2004, *59*, 69. (e) Codée, J. D. C., Litjens, R. E. J. N., Van den Bos, L. J., Overkleeft, H. S., Van der Marel, G. A. *Chem. Soc. Rev.* 2005, *34*, 769.
2. Gin, D. *J. Carbohydr. Chem.* 2002, *21*, 645–665.
3. Yu, B., Tao, H. *Tetrahedron Lett.* 2001, *42*, 2405–2407. (b) Yu, B., Sun, J. *Chem. Comm.* 2010, *46*, 4668–4679.
4. Mandal, P. K., Misra, A. K., *Synlett* 2007, 1207–1210.
5. (a) Motawia, M. S., Marcussen, J., Moeller, B. L. *J. Carbohydr. Chem.* 1995, *14*, 1279–1294. (b) Oshitari, T., Shibasaki, M., Yoshizawa, T., Tomita, M., Takao, K., Kobayashi, S. *Tetrahedron* 1997, *53*, 10993–11006.
6. Dinkelaar, J., Witte, M. D., Van den Bos, L. J., Overkleeft, H. S., Van der Marel, G. A. *Carb. Res.*, 2006, *341*, 1723–1729.
7. Yu, H., Ensley, H. E. *Tetrahedron Lett.* 2003, *44*, 9363–9366.

11 Highly Diastereoselective Construction of L-Heptosides by a Sequential Grignard Addition/Fleming-Tamao Oxidation

Maxime Durka
Kevin Buffet
Tianlei Li
Abdellatif Tikad
*Bas Hagen**
Stéphane P. Vincent†

CONTENTS

* Checker under supervision of G. A. van der Marel: marel_g@chem.leidenuniv.nl.
† Corresponding author: stephane.vincent@fundp.ac.be.

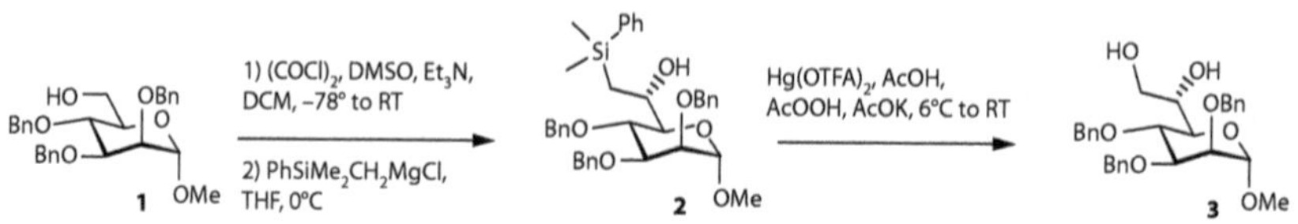

SCHEME 11.1

Lipopolysaccharide (LPS) is a key component of the outer membrane of Gram-negative bacteria.[1] This complex molecule plays key roles in the mortality of many infectious diseases as well as in the virulence of numerous human pathogens.[2] LPS consists of three main substructures: lipid A, the oligosaccharide core, and the O-antigen. The oligosaccharide core unit can itself be divided into two parts: the inner core which is formed by at least one molecule of 3-deoxy-α-D-*manno*-oct-2-ulopyranosonic acid (Kdo) and two molecules of L-*glycero*-α-D-*manno*-heptopyranose (L-heptose), and the outer core which is composed of hexoses.[3,4] The synthesis of heptosides has thus attracted much attention for the synthetic construction of bacterial oligosaccharides,[5–8] for the preparation of immunogenic glycocon-jugates[9,10] and for the development of novel antibacterial agents.[11–14]

Several synthetic procedures have been developed for the construction of L-heptoses, either through a homologation of a mannoside,[4] by a Wittig (or Horner-Wadworth-Emmons) reaction[15] or by reconstitution of the heptose skeleton through a central asymmetric aldol reaction.[16] For the synthesis of fluoro-heptoside, we also explored the stereoselective epoxide formation followed by a Cesium acetate opening.[11] The two strategies most commonly used to date are a D-selective dihy-droxylation of the appropriate alkene[17] followed, if L-heptosides are required, by a Mitsunobu inversion[5,18] and the L-selective addition of Grignard reagents.[6,7,9,19,20]

Due to its high L-diastereoselectivity, the most appealing strategy for us was the sequence developed by van Boom and collaborators[7,19,21] which consists of the addi-tion of the silylated Grignard reagent $PhMe_2SiCH_2MgCl$ followed by a Fleming-Tamao oxidation (Scheme 11.1).[22,23] The sequence began with primary alcohol **1** that was converted to the corresponding aldehyde using a standard Swern oxidation protocol. The intermediate aldehyde was used for the next step without purification. This reaction gave the L-adduct **2** in 65% yield with a L/D diastereoselectivity supe-rior to 14:1 as reported in the literature.[7]

For the Fleming-Tamao step, we developed a novel procedure to avoid side-reac-tions during the scale-up. We found that the oxidation could be performed in pres-ence of mercury trifluoroacetate,[24] to give alcohol **3** in 76% yield (based on recovered starting material).

This procedure could be performed on a multigram scale and applied to the syn-thesis of inhibitors of the two first enzymes of the bacterial heptose biosynthesis[14] and heptosyltransferase WaaC.[13]

EXPERIMENTAL

GENERAL METHODS

All chemicals were purchased from Sigma, Aldrich, Fluka, or Acros and were used without further purification. Tetrahydrofuran (THF) was freshly distilled over sodium benzophenone, dichloromethane over P_2O_5. ^{1}H, ^{13}C, and ^{31}P NMR (nuclear magnetic resonance) spectra were recorded with a Jeol ECX-400 spectrometer. All compounds were characterized by ^{1}H, ^{13}C (22°C) NMR. Spectra were interpreted with the aid of ^{1}H-^{1}H and ^{1}H-^{13}C correlation experiments. Column chromatography was performed on silica gel Kieselgel Si 60 (40–63 μm).

METHYL 2,3,4-TRI-O-BENZYL-7-(PHENYLDIMETHYL)SILANE-7-

DEOXY-L-GLYCERO-α-D-MANNO-HEPTOPYRANOSIDE (2)

DMSO (123 μL, 1.72 mmol, 1.6 eq.) was added dropwise (1 drop every ~15 seconds)[*] in 30 minutes at –78°C, under argon, to a solution of oxalyl chloride (127 μL, 1.51 mmol, 1.4 eq.), in dry CH_2Cl_2 (16 mL). After additional 30 minutes, a solution of primary alcohol 1[25] (500 mg, 1.08 mmol, 1 eq.) in dry CH_2Cl_2 (5 mL) was added dropwise in 12 minutes. After 1 h at –78°C, NEt$_3$ (450 μL, 3.23 mmol, 3 eq.) was added dropwise (over 2 minutes) and the reaction mixture was allowed to warm to room temperature. After 1 h, the solution was washed with sat. NaHCO$_3$ (10 mL), and the aqueous phases were extracted with CH_2Cl_2 (3 × 25 mL). The combined organic phases were washed with brine (2 × 75 mL), dried over MgSO$_4$, filtered, concentrated in vacuo, co-evaporated with toluene (3 × 10 mL), and dried under vacuum for 2 h.[†] The resulting crude aldehyde was used without further purification or can be stored at –18°C under argon for at least 3 days. ^{1}H-NMR (400 MHz, CDCl$_3$) δ = 9.73 (s, 1H, H-6), 7.34–7.26 (m, 15H, H^{Ar}), 4.84 (d, J = 2.7 Hz, 1H, H-1), 4.83 (AB, J = 10.7 Hz, 1H, CH$_2$Ph), 4.70 (s, 2H, CH$_2$Ph), 4.64 (AB, J = 10.7 Hz, 1H, CH$_2$Ph), 4.60 (s, 2H, CH$_2$Ph), 4.08 (d, J = 9.2 Hz, 1H, H-5), 4.04 (t, J = 8.9 Hz, 1H, H-4), 3.93 (dd, J = 3.0 Hz, J = 8.0 Hz, 1H, H-3), 3.75 (t, J = 2.8 Hz, 1H, H-2), 3.36 (s, 3H, H^{Me}).

(Phenyldimethylsilyl)methyl chloride (482 μL, 2.69 mmol, 2.5 eq.) was added dropwise under argon to a suspension of dry magnesium turnings (69.7 mg, 2.91 mmol, 2.7 eq.) in dry THF (12 mL). Activation of magnesium with 1,2-dibromoethane (approx. 0.03 mL) was required to initiate the Grignard reaction. The mixture was refluxed for 40 minutes (magnesium turning almost totally disappeared). The Grignard reagent thus obtained was cooled to room temperature and transferred under argon by cannula to a dry flask and cooled to 0°C. A solution of the aforementioned aldehyde (1.08 mmol, 1 eq.) in dry THF (5 mL) was added dropwise to the cold Grignard reagent. After stirring for 1 h at 0°C, the mixture was allowed to

---

[*] After each addition, the solution becomes cloudy and recovers its original homogeneity after several seconds; the DMSO drops must be added only after the cloudiness disappears.

[†] This aldehyde decomposes when left too long at room temperature, even under argon or under vacuum.

warm to room temperature and stirred for 15 h. The reaction was cooled to 0°C and quenched by slow addition of cold water (5 mL), filtered through a pad of Celite, the filtrate was diluted with a saturated solution of NH_4Cl (25 mL), and extracted with CH_2Cl_2 (3 × 30 mL). The organic layer was washed successively with a solution of diluted bleach (2%, 75 mL) and brine (2 × 75 mL). The organic layer was dried over $MgSO_4$, filtered, and concentrated under reduced pressure. Purification of the residue by flash chromatography (9:1 cyclohexane–EtOAc) afforded the product **2** (427 mg, 65%) as a colorless oil. $[\alpha]_D$ +13.1 [(*c* 1, $CHCl_3$,); ref.[26] +11.5 (*c* 1, $CHCl_3$,)]; (for less detailed NMR data see literature)[26] ^{1}H-NMR (400 MHz, $CDCl_3$) δ = 7.59–7.50 (m, 2H, H^{Ar}), 7.46–7.23 (m, 16H, H^{Ar}), 7.23–7.15 (m, 2H, H^{Ar}), 4.89 (d, *J* = 10.5 Hz, 1H, CH$_2$^{Bn}), 4.78 (d, *J* = 12.3 Hz, 1H, CH$_2$^{Bn}), 4.71 (d, *J* = 1.6 Hz, 1H, H-1), 4.67 (d, *J* = 12.4 Hz, 1H, CH$_2$Ph), 4.63 (s, 2H, CH$_2$Ph), 4.60 (d, *J* = 10.6 Hz, 1H, CH$_2$Ph), 4.10 (dd, *J* = 7.2 Hz, *J* = 14.0 Hz, 1H, H-6), 4.05 (t, 1H, *J* = 9.4 Hz, H-4), 3.83 (dd, *J* = 9.4 Hz, *J* = 3.0 Hz, 1H, H-3), 3.80–3.73 (m, 1H, H-2), 3.32 (d, *J* = 9.6 Hz, 1H, H-5), 3.25 (s, 3H, H^{Me}), 1.37 (dd, *J* = 14.8 Hz, *J* = 10.8 Hz, 1H, H-7a), 0.92 (dd, *J* = 14.8 Hz, *J* = 3.9 Hz, 1H, H-7b), 0.36 and 0.35 (2 s, 6H, H^{SiMe}); ^{13}C-NMR (101 MHz, $CDCl_3$): d = 139.4, 138.7, 138.48, 138.46 (4 × C$_q$^{Ar}), 133.8, 128.9, 128.5, 128.4, 128.3, 127.9, 127.8, 127.76, 127.7, 127.6 (20 × CH^{Ar}), 99.6 (C-1), 80.4, 75.6, 75.3, 74.9 (C-2, C-3, C-4, C-5), 75.5, 72.9, 72.3 (3 × CH$_2$Ph), 67.3 (C-6), 54.8 (C^{Me}), 21.9 (C-7), –1.9 (C^{SiMe}), –2.3(C^{SiMe}); MS (ESI+): *m/z*: 635.3 (100%) [M + Na]$^+$. Calcd. for $C_{37}H_{44}O_6Si$: C 72.52, H 7.24; found: C 72.49, H 7.36.

METHYL 2,3,4-TRI-*O*-BENZYL-L-*GLYCERO*-α-D-*MANNO*-HEPTOPYRANOSIDE (3)

AcOK (0.160 g, 1.63 mmol, 10 eq.) was added under argon to a solution of dry **2** (0.100 g, 0.163 mmol, 1 eq.) in 1 mL of glacial AcOH, and the solution was sonicated until all the solids dissolved (~5 minutes). Mercury trifluoroacetate (0.278 g, 0.653 mmol, 4 eq.) was added and the solution was sonicated for 5 minutes. The solution was cooled to 6°C and, after 10 minutes at 6°C, 40% peracetic acid in AcOH (135 μL, 0.816 mmol, 5 eq.) was added dropwise (1 drop every 20 seconds). The reaction was carried out under exclusion of light. The solution was stirred at 6°C for 30 minutes and for 4 hours at room temperature. After that time, in some cases, the reaction did not go to completion. The reaction could be monitored by thin-layer chromatography (TLC) (pentane/EtOAc, 3:2 v/v). The solution was diluted with CH_2Cl_2 (15 mL) and washed successively with a saturated solution of $NaHCO_3$ (10 mL) and water (10 mL). Aqueous phases were extracted with CH_2Cl_2 (3 × 10 mL). Combined organic phases were dried over $MgSO_4$ and concentrated under reduced pressure. Chromatography (cyclohexane → 6:4 cyclohexane–EtOAc) afforded the desired product **3**[23] as a colorless oil (54 mg, 66%, accompanied by 10 mg of starting material, 10%).

^{1}H-NMR (CDCl$_3$, 400 MHz): δ 7.36–7.25 (m, 15H, H$_{arom}$), 4.97 (AB, J_{AB} = 10.8 Hz, 1H, CH$_2$Ph), 4.78 (AB, J_{AB} = 12.4 Hz, 1H, CH$_2$Ph), 4.72–4.63 (m, 5H, H-1, 4CH$_2$Ph), 4.14 (t, J = 9.9 Hz, 1H, H-4), 3.97 (m, 1H, H-6), 3.87 (dd, J = 3.0 Hz, J = 9.4 Hz, 1H, H-5), 3.82–3.77 (m, 2H, H-2, H-7a), 3.71–3.65 (m, 1H, H-7b), 3.60 (dd, J = 1.4 Hz, 9.6 Hz, 1H, H-3), 3.27 (s, 3H, OCH$_3$), 2.39 (d, J = 9.9 Hz, 1H, OH), 2.16 (dd, J = 2.8 Hz, 9.6 Hz, 1H, OH). ^{13}C-NMR (CDCl$_3$, 100 MHz): δ 138.47, 138.43, 138.26 (3C$_{q\ arom}$), 128.51, 128.12, 127.93, 127.86, 127.69 (15C$_{arom}$), 99.65 (C-1), 80.15, 75.41 (CH$_2$Ph), 74.59, 74.41 (C-2, C-4), 73.08 (CH$_2$ Ph), 72.66 (C-3), 72.31 (CH$_2$ Ph), 69.46, 65.16 (C-5 and C-6), 54.97 (OCH$_3$). α$_D$ (CHCl$_3$, c = 1, 20°C): +24.5 (literature: +30.5,[26] +21,[27] and +25[28]); MS (ESI+): m/z: 517.2 (100%) [M + Na]$^+$; HR-MS (ESI+): Calcd. for C$_{29}$H$_{34}$O$_7$Na 517.2197; found: 517.2221.

ACKNOWLEDGMENTS

This work was funded by FNRS (FRFC grant 2.4.625.08.F and "Crédit au chercheur" 1.5.167.10). We are also grateful to Mutabilis S.A. for a long-standing collaboration.

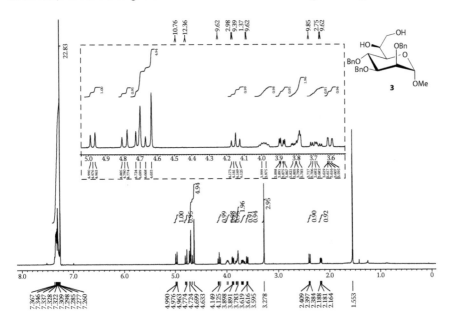

^{1}H-NMR spectra of the title compound **3**.

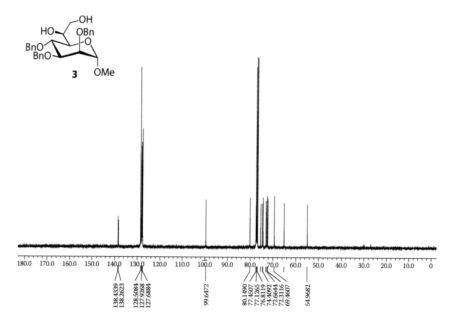

^{13}C-NMR spectra of the title compound **3**.

REFERENCES

1. Gronow, S.; Brade, H. *J. Endotoxin. Res.* 2001, *7*, 3–23; Silipo, A.; Molinaro, A.; Ierano, T.; De Soyza, A.; Sturiale, L.; Garozzo, D.; Aldridge, C.; Corris, P. A.; Khan, C. M. A.; Lanzetta, R.; Parrilli, M. *Chem. Eur. J.* 2007, *13*, 3501–3511; Komaniecka, I.; Choma, A.; Lindner, B.; Holst, O. *Chem. Eur. J.* 2010, *16*, 2922–2929.
2. Rietschel, E. T.; Kirikae, T.; Schade, F. U.; Mamat, U.; Schmidt, G.; Loppnow, H. *FASEB J.* 1994, *8*, 217–225.
3. Hansson, J.; Oscarson, S. *Curr. Org. Chem.* 2000, *4*, 535–564; Kosma, P. *Curr. Org. Chem.* 2008, *12*, 1021–1039.
4. Oscarson, S. in *Synthesis of Oligosaccharides of Bacterial Origin Containing Heptoses, Uronic Acids and Fructofuranoses as Synthetic Challenges, Vol. 186*, 1997, pp. 171–202; Oscarson, S. *Carbohydr. Polym.* 2001, *44*, 305–311.
5. Crich, D.; Banerjee, A. *J. Am. Chem. Soc.* 2006, *128*, 8078–8086.
6. Kubo, H.; Ishii, K.; Koshino, H.; Toubetto, K.; Naruchi, K.; Yamasaki, R. *Eur. J. Org. Chem.* 2004, *6*, 1202–1213.
7. Boons, G. J. P. H.; Overhand, M.; van der Marel, G. A.; van Boom, J. H. *Carbohydr. Res.* 1989, *192*, C1–C4.
8. Boons, G. J. P. H.; Overhand, M.; van der Marel, G. A.; van Boom, J. H. *Recl. Trav. Chim. Pays-Bas* 1992, *111*, 144–148.
9. Bernlind, C.; Bennett, S.; Oscarson, S. *Tetrahedron: Asymmetry* 2000, *11*, 481–492.
10. Bernlind, C.; Oscarson, S. *Carbohydr. Res.* 1997, *297*, 251–260.
11. Dohi, H.; Périon, R.; Durka, M.; Bosco, M.; Roué, Y.; Moreau, F.; Grizot, S.; Ducruix, A.; Escaich, S.; Vincent, S. P. *Chem. Eur. J.* 2008, *14*, 9530–9539.
12. (a) Zamyatina, A.; Gronow, S.; Oertelt, C.; Puchberger, M.; Brade, H.; Kosma, P. *Angew. Chem. Int. Ed.* 2000, *39*, 4150–4153. (b) Graziani, A.; Amer, H.; Zamyatina, A.; Hofinger, A.; Kosma, P. *Tetrahedron: Asymmetry* 2007, *18*, 115–122. (c) Read, J. A.; Ahmed, R. A.; Tanner, M. E. *Org. Lett.* 2005, *7*, 2457–2460.

13. Durka, M.; Buffet, K.; Iehl, J.; Holler, M.; Nierengarten, J.-F.; Vincent, S. P. *Chem. Eur. J.* 2012, *18*, 641–651.
14. Durka, M.; Tikad, A.; Périon, R.; Bosco, M.; Andaloussi, M. Floquet, S.; Malacain, E.; Moreau, F.; Oxoby, M.; Gerusz, V.; Vincent, S. P. *Chem. Eur. J.* 2011, *17*, 11305–11313.
15. (a) Brimacombe, J. S.; Kabir, A. K. M. *Carbohydr. Res.* 1986, *150*, 35–51; (b) Brimacombe, J. S.; Kabir, A. K. M. S. *Carbohydr. Res.* 1988, *179*, 21–30. (c) Güzlek, H.; Graziani, A.; Kosma, P. *Carbohydr. Res.* 2005, *340*, 2808–2811.
16. O'Hara, T.; Adibekian, A.; Esposito, D.; Stallforth, P.; Seeberger, P. H. *Chem. Commun.* 2010, *46*, 4106–4108.
17. van Straten, N. C. R.; Kriek, N. M. A. J.; Timmers, C. M.; Wigchert, S. C. M.; van der Marel, G. A.; van Boom, J. H. *J. Carbohydr. Chem.* 1997, *16*, 947–966.
18. (a) Crich, D.; Banerjee, A. *Org. Lett.* 2005, *7*, 1395–1398; (b) Dong, L.; Roosenberg, J. M.; Miller, M. J. *J. Am. Chem. Soc.* 2002, *124*, 15001–15005.
19. Boons, G. J. P. H.; van der Marel, G. A.; Poolman, J. T.; van Boom, J. H. *Recl. Trav. Chim. Pays-Bas* 1989, *108*, 339–343.
20. (a) Dasser, M.; Chretien, F.; Chapleur, Y. *J. Chem. Soc. Perkin Trans.* 1990, 3091–3094. (b) Dziewiszek, K.; Banaszek, A.; Zamojski, A. *Tetrahedron Lett.* 1987, *28*, 1569–1572. (c) Kim, M.; Grzeszczyk, B.; Zamojski, A. *Tetrahedron Lett.* 2000, *56*, 9319–9337. (d) Bernlind, C.; Oscarson, S. *J. Org. Chem.* 1998, *63*, 7780–7788. (e) Jorgensen, M.; Iversen, E. H.; Madsen, R. *J. Org. Chem.* 2001, *66*, 4625–4629.
21. Boons, G. J. P. H.; Overhand, M.; van der Marel, G. A.; van Boom, J. H. *Angew. Chem. Int. Ed.* 1989, *28*, 1504–1506.
22. (a) Fleming, I.; Sanderson, P. E. J. *Tetrahedron Lett.* 1987, *28*, 4229–4232. (b) Boons, G. J. P. H.; Elie, C. J. J.; van der Marel, G. A.; van Boom, J. H. *Tetrahedron Lett.* 1990, *31*, 2197–2200.
23. Boons, G. J. P. H.; van der Marel, G. A.; van Boom, J. H. *Tetrahedron Lett.* 1989, *30*, 229–232.
24. Ley, S. V.; Abad-Somovilla, A.; Anderson, J. C.; Ayats, C.; Banteli, R.; Beckmann, E.; Boyer, A.; Brasca, M. G.; Brice, A.; Broughton, H. B.; Burke, B. J.; Cleator, E.; Craig, D.; Denholm, A. A.; Denton, R. M.; Durand-Reville, T.; Gobbi, L. B.; Gobel, M.; Gray, B. L.; Grossmann, R. B.; Gutteridge, C. E.; Hahn, N.; Harding, S. L.; Jennens, D. C.; Jennens, L.; Lovell, P. J.; Lovell, H. J.; de La Puente, M. L.; Kolb, H. C.; Koot, W. J.; Maslen, S. L.; McCusker, C. F.; Mattes, A.; Pape, A. R.; Pinto, A.; Santafianos, D.; Scott, J. S.; Smith, S. C.; Somers, A. Q.; Spilling, C. D.; Stelzer, F.; Toogood, P. L.; Turner, R. M.; Veitch, G. E.; Wood, A.; Zumbrunn, C. *Chem. Eur. J.* 2008, *14*, 10683–10704.
25. Borén, H. B.; Eklind, K.; Garegg, P. J.; Lindberg, B.; Pilotti, A. *Acta Chem. Scand.* 1972, *26*, 4143.
26. Boons, G. J. P. H.; Steyger, R.; Overhand, M.; van der Marel, G. A.; van Boom, J. H. *J. Carbohydr. Chem.* 1991, *10*, 995–1007.
27. Stewart, A.; Bernlind, C.; Martin, A.; Oscarson, S.; Richards, J. C.; Schweda, E. K. H. *Carbohydr. Res.* 1998, *313*, 193–202.
28. Khare, N. K.; Sood, R. K.; Aspinall, G. O. *Can. J. Chem.* 1994, *72*, 237–246.

12 Synthesis of Fluorinated *Exo*-glycals Mediated by Selectfluor

Guillaume Eppe
Audrey Caravano
*Jérôme Désiré**
Stéphane P. Vincent†

CONTENTS

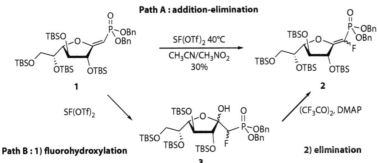

SCHEME 12.1 Synthesis of fluorinated *exo*-glycal 2 (SF(OTf)₂ = selectfluor bistriflate form).

* Checker under supervision of Y. Blériot: yves.bleriot@univ-poitiers.fr.
† Corresponding author: stephane.vincent@fundp.ac.be.

Fluorinated carbohydrates constitute a very important class of mechanistic probes for glycosyl processing enzymes.[1,2] In the mechanistic studies of retaining glycosidases, 2-fluoro-glycosides have become key tools to evidence a transient covalent process as well as to identify the catalytic nucleophiles.[3–5] Fluorinated sugars have also been successfully exploited as mechanistic probes for glycosyltransferases.[6–9] *Exo*-glycals are anomeric enol-ethers derived from carbohydrates in which the double bond is exocyclic. This class of molecules has found many applications as synthetic intermediates.[10] In addition, some *exo*-glycals have been designed as transition-state analogues of enzymatic glycosyl transfers and were shown to be extremely potent sialyltransferase inhibitors.[11] Interestingly, *exo*-glycals derived from galactofuranose (Gal*f*) displayed unique time-dependent inactivation properties against UDP-galactopyranose mutase (UGM), an essential enzyme implied in the biosynthesis of the cell wall of important pathogens.[12] Importantly, we could show that the corresponding fluorinated *exo*-glycal improved the properties of these inhibitors.[13] Beyond carbohydrate chemistry, phosphonates and their fluorinated analogues constitute a very important class of molecules that have found applications in both material and pharmaceutical sciences.[14,15]

Here we describe a method for the synthesis of fluorinated and phosphonylated *exo*-glycals.[13] The synthetic sequence we developed is a Selectfluor mediated fluorination/elimination sequence on phosphonylated *exo*-glycals, which is a new entry into the carbohydrate fluorination chemistry. A survey of the literature describing the synthesis of fluorinated enol-ethers shows that there is no other procedure reported to date for preparing tetrasubstituted olefins bearing a fluoro, a phosphonyl, an alkoxy, and an alkyl group. Recently, novel reactions of carbohydrate fluorinations have been developed thanks to the development of a new generation of fluorinating reagents such as Selectfluor (SF).[16,17] This reagent appeared to be the best fluorinating molecule for the preparation of 2-fluoro-glycosides.[17]

The starting tetrasilylated *exo*-glycal **1** can easily be prepared in two steps from tetrasilylated γ-galactonolactone.[12,18] Since fluorination of glycal **1** gives the desired product in only 30% yield we developed the preparation of the target molecules **2** through a two-step sequence: (1) synthesis of a fluorinated and phosphonylated intermediate ketol **3**, and (2) application of an elimination procedure[19,20] to generate product **2**. For the first step, acetonitrile is the solvent of choice, either pure or mixed with anhydrous nitromethane. The best yields were obtained by avoiding the usual aqueous workup, and diluting the crude reaction mixture in dichloromethane and exposing it to silica gel. Noteworthy, we could sometimes observe the partial formation of the desired *exo*-glycal **2**.* Therefore, to improve the yields and simplify the intermediate silica gel purification, we collected the two lactols **3** along with the *exo*-glycals **2** (if they are observed). Non-standard conditions were then optimized for the preparation of *exo*-glycal **2**. Although pyridine and triethylamine are the usual bases for this elimination, an excess of DMAP gave, by far, the best results. The two fluorinated *exo*-glycals **2** were thus obtained in 46% overall yield from **1**, in a 77/23 (Z/E) ratio.

---

* It is possible that molecular sieves play a role in the formation of fluorinated *exo*-glycals. The amount observed may thus vary from one experiment to another. The silica gel purification protocol described in this manuscript takes this into account.

EXPERIMENTAL

MATERIALS AND PROCEDURES

All chemicals were purchased from Sigma, Aldrich, or Fluka and were used without further purification. Tetrahydrofuran was freshly distilled over sodium benzophenone, dichloromethane over P_2O_5, and acetonitrile and nitromethane over CaH_2. ^1H, ^{13}C, and ^{31}P nuclear magnetic resonance (NMR) spectra were recorded with JEOL (JNM EX-400) spectrometers. Molecular sieves (4Å Fluka Analytical, Lot 445816/1) were heated at 350°C for 24 h under vacuum before use. ^1H, ^{13}C, and ^{31}P NMR spectra were interpreted with the aid of ^1H-^1H and ^1H-^{13}C correlation experiments. Chemical shifts are reported relative to residual solvent signals (CDCl$_3$: ^1H: δ = 7.27; ^{13}C: δ = 77.23) or external reference (85% H_3PO_4 in H_2O: ^{31}P: δ = 0.0). Specific optical rotations were measured on a Perkin Elmer 241 Polarimeter in a 1 dm cell. Melting points were determined with a Büchi 535 apparatus. Column chromatographies were performed on silica gel Kieselgel Si 60 (40–63 µm).

[1(1')Z]-2,3,5,6-TETRA-*O*-*TERT*-BUTYLDIMETHYLSILYL-1-
(DIBENZYLOXYPHOSPHORYL)METHYLIDENE-D-GALACTOFURANOSE (1)

Butyl lithium (6.9 mL, 17.3 mmol, of a 2.5 M solution in hexane) was added under argon atmosphere to a cooled (–70°C) solution of dibenzyl methylphosphonate (4.96 g, 18.0 mmol) in THF (36 mL) followed, after 20 min, by a solution of 2,3,5,6-tetra-*O*-*tert*-butyldimethylsilyl-D-galactono-1,4-lactone[12] (4.56 g, 7.19 mmol) in THF (7 mL). The temperature was maintained at –70°C for 10 min, and then the mixture was allowed to reach -40°C over a 1 h period. The solution was diluted with 1 M phosphate buffer pH = 7 (170 mL) and extracted with CH_2Cl_2 (2 × 350 mL). The combined organic phase was dried over MgSO$_4$, filtered, and concentrated, and the residue was dried in vacuo overnight. The crude product was dissolved in THF (50 mL) at 0°C, and pyridine (5.81 mL, 71.9 mmol) and trifluoroacetic anhydride (5.0 mL, 36.0 mmol) were added. After 4 h at 0°C, the reaction was quenched by addition of saturated aqueous NaHCO$_3$, and the mixture was extracted with EtOAc (350 mL). The organic phase was dried over MgSO$_4$, filtered, and concentrated, and chromatography (5:1 cyclohexane–EtOAc) afforded **1** (4.55 g, 71%) as a colorless syrup.

$[\alpha]_D^{21}$ = +6.4 (c = 1.1, in CHCl$_3$); ^1H NMR (400 MHz, CDCl$_3$): δ = 7.38–7.30 (m, 10H; H arom.), 5.09 (ABX, J(H,P) = 7.4 Hz, J(H,H) = 12.0 Hz, 2H; CH$_2$Ph), 5.02 (ABX, J(H,P) = 7.9 Hz, J(H,H) = 12.0 Hz, 2H; CH$_2$Ph), 4.61 (dd, J(1',2) = 1.1 Hz, J(1',P) = 10.5 Hz, 1H; H-1'), 4.47 (dt, J(1',2) = 1.1 Hz, J(2,P) = J(2,3) = 3.9 Hz, 1H; H-2), 4.41 (t, J(3,4) = J(4,5) = 3.9 Hz, 1H; H-4), 4.33 (dt, J(2,3) = J(3,4) = 3.9 Hz, J(3,P) = 0.6 Hz, 1H; H-3), 3.92 (dt, J(5,6a) = 7.0 Hz, J(5,6b) = 4.6 Hz, 1H; H-5), 3.77 (ABX, J(5,6a) = 7.0 Hz, J(6a,6b) = 10.2 Hz, 1H; H-6a), 3.65 (ABX, J(5,6b) = 4.6 Hz, J(6a,6b) = 10.2 Hz, 1H; H-6b), 0.94 (s, 9H; Si-*t*Bu), 0.92 (s, 9H; Si-*t*Bu), 0.90 (s, 9H; Si-*t*Bu), 0.88 (s, 9H; Si-*t*Bu), 0.16 (s, 6H; 2Si-Me), 0.13 (s, 3H; Si-Me), 0.12 (s, 3H; Si-Me), 0.11 (s, 3H; Si-Me), 0.09 (s, 3H; Si-Me), 0.08 (s, 3H; Si-Me), 0.02 ppm (s, 3H; Si-Me); ^{13}C NMR (100 MHz, CDCl$_3$): δ = 173.03 (d, J(1,P) = 2.6 Hz; C-1), 136.93 (d, J(C,P) = 7.3 Hz; C^q arom.), 136.91 (d, J(C,P) = 7.5 H; C^q arom.), 128.23–127.57

(10 × CH arom.), 90.27 (C-4), 81.78 (d, $J(1',P) = 195.7$ Hz; C-1'), 81.05 (d, $J(2,P) =$ 13.9 Hz; C-2), 76.39 (C-3), 72.57 (C-5), 66.63 (d, $J(C,P) = 4.5$ Hz; CH_2Ph), 66.60 (d, $J(C,P) = 5.0$ Hz; CH_2Ph), 64.39 (C-6), 25.93 ($Si-C(CH_3)_3$), 25.82 ($Si-C(CH_3)_3$), 25.66 (2 $Si-C(CH_3)_3$), 18.28 ($Si-C(CH_3)_3$), 18.05 ($Si-C(CH_3)_3$), 17.83 ($Si-C(CH_3)_3$), 17.72 ($Si-C(CH_3)_3$), −3.82 (Si-Me), −4.02 (Si-Me), −4.34 (Si-Me), −4.38 (Si-Me), −4.41 (Si-Me), −4.81 (Si-Me), −5.40 (Si-Me), −5.42 ppm (Si-Me); MS (DCI-NH_3): m/z: 893 [M + H]$^+$; calcd. for $C_{45}H_{81}O_8PSi_4$: C 60.49 H 9.14; found: C 60.36 H 9.27.

[1(1′)Z]-1-Deoxy-1-(dibenzyloxyphosphoryl)fluoromethylidene-2,3,5,6-tetra-O-tert-butyldimethylsilyl-D-galactofuranose [(Z)-2] and [1(1′)E]-1-Deoxy-1-(dibenzyloxyphosphoryl)fluoromethylidene-2,3,5,6-tetra-O-tert-butyldimethylsilyl-D-galactofuranose [(E)-2]

Step 1: Fluorohydroxylation

A mixture of *exo*-glycal **1** (0.25 g, 0.28 mmol) and powdered molecular sieves 4Å (0.66g) in a mixture of anhydrous MeCN (10 mL) and nitromethane (10 mL) was stirred for 3 h under argon. The mixture was cooled to 0°C, Selectfluor (bistriflate, 0.533 g, 1.22 mmol) was added, and the solution was stirred overnight at 30°C. EtOAc was added, and the mixture was filtered through a pad of Celite®.* The filtrate was diluted with CH_2Cl_2 (0.2 mL), and silica gel (~200 mg) was added. After concentration, purification by silica gel chromatography (15:1 PE–EtOAc) gave a colorless mixture of two lactols **3A** and **3B** R_f (15:1 PE–EtOAc): 0.38 and 0.27 for **3A** and **3B**. ^{31}P NMR of the collected mixture (101 MHz, $CDCl_3$): $\delta = 18.22$ (d, $J(P,F) = 76.8$ Hz; lactol major diastereomer), 16.52 ppm (d, $J(P,F) = 71.8$ Hz; lactol minor diastereomer).

Step 2: Elimination

The foregoing mixture of compounds **3A**, **3B**, **(E)-2**, and **(Z)-2** (163 mg, 0.188 mmol) was dissolved in a mixture of anhydrous CH_2Cl_2/pyridine (6/4, 5.0 mL) under argon atmosphere. The solution was cooled to 0°C, and DMAP (0.459 g, 3.76 mmol) and trifluoroacetic anhydride (0.270 mL, 1.88 mmol) were successively added. The solution was stirred for 24 hours at 35°C, and the solvents were evaporated. The residue was diluted with CH_2Cl_2 (10 mL), and the organic layer was washed with water, dried over $MgSO_4$, filtered, concentrated, and material in the residue was purified by silica gel chromatography (12:1 PE–EtOAc) to give compound **(Z)-2** (85 mg, 50%) and compound **(E)-2** (25 mg, 15%, overall yield 46% for the two steps).

(Z)-2: $[\alpha]_D^{18} = -7.3$ (c = 1.0, in $CHCl_3$); ^1H NMR (400 MHz, $CDCl_3$): $\delta =$ 7.38–7.25 (m, 10H; H arom.), 5.13 (ABX, $J(H,P) = 6.5$ Hz, $J(H,H) = 11.7$Hz, 2H; 2 C*H*Ph), 5.08 (ABX, $J(H,P) = 7.7$ Hz, $J(H,H) = 11.7$ Hz, 1H; C*H*Ph), 5.01 (ABX, $J(H,P) = 7.0$ Hz, $J(H,H) = 11.7$ Hz, 1H; C*H*Ph), 4.74 (d, 1H; H-3), 4.37 (dd, $J(3,4) = 1.6$ Hz, $J(4,5) = 9.1$ Hz, 1H; H-4), 4.23 (d, $J(2,3) =$

---

* If already present in the crude reaction mixture as minor side products, the *exo*-glycals have the following R_f (cyclohexane/ethylacetate 10/1): 0.31 (**(E)-2**), 0.20 and 0.25 (lactols diastereoisomers), 0.11 (**(Z)-2**).

0.7 Hz, 1H; H-2), 3.97 (td, $J(4,5) = 9.1$ Hz, $J(5,6a,b) = 4.0$ Hz, 1H; H-5), 3.70 (AX, $J(5,6a,b) = 4.0$ Hz, 2H; H-6a and H-6b), 0.90 (s, 9H; Si-*t*Bu), 0.88 (s, 9H; Si-*t*Bu), 0.85 (s, 9H; Si-*t*Bu), 0.83 (s, 9H; Si-*t*Bu), 0.15 (s, 3H; Si-Me), 0.11 (s, 3H; Si-Me), 0.10 (s, 3H; Si-Me), 0.07 (s × 2, 6H; 2 Si-Me), −0.01 ppm (s, 3H; Si-Me); ^{13}C NMR (100 MHz, CDCl$_3$): δ = 160.85 (dd, $J_{1\text{-P}} = 21.6$ Hz, $J_{1\text{-F}} = 37.3$ Hz; C-1), 136.03 (d, $J_{\text{C-P}} = 8.3$ Hz; 2C^q arom.), 133.57 (dd, $J_{1'\text{-P}} = 232.4$ Hz, $J_{1'\text{-F}} = 241.4$ Hz; C-1'), 128.53–127.61 (10 C arom.), 95.91 (C-4), 76.75 (C-3), 76.10 (C-2), 72.35 (C-5), 67.80 (d, $J_{\text{C-P}} = 4.5$ Hz; CH_2Ph), 67.71 (d, $J_{\text{C-P}} = 5.1$ Hz; CH_2Ph), 66.50 (C-6), 26.11 (Si-C(CH$_3$)$_3$), 25.89 (Si-C(CH$_3$)$_3$), 25.61 (Si-C(CH$_3$)$_3$), 25.55 (Si-C(CH$_3$)$_3$), 18.56 (Si-C(CH$_3$)$_3$), 18.07 (Si-C(CH$_3$)$_3$), 17.93 (Si-C(CH$_3$)$_3$), 17.75 (Si-C(CH$_3$)$_3$), −4.55 (Si-Me), −4.66 (Si-Me), −4.70 (Si-Me), −4.88 (Si-Me), −5.14 (Si-Me), −5.17 (Si-Me), −5.22 (Si-Me), −5.27 ppm (Si-Me); ^{31}P NMR (101 MHz, CDCl$_3$): δ = 7.35 ppm (d, $J(P,F) = 94.1$ Hz); ^{19}F NMR (235 MHz, CDCl$_3$): δ = −174.12 ppm (d, $J(P,F) = 95.4$ Hz); MS (DCI-NH$_3$): m/z (%) : 911 (20) [M + H]$^+$, 928 (100) [M + NH$_4$]$^+$; HRMS: m/z: calcd. for C$_{45}$H$_{81}$O$_8$FSi$_4$P: 911.4730; found: 911.4722; R$_f$ (PE/AcOEt 12/1): 0.20.

(*E*)-2: [α]$_D^{18}$ = −40.4 (c = 1.0, in CHCl$_3$); ^{1}H NMR (400 MHz, CDCl$_3$): δ = 7.40–7.24 (m, 10H; H arom.), 5.16 (s, 1H; H-3), 5.10 (ABX, $J(H,P) = 6.8$ Hz, $J(H,H) = 11.9$ Hz, 1H; C*H*Ph), 5.06 (ABX, $J(H,P) = 6.2$ Hz, $J(H,H) = 11.7$ Hz, 2H; 2 C*H*Ph), 5.02 (ABX, $J(H,P) = 6.8$ Hz, $J(H,H) = 11.7$ Hz, 1H; C*H*Ph), 4.39 (d, $J(4,5) = 9.9$ Hz, 1H; H-4), 4.15 (d, $J(2,3) = 3.3$ Hz, 1H; H-2), 4.09 (td, $J(4,5) = 9.9$ Hz, $J(5,6a,b) = 3.5$ Hz, 1H; H-5), 3.72 (AX, $J(5,6a,b) = 3.5$ Hz, 2H; H-6a and H-6b), 0.90 (s×2, 18H; Si-*t*Bu), 0.86 (s, 9H; Si-*t*Bu), 0.84 (s, 9H; Si-*t*Bu), 0.23 (s, 3H; Si-Me), 0.21 (s, 3H; Si-Me), 0.12 (s, 3H; Si-Me), 0.09 (s, 3H; Si-Me), 0.08 (s × 2, 6H; 2 Si-Me), 0.07 (s, 3H; Si-Me), 0.02 ppm (s, 3H; Si-Me); ^{13}C NMR (100 MHz, CDCl$_3$): δ = 159.15 (dd, $J(1,P) = 4.4$ Hz, $J(1,F) = 48.4$ Hz; C-1), 135.82 (d, $J(C,P) = 7.4$ Hz; C^q arom.), 135.79 (d, $J(C,P) = 8.8$ Hz; C^q arom.), 131.30 (dd, $J(1',P) = 247.6$ Hz, $J(1',F) = 254.9$ Hz; C-1'), 128.45–127.67 (10 C arom.), 94.52 (C-4), 77.27 (d, $J(3,P) = 2.6$ Hz; C-3), 76.73 (C-2), 72.30 (C-5), 67.90 (d, $J(C,P) = 5.2$ Hz; CH_2Ph), 67.64 (d, $J(C,P) = 4.2$ Hz; CH_2Ph), 64.00 (C-6), 26.10 (Si-C(CH$_3$)$_3$), 25.87 (Si-C(CH$_3$)$_3$), 25.80 (Si-C(CH$_3$)$_3$), 25.55 (Si-C(CH$_3$)$_3$), 18.56 (Si-C(CH$_3$)$_3$), 18.24 (Si-C(CH$_3$)$_3$), 17.95 (Si-C(CH$_3$)$_3$), 17.79 (Si-C(CH$_3$)$_3$), −4.65 (Si-Me), −4.70 (Si-Me), −4.73 (3 Si-Me), −4.75 (Si-Me), −5.27 ppm (2 Si-Me); ^{31}P NMR (101 MHz, CDCl$_3$): δ = 8.37 ppm (d, $J(P,F) = 76.8$ Hz); ^{19}F NMR (235 MHz, CDCl$_3$): δ = −163.66 ppm (d, $J(P,F) = 80.7$ Hz); MS (DCI-NH$_3$): m/z (%): 911 (100) [M + H]$^+$, 928 (20) [M + NH$_4$]$^+$; HRMS: m/z: calcd. for C$_{45}$H$_{81}$O$_8$FSi$_4$P: 911.4730; found: 911.4732; R$_f$ (PE/AcOEt 12/1): 0.46.

ACKNOWLEDGMENTS

This work was funded by FNRS (FRFC grant 2.4.625.08.F and "Crédit au chercheur" 1.5.167.10 and F.R.I.A., Fonds pour la Formation à la Recherche dans l'Industrie et dans l'Agriculture).

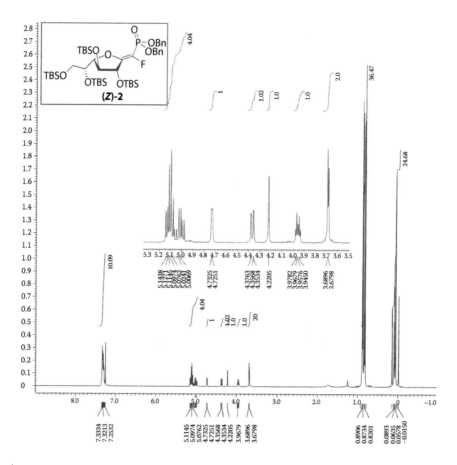

¹H NMR spectrum for title compound (Z)-2.

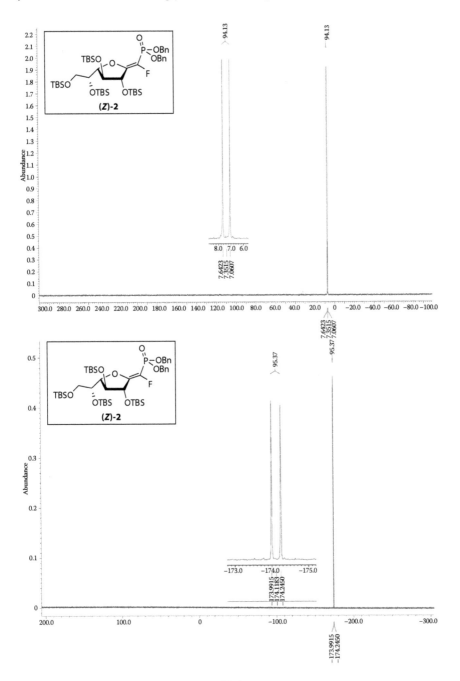

^{31}P and ^{31}F NMR spectra for title compound (**Z**)-**2**.

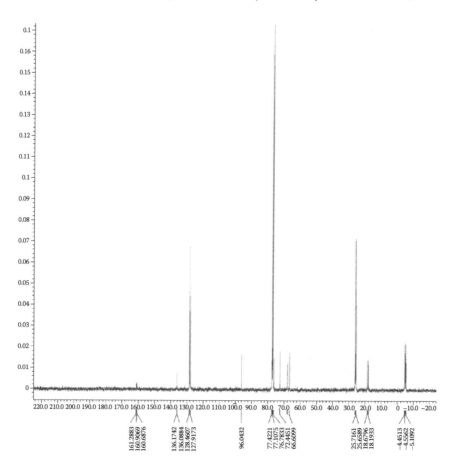

^{13}C NMR spectrum for title compound (**Z**)-**2**.

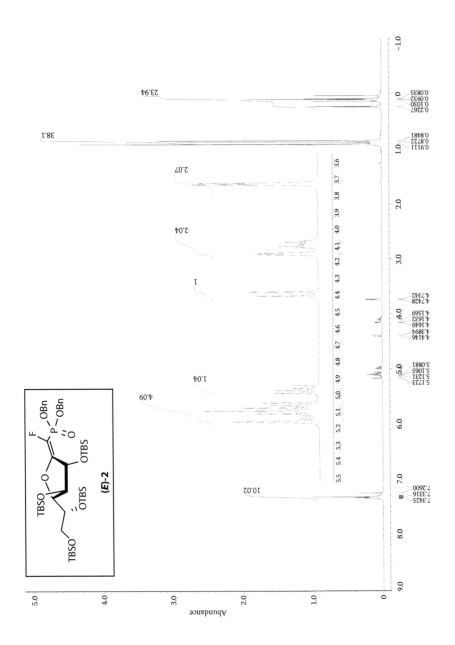

^{1}H NMR spectrum for title compound (*E*)-**2**.

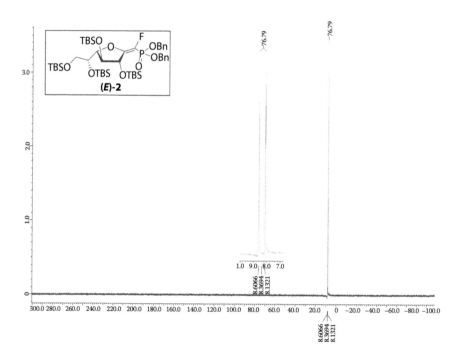

^{31}P NMR spectrum for title compound (*E*)-**2**.

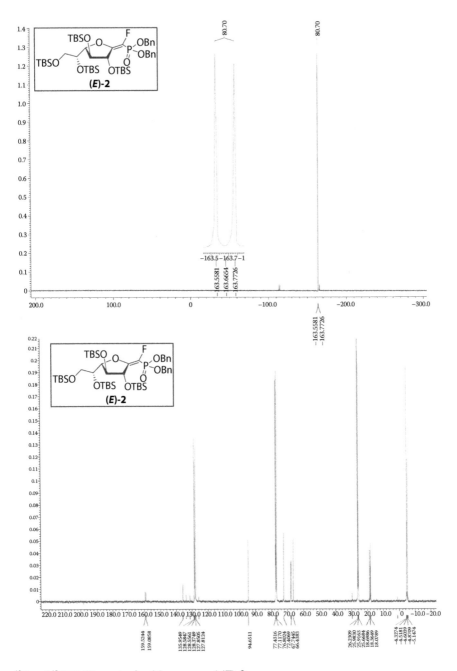

¹³C and ¹⁹F NMR spectra for title compound **(E)-2**.

REFERENCES

1. Zechel, D. L.; Withers, S. G. *Acc. Chem. Res.* 2000, *33*, 11.
2. Pongdee, R.; Liu, H.-W. *Bioorg. Chem.* 2004, *32*, 393–437.
3. Withers, S. G. *Carbohydr. Polym.* 2001, *44*, 325–337.
4. Zechel, D. L.; Withers, S. G. *Comprehensive Natural Products Chemistry*, eds. Barton, D.; Nakanishi, K.; Poulter, C. D. 1999, pp. 279–314.
5. Vocadlo, D. J.; Withers, S. G. *Carbohydr. Res.* 2005, *340*, 379–388.
6. Burkart, M. D.; Vincent, S. P.; Düffels, A.; Murray, B. W.; Ley, S. V.; Wong, C.-H. *Bioorg. Med. Chem.* 2000, *8*, 1937–1946.
7. Burkart, M. D.; Vincent, S. P.; Wong, C.-H. *Chem. Commun.* 1999, 1525–1526.
8. Dohi, H.; Périon, R.; Durka, M.; Bosco, M.; Roué, Y.; Moreau, F.; Grizot, S.; Ducruix, A.; Escaich, S.; Vincent, S. P. *Chem. Eur. J.* 2008, *14*, 9530–9539.
9. Chiu, C. P. C.; Watts, A. G.; Lairson, L. L.; Gilbert, M.; Lim, D.; Wakarchuk, W. W.; Withers, S. G.; Strynadka, N. C. J. *Nat. Struct. Mol. Biol.* 2004, *11*, 163–170.
10. Taillefumier, C.; Chapleur, Y. *Chem. Rev.* 2004, *104*, 263–292.
11. Müller, B; Schaub, C.; Schmidt, R. R. *Angew. Chem. Int. Ed.* 1998, *37*, 2893–2896.
12. Caravano, A.; Vincent, S. P.; Sinaÿ, P. *Chem. Commun.* 2004, 1216–1217.
13. Caravano, A.; Dohi, H.; Sinaÿ, P.; Vincent, S. P. *Chem. Eur. J.* 2006, *11*, 3114–3123.
14. Van der Jeught, S.; Stevens, C. V. *Chem. Rev.* 2009, *109*, 2672–2702.
15. Radwan-Olszewska, K.; Palacios, F.; Kafarski, P. *J. Org. Chem.*, 2011, *76*, 1170–1173.
16. Vincent, S. P.; Burkart, M. D.; Tsai, C.-Y.; Zhang, Z.; Wong, C.-H. *J. Org. Chem.*, 1999, *64*, 5264–5279.
17. Nyffeler, P. T.; Duron, S. G.; Burkart, M. D.; Vincent, S. P.; Wong, C.-H. *Angew. Chem. Int. Ed.* 2005, *44*, 192–212.
18. Caravano, A.; Mengin-Lecreulx, D.; Brondello, J. -M.; Vincent, S. P.; Sinaÿ, P. *Chem. Eur. J.* 2003, *23*, 5888–5898.
19. Yang, W.-B.; Chang, C.-F.; Wang, S.-H.; Teo, C.-F.; Lin, C.-H. *Tetrahedron Lett.* 2001, *42*, 4657–4660.
20. Yang, W.-B.; Wu, C.-Y.; Chang, C.-C.; Wang, S.-H.; Teo, C.-F.; Lin, C.-H. *Tetrahedron Lett.* 2001, *42*, 6907–6910.

13 A Direct and Stereospecific Approach to the Synthesis of α-Glycosyl Thiols

*Roger A. Ashmus**
Lei Zhang
Xiangming Zhu†

CONTENTS

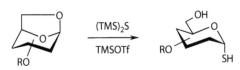

SCHEME 13.1

Thioglycosides, in which the anomeric oxygen is replaced by sulfur, receive considerable attention due to altered stability to chemical and enzymatic hydrolysis while their solution conformation and biological activities are comparable to native counterparts.[1] Consequently, much effort has been devoted in the past two decades to the synthesis of thioglycosides, including thiosaccharides and S-glycoconjugates, in

---

* Checker under supervision of T. Lowary: tlowary@ualberta.ca.
† Corresponding author: xiangming_zhu@hotmail.com.

order to provide valuable compounds for biological studies.[2] For instance, carbohydrate epitopes of conjugate vaccines have been modified to contain S-linked residues and the resulting S-linked immunogens generated an antigen-specific immune response that exceeds the response to the native oligosaccharides.[3]

Currently, glycosyl thiols or their precursors, such as anomeric thioacetates, which can be S-deacetylated in situ to generate the desired glycosyl thiols, are the key building blocks for the construction of thioglycosides.[2] The great advantage of using glycosyl thiols is that both α- and β-glycosyl thiols are quite stable and do not mutarotate under basic conditions,[4] hence the anomeric stereochemistry is maintained during the glycosylation process. Given the great value of thioglycosides in biological studies and the wide occurrence of α-linkage in various glycoconjugates,[5] there is a high demand for procedures for the stereoselective synthesis of α-glycosyl thiols that can be used to construct α-thioglycosides. We report herein a direct and stereospecific approach for the synthesis of α-glycosyl thiols.

We envisioned that 1,6-anhydrosugars could serve as glycosylating agents for the synthesis of α-glycosyl thiols because β-attack on 1,6-anhydrosugars by any nucleophiles is not possible in an S_N2-type pathway. Also, we anticipated that bis(trimethylsilyl)sulfide could be used as the required sulfur nucleophile to directly introduce the sulfhydryl group. In view of the acid lability of the trimethylsilyl group, it could be cleaved in situ under glycosidation conditions. We performed preliminary experiments with 1,6-anhydrogalactose **1**[6] (Table 13.1). Treatment of **1** with 1.4 equivalent of bis(trimethylsilyl)sulfide in the presence of 0.8 equivalent of TMSOTf at room temperature failed to provide any glycosyl thiol. However, when the reaction mixture was heated to 50°C, **5** was formed as the sole anomer in high yield (Table 13.1, entry 1). The reaction was very clean as indicated by thin-layer chromatography (TLC), and the anomeric configuration of the product could be readily determined by nuclear magnetic resonance (NMR) spectroscopy.

Encouraged by this result, a series of protected 1,6-anhydrosugars were then subjected to the above ring-opening conditions (Table 13.1). The results, summarized in Table 13.1, indicated that under the above reaction conditions, 1,6-anhydrosugars **(1–4)**[7,8] can be converted effectively into α-glycosyl thiols **(5–8)** in a stereospecific way. All the 1,6-anhydrosugars are known compounds, but the products α-thiols, to our knowledge, are new structures and are fully characterized by NMR and mass spectroscopies as described in the next section.

EXPERIMENTAL

GENERAL METHODS

Reactions were performed in oven-dried glassware under argon using dry solvents. Solvents were evaporated under reduced pressure at <40°C. All reactions were monitored by TLC on silica gel 60 F_{254}-coated aluminium sheets and the spots were visualized by ultraviolet (UV) or by charring with 8% H_2SO_4 in methanol. Flash chromatography was performed with the appropriate solvent system using 46–60

TABLE 13.1
Synthesis of α-Glycosyl Thiols

Entry	Substrate	Reaction Time	Product	Yield (%)[a]	α/β ratio
1	**1**[6]	9 h	**5**	79	α only
2	**2**[7]	10 h	**6**	88	α only
3	**3**[8]	7 h	**7**	81	α only
4	**4**[8]	7 h	**8**	83	α only

[a]Isolated yield following chromatography.

μm silica gel. Optical rotations were measured at 20°C. ^{1}H NMR (400 or 500 MHz) spectra were measured for solutions in $CDCl_3$ with tetramethylsilane as internal standard. ^{13}C NMR spectra were recorded at 100 or 125 MHz using $CDCl_3$ as solvent, and the signals were assigned with the aid of Distortionless Enhancement by Polarisation Transfer (DEPT), Heteronuclear Single Quantum Correlation (HSQC) experiments. Yields refer to chromatographically pure compounds.

SYNTHESIS OF α-GLYCOSYL THIOLS: GENERAL PROCEDURE

To a solution of the 1,6-anhydrosugar (1.0 mmol) and bis(trimethylsilyl)sulfide (1.4 mmol) in CH_2Cl_2 (10 mL) was added TMSOTf (0.8 mmol) at 0°C. The mixture was refluxed for a period of time indicated in Table 13.1, poured into aqueous NaHCO$_3$, and extracted with EtOAc. The organic layer was washed successively with water and brine, dried over Na$_2$SO$_4$, and concentrated in vacuo to give a residue which was purified by flash column chromatography to afford the corresponding α-glycosyl thiol.

2,3,4-Tri-*O*-benzyl-1-thio-α-d-galactopyranose (5)

Purified by flash column chromatography (cyclohexane/EtOAc, 8:1→5:1) to give the title compound as a colorless syrup: [α]$_D$ = +93.6 (*c* 0.8, CHCl$_3$); ^{1}H NMR (500 MHz, CDCl$_3$): δ 7.38–7.24 (m, 15H, ArH), 5.84 (t, 1H, *J* 5.0, 4.5 Hz, H-1), 4.93 and 4.65 (AB peak, 2H, *J* 11.5 Hz, PhC*H*$_2$), 4.84 and 4.69 (AB peak, 2H, *J* 12.0 Hz, PhC*H*$_2$), 4.72 and 4.69 (AB peak, 2H, *J* 11.0 Hz, PhC*H*$_2$), 4.25 (dd, 1H, *J* 5.0, 9.5 Hz, H-2), 4.15 (t, 1H, *J* 5.5, 6.0 Hz, H-5), 3.89 (br s, 1H, H-4), 3.81 (dd, 1H, *J* 2.5, 9.5 Hz, H-3), 3.72 (dd, 1H, *J* 6.5, 11.5 Hz, H-6$_a$), 3.54–3.49 (m, 1H, H-6$_b$), 1.83 (d, 1H, *J* 4.0 Hz, SH); ^{13}C NMR (125 MHz, CDCl$_3$): δ 138.4, 138.0, 137.8, 128.6, 128.5, 128.44, 128.39, 128.0, 127.9, 127.8, 127.7, 127.6 (aromatic carbons), 79.4 (C-1), 78.6 (C-3), 75.9 (C-2), 74.6 (C-4), 74.4, 73.6, 72.6 (3 × PhCH$_2$), 71.6 (C-5), 62.0 (C-6). ESI-MS m/z 489.2 [M + Na]$^+$. ESI-HRMS calcd. for C$_{27}$H$_{30}$NaO$_5$S [M + Na]$^+$ 489.1712, found: 489.1712.

2,3,4-Tri-*O*-benzyl-1-thio-α-d-glucopyranose (6)

Purified by flash column chromatography (cyclohexane/EtOAc, 8:1→5:1) to give the title compound as a colorless syrup: [α]$_D$ = + 95.1 (*c* 1.0, CHCl$_3$); ^{1}H NMR (400 MHz, CDCl$_3$): δ 7.31–7.22 (m, 15H, ArH), 5.62 (t, 1H, *J* 5.2, 4.8 Hz, H-1), 4.88 and 4.74 (AB peak, 2H, *J* 10.8 Hz, PhC*H*$_2$), 4.81 and 4.58 (AB peak, 2H, *J* 11.0 Hz, PhC*H*$_2$), 4.66 and 4.57 (AB peak, 2H, *J* 11.8 Hz, PhC*H*$_2$), 4.02 (m, 1H, H-5), 3.82 (t, 1H, *J* 9.2, 8.8 Hz, H-3), 3.68 (m, 3H, H-2, H-6$_a$, H-6$_b$), 3.48 (t, 1H, *J* 10.0, 9.2 Hz, H-4), 1.84 (d, 1H, *J* 4.8 Hz, SH); ^{13}C NMR (100 MHz, CDCl$_3$): δ 138.7, 138.2, 137.7, 128.66, 128.65, 128.6, 128.22, 128.20, 128.17, 128.13, 128.09, 127.8 (aromatic carbons), 81.8 (C-3), 79.5 (C-2), 78.9 (C-1), 77.1 (C-4), 75.9, 75.2, 72.6 (3 × PhCH$_2$), 72.0 (C-5), 61.9 (C-6). ESI-MS m/z 489.2 [M + Na]$^+$. ESI-HRMS calcd. for C$_{27}$H$_{30}$NaO$_5$S [M + Na]$^+$ 489.1712, found: 489.1698.

2,3-Di-*O*-allyl-4-*O*-benzyl-1-thio-α-d-glucopyranose (7)

Purified by flash column chromatography (cyclohexane/EtOAc, 8:1→4:1) to give the title compound as a colorless syrup: [α]$_D$ = + 117 (*c* 0.2, CHCl$_3$); ^{1}H NMR (500 MHz, CDCl$_3$): δ 7.35–7.25 (m, 5H, ArH), 6.00–5.88 (m, 2H, 2 × CH$_2$C*H*CH$_2$), 5.71 (t, 1H, *J* 5.2, 4.8 Hz, H-1), 5.31 (m, 2H, CH$_2$CHC*H*$_2$), 4.88 and 4.66 (AB peak, 2H, *J* 10.5, 11.0 Hz, PhC*H*$_2$), 4.40 (dd, 1H, *J* 5.5, 12.0 Hz, C*H*$_2$CHCH$_2$), 4.28 (dd, 1H, *J* 5.5, 12.0 Hz, C*H*$_2$CHCH$_2$), 4.05 (dt, 1H, *J* 3.5, 10.0 Hz H-5), 3.79–3.68 (overlapped, 3H, H-3, H-4, H-6$_a$), 3.62 (dd, 1H, *J* 5.9, 9.5 Hz, H-2), 3.49 (t, 1H, *J* 9.5 Hz, H-6$_b$),

1.87 (d, 1H, J 4.5 Hz, SH); ^{13}C NMR (100 MHz, CDCl$_3$): δ 138.1, 135.1, 134.3 (2 × CH$_2$CHCH$_2$), 128.50, 128.45, 128.1, 127.9, 127.8 (aromatic carbons), 117.6, 116.7 (2 × CH$_2$CHCH$_2$), 81.3 (C-1), 79.1 (C-3), 78.8 (C-2), 77.0 (C-4), 75.1 (PhCH$_2$), 74.4, 71.8 (2 × CH$_2$CHCH$_2$), 71.5 (C-5), 61.8 (C-6). ESI-MS m/z 389.1 [M + Na]$^+$. ESI-HRMS calcd. for C$_{19}$H$_{26}$NaO$_5$S [M + Na]$^+$ 389.1399, found: 389.1418.

2,3-DI-O-ALLYL-4-O-BENZYL-1-THIO-α-D-GALACTOPYRANOSE (8)

Purified by flash column chromatography (cyclohexane/EtOAc, 8:1→4:1) to give the title compound as a colorless syrup: $[\alpha]_D$ = + 76.2 (c 2.0, CHCl$_3$); ^{1}H NMR (500 MHz, CDCl$_3$): δ 7.39–7.32 (m, 5H, ArH), 6.02–5.91 (m, 2H, 2 × CH$_2$CHCH$_2$), 5.87 (t, 1H, J 4.5 Hz, H-1), 5.36 (m, 2H, CH$_2$CHCH$_2$), 5.22 (m, 2H, CH$_2$CHCH$_2$), 4.98 and 4.96 (AB peak, 2H, J 11.5 Hz, PhCH$_2$), 4.33–4.17 (m, 4H, 2 × CH$_2$CHCH$_2$), 4.20 (overlapped m, 1H, H-5), 4.12 (dd, 1H, J 5.5, 10.0 Hz, H-2), 3.91 (br s, 1H, H-4), 3.76 (dd, J 6.5, 11.5Hz H-6$_a$), 3.68 (dd, 1H, J 2.5, 9.5 Hz, H-3), 3.56 (m, 1H, H-6$_b$), 1.83 (d, 1H, J 4.0 Hz, SH); ^{13}C NMR (125 MHz, CDCl$_3$): δ 138.1, 134.9, 134.5 (2 × CH$_2$CHCH$_2$), 128.6, 128.51, 128.0 (aromatic carbons), 117.3, 116.6 (2 × CH$_2$CHCH$_2$), 79.5 (C-1), 78.3 (C-3), 75.6 (C-2), 74.49 (C-4), 74.45 (PhCH$_2$), 72.2, 71.7 (2 × CH$_2$CHCH$_2$), 71.6 (C-5), 62.1 (C-6). ESI-MS m/z 389.1 [M + Na]$^+$. ESI-HRMS calcd. for C$_{19}$H$_{27}$O$_5$S [M + H]$^+$ 367.1579, found: 367.1562.

ACKNOWLEDGMENTS

This work was supported by Research Frontier Program of Science Foundation Ireland and the Natural Science Foundation of Zhejiang Province (R4110195).

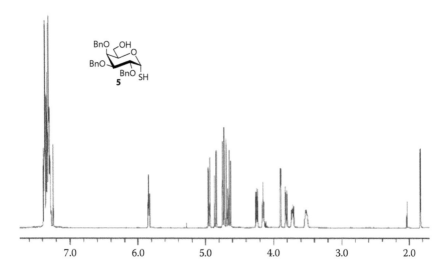

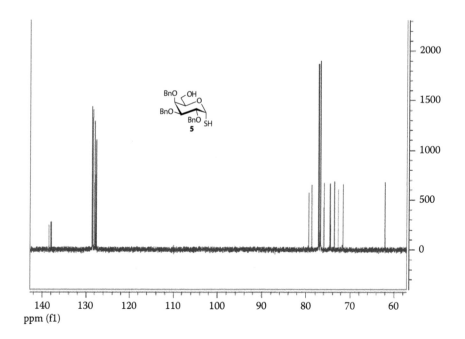

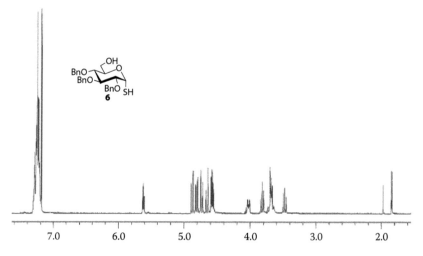

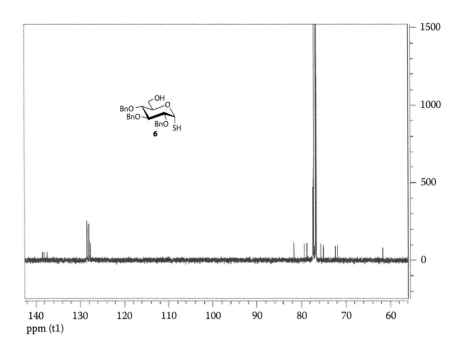

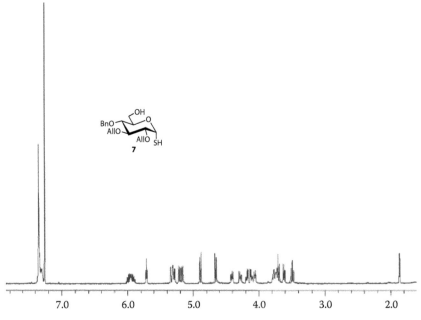

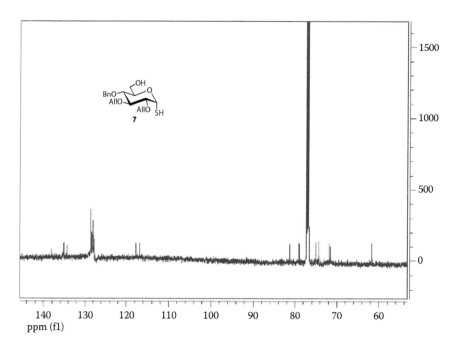

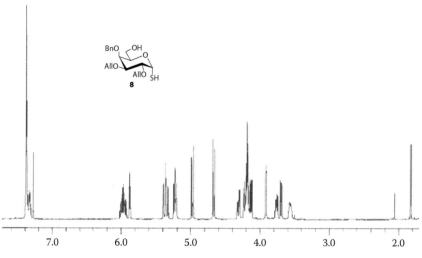

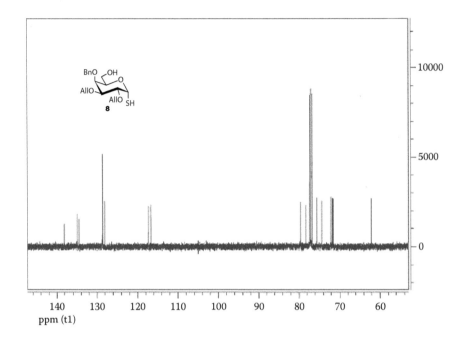

ppm (t1)

REFERENCES

1. (a) Driguez, H. *Chembiochem* 2001, *2*, 311–318. (b) Knapp, S.; Gonzalez, S.; Myers, D. S.; Eckman, L. L.; Bewley, C. A. *Org. Lett.* 2002, *4*, 4337–4339. (c) Jahn, M.; Marles, J.; Warren, R. A. J.; Withers, S. G. *Angew. Chem. Int. Ed.* 2003, *42*, 352–354. (d) Uhrig, M. L.; Manzano, V. E.; Varela, O. *Eur. J. Org. Chem.* 2006, 162–168.
2. Pachamuthu, R.; Schmidt, R. R. *Chem. Rev.* 2006, *106*, 160–187.
3. (a) Rich, J. R.; Bundle, D. R. *Org. Lett.* 2004, *6*, 897–900. (b) Bundle, D. R.; Rich, J. R.; Jacques, S.; Yu, H. N.; Nitz, M.; Ling, C. C. *Angew. Chem. Int. Ed.* 2005, *44*, 7725–7729.
4. MacDougall, J. M.; Zhang, X. D.; Polgar, W. E.; Khroyan, T. V.; Toll, L.; Cashman, J. R. *J. Med. Chem.* 2004, *47*, 5809–5815.
5. (a) Vankar, Y. D.; Schmidt, R. R. *Chem. Soc. Rev.* 2000, *29*, 201–216. (b) Kinjo, Y.; Wu, D.; Kim, G.; Xing, G.; Poles, M. A.; Ho, D. D.; Wong, C. H.; Kronenberg, M. *Nature* 2005, *102*, 1351–1356.
6. Cruzado, M. C.; Martin-Lomas, M. *Tetrahedron Lett.* 1986, *27*, 2497–2500.
7. Ruckel, E. R.; Schuerch, C. *J. Org. Chem.* 1966, *31*, 2233–2239.
8. Zhu, X.; Dere, R. T.; Jiang, J.; Zhang, L.; Wang, X. *J. Org. Chem.* 2011, *76*, 10187–10197.

14 Efficient Microwave-Assisted Synthesis of 1′,2,3,3′,4,4′,6-Hepta-O-benzyl-sucrose and 1′,2,3,3′,4,4′-Hexa-O-benzylsucrose

*M. Teresa Barros**
Krasimira T. Petrova
Paula Correia-da-Silva
Ana-Paula Esteves†

CONTENTS

---

* Corresponding author: mtbarros@dq.fct.unl.pt.
† Checker: aesteves@quimica.uminho.pt.

107

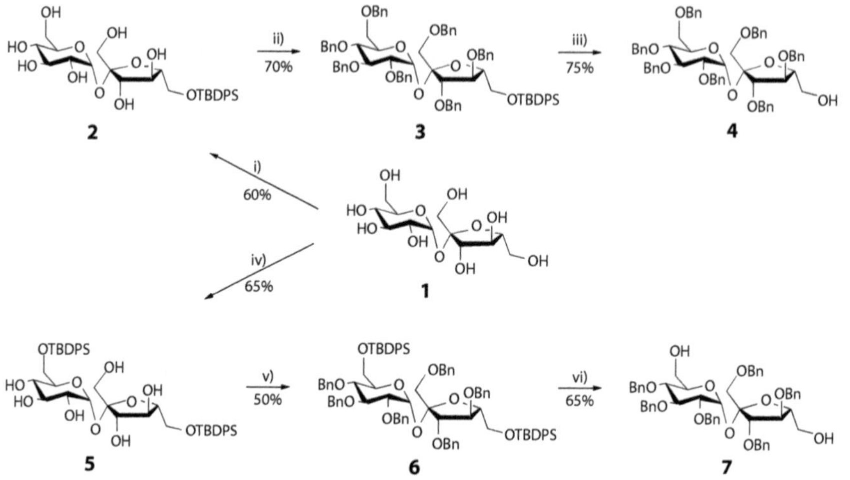

SCHEME 14.1   (i) 1.1 equiv. TBDPSCl, 4-DMAP, py, MW (90°C, 300 W), 5 min, 60%; (ii) NaH/BnBr, DMF, MW (145°C, 300 W), 5 min, 70%; (iii) TBAF, THF, MW (65°C, 300 W), 5 min, 75%; (iv) 2.2 equiv. TBDPSCl, 4-DMAP, py, MW (90°C, 300 W), 20 min, 65%; (v) NaH/BnBr, DMF, MW (145°C, 300 W), 5 min, 50%; (vi) TBAF, THF, MW (65°C, 300 W), 5 min, 65%.

Methods used to produce and purify carbohydrate derivatives are often tedious and complex,[1] which makes the synthesis of selectively protected sucrose derivatives laborious. Clearly, there is a need for simpler procedures and faster and environmentally friendlier heating methods, and these can be provided by the use of microwave techniques. Protocols presented in this chapter (Scheme 14.1) allow significant reduction of time and energy compared with other routes[2–4] and have potential for automatization of tedious multi-step syntheses of useful sucrose derivatives.

The first step to the target compound **4** consisted of a regioselective silylation of the 6′-hydroxyl group of sucrose (**1**) using 1.1 equiv. of *tert*-butyldiphenylchlorosilane (TBDPSCl) in dry pyridine. The reaction was completed in 5 min under microwave irradiation (90°C, 300 W). The monosilylated sucrose **2** was benzylated using BnBr/NaH reagent in DMF leading to compound **3** in 5 min (145°C, 300 W). Selective deprotection of the silyl group, using TBAF (tetrabutylammonium fluoride), in THF led to 1′,2,3,3′,4,4′,6-hepta-*O*-benzyl-sucrose (**4**) in a good yield (75% or 32% overall from sucrose).

Similar strategy has been applied to obtain 1′,2,3,3′,4,4′-hexa-*O*-benzylsucrose (**7**). In the first step, TBDPSCl (2.2 equiv.) was used to obtain 6,6′-di-*O*-TBDPS-sucrose **5**, which was subsequently benzylated to obtain **6**. The primary positions were then selectively deprotected using TBAF, to give compound **7** in good yield (65% or 21% overall from sucrose).

Microwave-enhanced synthesis has been applied in almost all areas of chemistry,[5] but carbohydrate chemistry has suffered a certain delay in adopting this technology, as documented by the limited number of applications.[6,7] In particular, it had previously not been applied to sucrose chemistry, and the general opinion is that the method is hampered by competitive degradation of sucrose because of its thermal instability.[8] Herein, a method to overcome these limitations and to apply highly efficient and fast synthetic protocols for the selective functionalization of sucrose under microwave irradiation is presented. It is to note that in the case of sucrose functionalization, elimination of solvent is not recommended as it led to significant overheating and lower yields.

The key point to successful synthesis under microwave irradiation is the use of equipment specially designed for chemical laboratories. Monomodal microwave equipment has overcome the uncertainties associated with the use of domestic microwave ovens. These reactors offer much more precise temperature and pressure control, and the software provides simplified process monitoring, which results in accurate, reproducible reaction conditions.[9] The energy transfer in a microwave-assisted reaction is rapid, the decomposition of the substances can be avoided, and high yields can be obtained in short reaction times by proper programming of the temperature. When the desired temperature is reached, the power is automatically reduced and maintained by the software during the reaction period.

EXPERIMENTAL

GENERAL METHODS

Reagents and solvents were purified before use.[10] Solvents used as reaction media (pyridine, DMF, and THF) were freshly distilled. The reactions under microwave irradiation were performed in open flasks equipped with temperature control sensor and magnetic stirring using a monomodal microwave reactor MicroSynth Labstation (MileStone Inc.).[9] This synthesizer has a single-mode cavity with temperature and pressure control. In this way, the temperature runaway and explosion risk are avoided. It should be noticed that the reaction conditions are expressed as a function of the reaction temperature and not the magnetron power, which is the case for most microwave-assisted reactions published. Nuclear magnetic resonance (NMR) spectra were recorded at 400 MHz in $CDCl_3$ or DMSO-d_6, with chemical shift values (δ) in ppm downfield from TMS (0 ppm) or the residual solvent peak of DMSO (2.50 ppm). The signals were assigned with the aid of DEPT (Distortionless Enhancement by Polarization Transfer), COSY (COrrelation SpectroscopY), and HMQC (Heteronuclear Multiple Quantum Coherence) experiments. Optical rotations were measured at 20°C on an AA-1000 polarimeter (0.5 dm cell) at 589 nm. Melting points were determined with a capillary apparatus and are uncorrected. Fourier transform infrared (FTIR) spectra were recorded on Perkin-Elmer Spectrum BX apparatus in KBr dispersions.

6'-O-TERT-BUTYLDIPHENYLSILYLSUCROSE (2)[11]

To a solution of sucrose (1.000 g, 2.92 mmol) in dry pyridine (15 mL) was added 4-dimethylaminopyridine (0.05 g) and 1.1 mol equiv. tert-butyldiphenylsilyl chloride (0.884 g, 3.21 mmol). The reaction mixture was placed in the microwave oven and irradiated at 90°C and power of 300 W for 5 min. The reaction mixture was concentrated and chromatographed (ethyl acetate→100:100:1 ethyl acetate-acetone-water→10:10:1 ethyl acetate-acetone-water), to yield first 6,6'-di-O-tert-butyldiphenylsilyl-sucrose (0.238 g, 10%), which was identical in all respects to substance 5 described below.

Compound **2** was eluted next and obtained after concentration as a white solid (1.018 g, 60%). Crystallization from CH_3OH gave **2**, m.p. 190–192°C; Lit.[11] 192–195°C; $[\alpha]_D^{20}$ + 40.7 (c 1.1, CH_3OH); Lit.[12] $[\alpha]_D$ + 44 (c 1, CH_3OH). IR: ν_{max} (KBr): 3306 (O-H), 2930, 2856 (C-H, satd), 1471 (Si-Ar), 1428, 1378 (C-C-C), 1278 (Si-C), 1142 (C-C-O), 1112 (Si-Ar), 1054 (C-O-C), 824 (Ar) cm^{-1}. ^{1}H NMR (DMSO-d_6): δ 7.66 (m, 4H, Ar-H), 7.41 (m, 6H, Ar-H), 5.24 (d, J = 3.1 Hz, 1H, H-1), 3.90 (t, $J_{3'-4'}$ = $J_{4'-5'}$ = 8.5 Hz, 1H, H-4'), 3.79 (d, $J_{3'-4'}$ = 8.5 Hz, 1H, H-3'), 3.73 (m, 1H, H-5'), 3.63 (m, 1H, H-5), 3.44 (m, 7H, H-1',3,6,6'), 3.14 (dd, J_{2-3} = 9.5 Hz, J_{1-2} = 3.0 Hz, 1H, H-2), 3.07 (t, J = 9.3 Hz, 1H, H-4), 0.99 (s, 9H, CH_3). ^{13}C NMR (DMSO- d_6): δ 135.1, 133.2, 129.7, 127.8 (C_{Ar}), 104.4 (C-2'), 91.5 (C-1), 82.2, 76.8, 74.6, 73.1, 72.6, 71.7, 70.0 (C-2,3,3',4,4',5,5'), 65.7 (C-1'), 62.1, 60.8 (C-6,6'), 26.7 (CH_3), 18.9 ($SiC(CH_3)_3$). Anal. Calcd for $C_{28}H_{40}O_{11}Si$: C, 57.91; H, 6.94. Found: C, 58.13; H, 6.96.

1',2,3,3',4,4',6-Hepta-O-benzyl-6'-O-tert-butyldiphenylsilylsucrose (3)[13]

NaH (60% suspension in oil, 11.2 equiv., 0.270 g, 6.72 mmol) was added carefully at 0°C to a solution of 6'-O-tert-butyldiphenylsilylsucrose (**2**, 0.350 g, 0.60 mmol) in DMF (5 mL). After 20 min, benzyl bromide (1.437 g, 8.40 mmol, 14 equiv.) was added dropwise during 15 min. The mixture was placed in the microwave oven and irradiated at 145°C and power of 300 W for 5 min, when thin-layer chromatography (TLC) (10:10:1 ethyl acetate-acetone-water) showed that the starting material had been consumed. The reaction mixture was poured into cold H_2O (30 mL), the product was extracted with diethyl ether (4 × 20 mL), the combined organic layers were washed with H_2O (2 × 15 mL), dried over Na_2SO_4, and concentrated. Chromatography of the residue (5:1 hexane-ethyl acetate) gave first **3** (0.512 g, 70%) as a colorless oil. $[\alpha]_D^{20}$ + 31.7 (c 0.7, $CHCl_3$); Lit.[13] $[\alpha]_D^{20}$ + 30.9 (c 0.9, $CHCl_3$). IR: ν_{max} (CH_2Cl_2) 3052, 3030, 2926, 2858 (C-H, satd.), 1954, 1606 (Bn), 1591, 1496 (Si-Ar), 1452 (Bn), 1435, 1358 (C-C-C), 1311, 1261 (Si-C), 1203, 1092 (C-O-C), 916 (Si-Ar), 818 (Ar), 737 cm^{-1}. ^{1}H NMR ($CDCl_3$): δ 7.25 (m, 45 H, Ar-H), 5.76 (d, 1H, J_{1-2} = 3.4 Hz, H-1), 4.82 (d, 1H, J = 10.8 Hz, Ar-CH_2), 4.78 (d, 1H, J = 11.0 Hz, Ar-CH_2), 4.54 (m, 12H, H-1',3',6,6', Ar-CH_2), 4.29 (t, 1H, $J_{3'-4'}$ = $J_{4'-5'}$ = 8.5 Hz, H-4'), 4.25 (d, 1H, J = 11.2 Hz, Ar-CH_2), 4.05 (m, 1H, H-5'), 3.95 (m, 3H, H-5, Ar-CH_2), 3.86 (t, 1H, J_{2-3} = J_{3-4} = 9.2 Hz, H-3), 3.75 (d, 1H, J = 11.0 Hz, Ar-CH_2), 3.63 (t, 1H, J_{3-4} = J_{4-5} = 9.6 Hz, H-4), 3.53 (d, 1H, J = 11.0 Hz, Ar-CH_2), 3.46 (dd, 1H, J_{1-2} = 3.4 Hz, J_{2-3} = 9.6 Hz, H-2), 3.41 (m, 2H, Ar-CH_2), 3.27 (d, 1 H, J = 10.3 Hz, Ar-CH_2), 1.02 (s, 9 H, CH_3). ^{13}C NMR ($CDCl_3$): δ 139.0, 138.7, 138.4, 138.3, 138.2, 1387.0, 135.6, 133.4, 133.2 (C_q), 129.6, 128.3, 127.9, 127.7, 127.5 (C_{Ar}), 104.6 (C-2'), 89.8 (C-1), 84.2, 82.7, 82.0, 81.3, 79.9, 75.5, 74.7, 73.4, 73.3, 73.1, 72.4, 72.1, 71.2, 70.6 (C-2,3,3',4,4',5,5', 7 × OCH$_2$Ph and C-1'), 68.4, 65.0 (C-6,6'), 26.9 (CH_3), 19.3, ($SiC(CH_3)_3$). Anal. calcd. for $C_{77}H_{82}O_{11}Si$: C, 76.33; H, 6.82. Found: C, 76.40; H, 7.00. Eluted next was 1',2,3,3',4,4',6,6'-octa-O-benzyl-sucrose **8** (0.096 g, 15%). $[\alpha]_D^{20}$ +37.2 (c 1.0, $CHCl_3$). ^{1}H NMR ($CDCl_3$): δ 7.24 (m, 40H, Ar-H), 5.71 (d, 1H, J_{1-2} = 3.0 Hz, H-1), 4.88 (d, 1H, J = 10.8 Hz, Ar-CH_2), 4.81 (d, 1H, J = 10.8 Hz, Ar-CH_2), 4.66 (m, 6H, H-6,6', Ar-CH_2), 4.45 (m, 15H, H-1',3', Ar-CH_2), 4.17 (t, 1H, $J_{3'-4'}$ = $J_{4'-5'}$ = 9.2 Hz, H-4'), 4.11 (m, 1H, H-5), 4.06 (m, 1H, H-5'), 3.93 (t, 1H, J_{2-3} = J_{3-4} = 9.2 Hz, H-3), 3.69 (m, 4H, H-4, Ar-CH_2), 3.51 (m, 3H, H-2, Ar-CH_2), 3.36 (d, 1H, J = 9.9Hz, Ar-CH_2). ^{13}C NMR ($CDCl_3$): δ 138.9, 138.7, 138.4,

138.2, 138.1, 138.0 (C_q), 128.5, 128.3, 128.2, 128.0, 127.9, 127.9, 127.7, 127.6, 127.5, 127.4, 126.9 (C_{Ar}), 104.6 (C-2'), 89.9 (C-1), 83.9, 82.5, 81.9, 79.8, 79.6, 77.7, 75.5, 74.8, 73.4, 73.2, 73.0, 72.5, 72.2, 71.4, 70.6 (C-2,3,3',4,4',5,5', 8 × OCH_2Ph), 68.5, 65.3 (C-1',6,6'). Anal. calcd. for $C_{68}H_{70}O_{11}$: C, 76.81; H, 6.64. Found: C, 77.13; H, 6.85.

1',2,3,3',4,4',6-Hepta-O-benzylsucrose (4)[13]

Tetrabutyl ammonium fluoride (1M solution in THF, 0.60 mL, 0.60 mmol) was added, at 0°C, to a solution of compound **3** (0.638 g, 0.60 mmol) in dry THF (10 mL). The reaction mixture was placed in the microwave oven and irradiated at 65°C and power of 300 W for 5 min, when TLC (4:1 hexane–EtOAc) showed that the starting material had been consumed. The solvent was evaporated, and the residue was dissolved in dichloromethane (30 mL), washed with H_2O (2 × 10 mL), dried over Na_2SO_4, and concentrated. The residue was chromatographed (4:1 hexane-EtOAc) to give first 0.050 g (10%) of the recovered starting material **3** (1',2,3,3',4,4',6-hepta-O-benzyl-6'-O-tert-butyldiphenylsilylsucrose). Eluted next was the desired compound **4** (0.301 g, 75%, colorless oil). $[\alpha]_D^{20} + 44.3$ (c 0.8, $CHCl_3$); Lit.[13] $[\alpha]_D^{20} + 46.1$ (c 1.3, $CHCl_3$). IR: (CH_2Cl_2) 3031, 2920, 2867 (C-H, sat.), 1606 (Bn), 1541, 1496, 1450 (Bn), 1398, 1360 (C-C-C), 1315, 1260, 1207, 1078 (C-O-C), 897 (Ar), 735 cm^{-1}. ^{1}H NMR ($CDCl_3$): δ 7.30 (m, 35H, Ar-*H*), 5.54 (d, 1 H, $J_{1-2} = 3.2$ Hz, H-1), 4.86 (d, 1H, $J = 10.9$ Hz, CH_2-Ph), 4.80 (d, 1H, $J = 10.8$ Hz, CH_2-Ph), 4.75 (d, 1H, $J = 10.9$ Hz, CH_2-Ph), 4.62 (m, 7H, H-3',6,6', CH_2-Ph), 4.47 (m, 4H, H-1', CH_2-Ph), 4.36 (m, 2H, CH_2-Ph), 4.27 (d, 1H, $J = 11.9$ Hz, CH_2-Ph), 4.09 (t, 1H, $J_{3'-4'} = J_{4'-5'} = 10.6$ Hz, H-4'), 4.00 (m, 1H, H-5), 3.96 (t, 1H, $J_{2-3} = J_{3-4} = 9.6$ Hz, H-3), 3.72 (t, 2H, $J_{3-4} = J_{4-5} = 9.6$ Hz, H-4, CH_2-Ph), 3.64 (m, 1H, H-5'), 3.59 (d, 2H, $J = 11.0$ Hz, CH_2-Ph), 3.54 (dd, 1H, $J_{1-2} = 3.3$ Hz, $J_{2-3} = 9.7$ Hz, H-2), 3.47 (d, 1H, $J = 11.0$ Hz, CH_2-Ph). ^{13}C NMR ($CDCl_3$): δ 138.7, 138.4, 138.1, 137.9, 137.8 (C_q benzyl groups), 128.4, 127.7 (C_{Ar}), 103.8 (C-2'), 91.1 (C-1), 83.6, 81.7, 81.2, 79.5, 79.4, 77.3, 77.0, 76.7, 75.6, 74.9, 73.5, 73.4, 72.9, 72.5, 71.3, 71.2 (C-2,3,3',4,4',5,5', 7 × OCH_2Ph and C-1'), 67.9, 61.2 (C-6,6'). Anal. calcd. for $C_{61}H_{64}O_{11}$: C, 75.29; H, 6.63. Found: C, 75.40; H, 6.81.

6,6'-Di-O-tert-butyldiphenylsilylsucrose (5)[11]

To a solution of sucrose (1.000 g, 2.92 mmol) in dry pyridine (15 mL) was added 4-dimethylaminopyridine (0.05 g) and tert-butyldiphenylsilyl chloride (1.768 g, 6.42 mmol, 2.2 equiv.). The mixture was placed in the microwave oven and irradiated at 90°C and 300 W for 20 min. After concentration, column chromatography (EtOAc → EtOAc-acetone-water, 100:100:1 → 10:10:1) gave **5** (1.557 g, 65%) as a white solid. m.p. 175–180°C (from CH_3OH); Lit.[11] 212–213°C (the compound is dimorphous); $[\alpha]_D^{20} + 27.2$ (c 1.0, CH_3OH); Lit.[11] $[\alpha]_D + 26.0$ (c 1.0, CH_3OH). IR: v_{max} (KBr): 3385 (O-H), 2928, 2855 (C-H, satd), 1472 (Si-Ar), 1427, 1390, 1362 (C-C-C), 1266 (Si-C), 1113 (Si-Ar), 1011 (C-O-C), 824 (Ar) cm^{-1}. ^{1}H NMR (DMSO-d_6): δ 7.60 (m, 8H, Ar-*H*), 7.32 (m, 12H, Ar-*H*), 5.32 (d, $J_{1-2} = 3.2$ Hz, 1H, H-1), 3.95 (d, $J = 7.5$ Hz, 1H, H-6a), 3.77 (m, 3H, H-4', H-3', H-5'), 3.65 (d, $J = 9.7$ Hz, 1H, H-6b), 3.52 (m, 3H, H-6', H-3), 3.35 (m, 3H, H-1', H-5), 3.18 (dd, $J_{1-2} = 3.3$ Hz, $J_{2-3} = 9.5$ Hz, 1H, H-2),

3.04 (t, $J_{3-4} = J_{4-5} = 9.3$ Hz, 1H, H-4), 0.92 (s, 9H, CH_3), 0.90 (s, 9H, CH_3). ^{13}C NMR (DMSO-d_6): δ 135.7, 134.0, 133.7, 130.2, 128.3 (C_{Ar}), 105.5 (C-2′), 92.7 (C-1), 82.8, 77.6, 75.6, 73.7, 73.2, 72.4, 70.1 (C-2,3,3′,4,4′,5,5′), 66.2 (C-1′), 63.8, 62.5 (C-6,6′), 27.2 (CH_3), 19.5 (SiC(CH_3)$_3$). Anal. calcd. for $C_{44}H_{58}O_{11}Si_2$: C, 64.52; H, 7.14. Found: C, 64.67; H, 6.81. Eluted later was 6′-O-tert-butyldiphenylsilylsucrose (2, 0.170 g, 10%), which was identical to substance 2 described above.

1′,2,3,3′,4,4′-Hexa-O-benzyl-6,6′-di-O-tert-butyldiphenylsilylsucrose (6)[14]

NaH (60% suspension in oil, 0.094 g, 2.34 mmol, 9.6 equiv.) was added carefully at 0°C to a stirred solution of 5 (0.200 g, 0.244 mmol) in DMF (5 mL). After 20 min, benzyl bromide (0.501 g, 2.93 mmol, 12 equiv.) was added dropwise over 15 min. The reaction mixture was placed in the microwave oven and irradiated for 5 min at 145°C and power of 300 W. TLC (10:10:1 ethyl acetate-acetone-water) showed that the starting material had been consumed. The mixture was poured into cold H_2O (20 mL), and the product was extracted with diethyl ether (4 × 15 mL). The combined organic layers were washed with H_2O (2 × 10 mL), dried over Na_2SO_4, and concentrated. The residue was chromatographed (10:1 → 5:1 hexane-ethyl acetate) to give compound 6 as colorless oil. Yield 0.166 g (50%). $[α]_D$ + 30.1 (c 1.0, CHCl$_3$); Lit.[14] $[α]_D$ + 29.8 (c 1.0, CHCl$_3$). ^{1}H NMR (CDCl$_3$): δ 7.66 (m, 10H, Ar-H), 7.22 (m, 40H, Ar-H), 5.95 (d, 1H, J_{1-2} = 3.0 Hz, H-1), 4.91 (d, 1H, J = 10.8 Hz, CH_2-Ph), 4.84 (d, 1H, J = 10.5 Hz, CH_2-Ph), 4.72 (m, 3H, H-3′, CH_2-Ph), 4.60 (m, 2H, H-4′, CH_2-Ph), 4.46 (m, 5H, H-6,6′, CH_2-Ph), 4.33 (m, 3H, H-5, CH_2-Ph), 4.00 (m, 1H, CH_2-Ph), 3.91 (m, 2H, H-5′, CH_2-Ph), 3.85 (m, 1H, H-3), 3.74 (d, 1H, J = 10.8 Hz, CH_2-Ph), 3.57 (m, 4H, H-1′, 4, CH_2-Ph), 3.37 (dd, 1H, J_{1-2} = 3.3 Hz, J_{2-3} = 9.4 Hz, H-2), 1.06 (s, 9H, CH_3), 1.01 (s, 9H, CH_3). ^{13}C NMR (CDCl$_3$): δ 139.0, 138.3, 138.1, 138.0 (C_q benzyl groups), 129.6, 128.2, 127.8, 127.7, 127.6 (C_{Ar}), 104.4 (C-2′), 89.5 (C-1), 84.3, 82.2, 82.0, 80.7, 77.3, 76.7, 75.8, 74.7, 73.5, 72.6, 72.4, 71.9, 71.6, 69.5 (C-2,3,3′,4,4′,5,5′, 6 × OCH_2Ph and C-1′), 64.6, 62.3 (C-6,6′), 26.9 (CH_3), 19.3 (SiC(CH_3)$_3$). Anal. calcd. for $C_{86}H_{94}O_{11}Si_2$: C, 75.96; H, 6.97. Found: C, 75.65; H, 6.84. Eluted later were 1′,2,3,3′,4,4′,6-hepta-O-benzyl-6′-O-tert-butyldiphenylsilyl-sucrose (3, 0.074 g, 25%) and 1′,2,3,3′,4,4′,6,6′-octa-O-benzylsucrose (8, 0.026 g, 10%), identical to materials described above.

1′,2,3,3′,4,4′-Hexa-O-benzylsucrose (7)[15]

Tetrabutyl ammonium fluoride (1M solution in THF, 0.13 mL, 0.13 mmol) was added at 0°C to a solution of 6 (0.150 g, 0.11 mmol) in dry THF (4 mL). The mixture was placed in the microwave oven and irradiated for 5 min at 65°C and 300 W. TLC (4:1 hexane-EtOAc) showed that the starting material had been consumed. The solvent was evaporated and a solution of the residue in dichloromethane (10 mL) was washed with H_2O (2 × 5 mL), dried over Na_2SO_4 and concentrated. The residue was purified by flash column chromatography (eluent hexane-ethyl acetate, 4:1 → 1:1) to give 7 as colorless oil (0.063 g, 65%). $[α]_D^{20}$ + 41.5 (c 1.0, CHCl$_3$); Lit.[15] $[α]_D$ + 40.8 (c 1.0,

CHCl$_3$). ^{1}H NMR (CDCl$_3$): δ 7.27 (m, 30H, Ar-H), 5.49 (d, 1H, J_{1-2} = 2.9 Hz, H-1), 4.86 (d, 2H, J = 10.6Hz, Ar-CH_2), 4.76 (d, 1H, J = 10.9 Hz, Ar-CH_2), 4.64 (m, 6H, H-6,6′, Ar-CH_2), 4.46 (m, 3H, H-3′, Ar-CH_2), 4.32 (m, 2H, H-4′, Ar-CH_2), 4.14 (m, 1H, H-5), 3.98 (m, 2H, H-4, Ar-CH_2), 3.82 (m, 2H, H-3, Ar-CH_2), 3.64 (m, 1H, H-5′), 3.57 (m, 2H, H-1′), 3.50 (dd, 1H, J_{1-2} = 3.2 Hz, J_{2-3} = 9.6 Hz, H-2), 3.43 (m, 3H, H-4, Ar-CH_2). ^{13}C NMR (CDCl$_3$): δ 138.6, 138.3, 138.1, 138.0, 137.7 (C$_q$), 128.4, 127.9, 127.8, 127.7, 127.6 (C$_{Ar}$), 103.9 (C-2′), 90.7 (C-1), 83.5, 81.7, 81.0, 79.9, 79.5, 77.6 and 71.3 (C-2,3,3′,4,4′,5,5′), 75.6, 75.0, 73.4, 73.1, 73.0, 72.5 (C-1′, 6 × OCH$_2$Ph), 61.9, 61.0 (C-6,6′). Anal. calcd. for C$_{54}$H$_{58}$O$_{11}$: C, 73.45; H, 6.62. Found: C, 73.23; H, 6.84.

ACKNOWLEDGMENTS

This work has been supported by Fundação para a Ciência e a Tecnologia through Grant no. PEst-C/EQB/LA0006/2011.

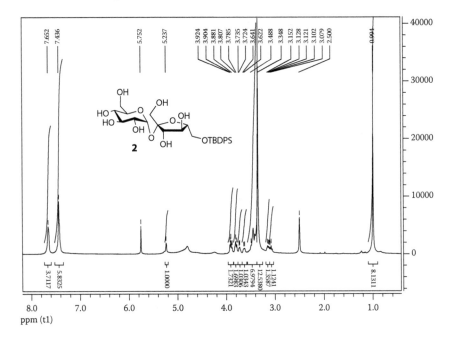

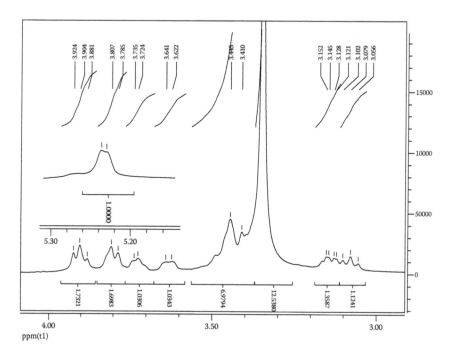

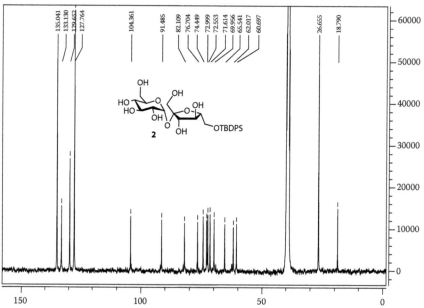

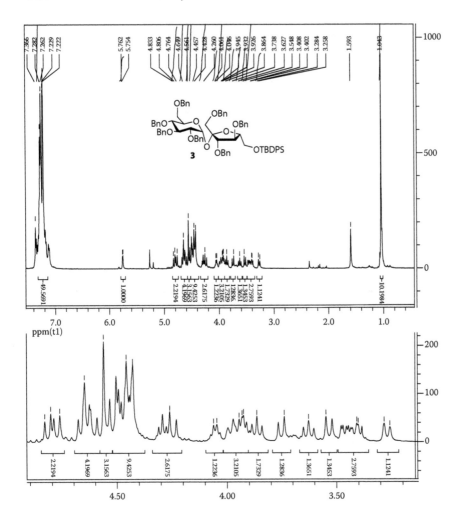

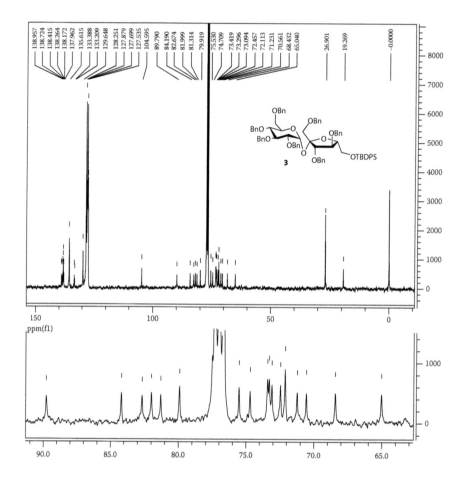

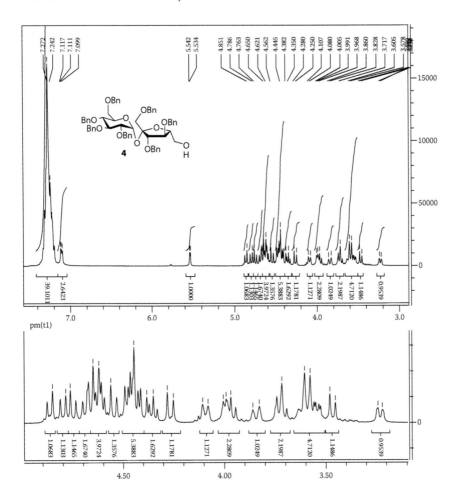

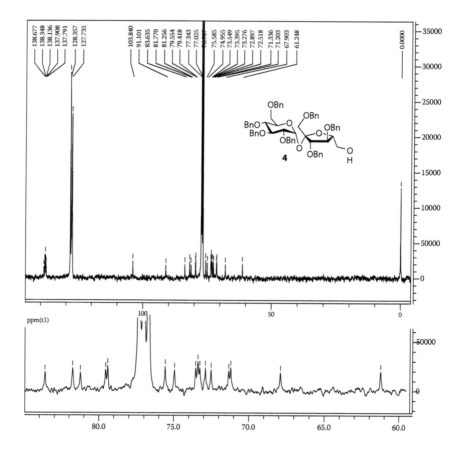

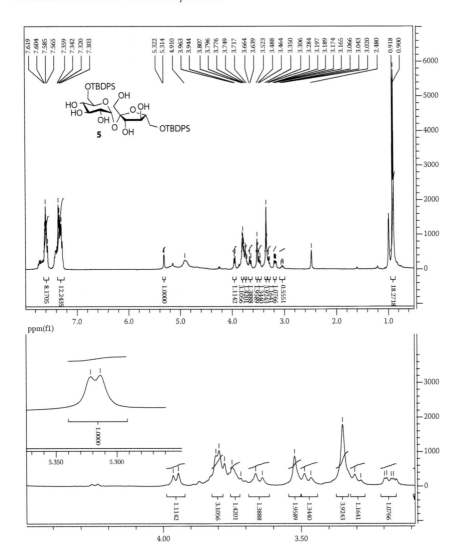

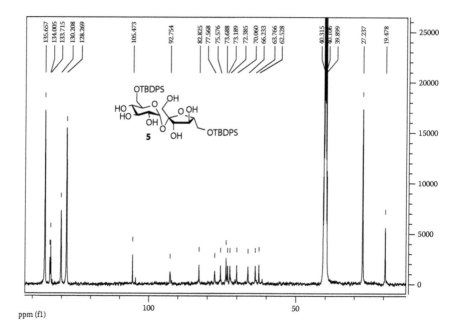

ppm (f1)

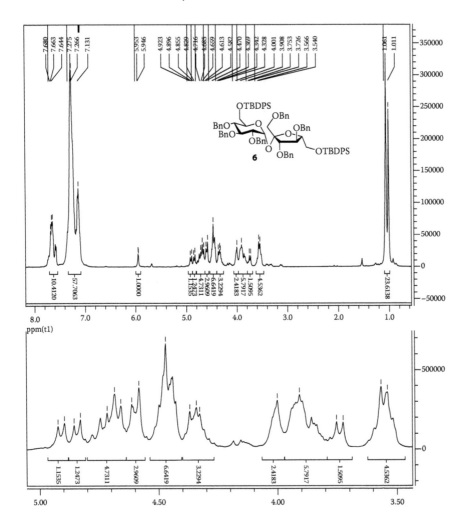

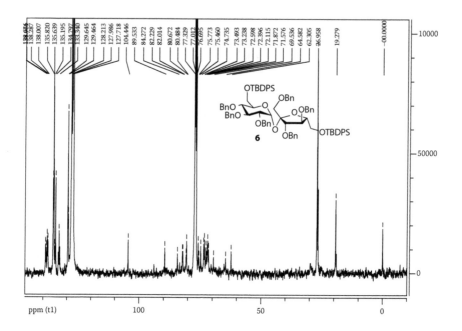

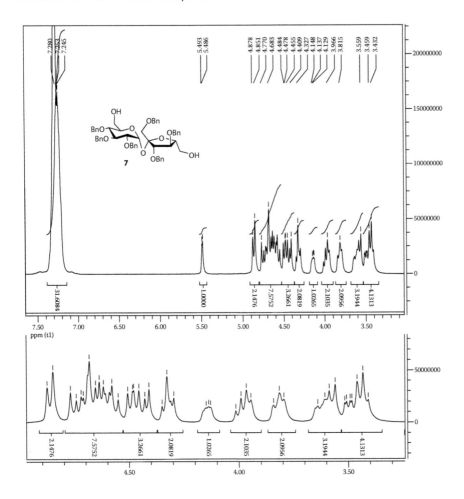

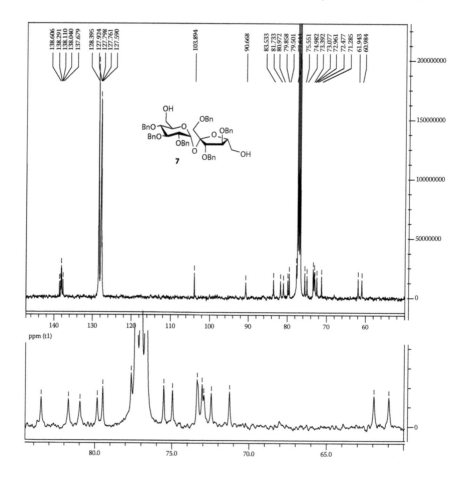

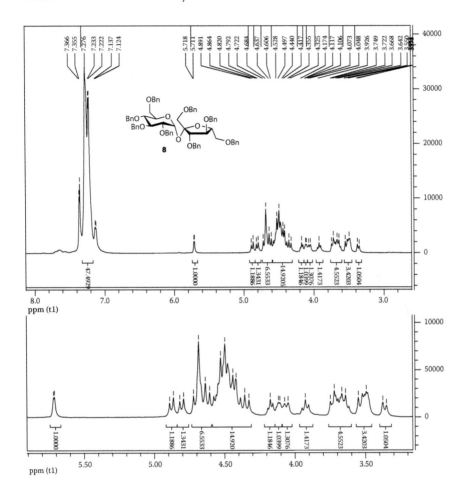

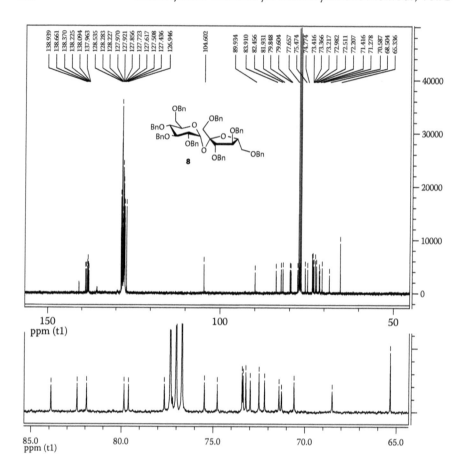

REFERENCES

1. Jarosz, S.; Mach, M. *Eur. J. Org. Chem.* 2002, 769–780.
2. Barros, M. T.; Sineriz, F. *Synthesis* 2002, *10*, 1407–1411.
3. Barros, M. T.; Maycock, C. D.; Sineriz, F.; Thomassigny, C. *Tetrahedron* 2000, *56*, 6511–6516.
4. Crucho, C. C.; Petrova, K. T.; Pinto, R. C.; Barros, M. T. *Molecules* 2008, *13*, 762–770.
5. Lidstrom, P.; Tierney, J.; Wathey, B.; Westman, J. *Tetrahedron* 2001, *57*, 9225–9283.
6. Corsaro, A.; Chiacchio, U.; Pistarà, V.; Romeo, G. *Curr. Org. Chem.* 2004, *8*, 511–538.
7. Soderberg, E.; Westman, J.; Oscarson, S. *J. Carbohydr. Chem.* 2001, *20*, 397–410.
8. Queneau, Y.; Jarosz, S.; Lewandowski, B.; Fitremann, J. *Adv. Carbohydr. Chem. Biochem.* 2007, *61*, 217–292.
9. Milestone: www.milestonesrl.com.
10. Perrin, D. D.; Armagedo, W. L. F.; Perrin, D. R. *Purification of Laboratory Chemicals*; 2nd ed.; Pergamon Press: New York, 1980.
11. Karl, H.; Lee, C. K.; Khan, R. *Carbohydr. Res.* 1982, *101*, 31–38.

12. Barros, M. T.; Maycock, C. D.; Rodrigues, P.; Thomassigny, C. *Carbohydr. Res.* 2004, *339*, 1373–1376.
13. Barros, M. T.; Petrova, K.; Ramos, A. M. *J. Org. Chem.* 2004, *69*, 7772–7775.
14. Gawet, A.; Jarosz, S. *J. Carbohydr. Chem.* 2010, *29*, 332–347.
15. Mach, M.; Jarosz, S.; Listkowski, A. *J. Carbohydr. Chem.* 2001, *20*, 485–493.

Section II

Synthetic Intermediates

15 Oligosaccharide-Salicylaldehyde Derivatives as Precursors of Water-Soluble, Biocompatible Anion Receptors

*Emiliano Bedini**
Gianpiero Forte
Ivone Carvalho†
Michelangelo Parrilli
Antonella Dalla Cort

CONTENTS

---

* Corresponding author: ebedini@unina.it.
† Checker: carronal@usp.br

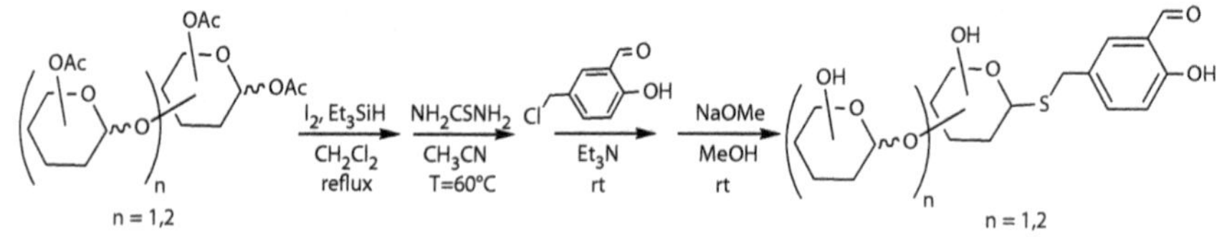

SCHEME 15.1

Selective recognition of anions in water is important in supramolecular and analytical chemistry due to their involvement in environmental pollution and in many biochemical processes.[1] It requires development of water-soluble receptors which are able to overcome the strong competition of water molecules.[2] Many compounds have been synthesized for this purpose.[3] In general, to make organic receptors soluble in aqueous media, ionic groups (carboxylate, sulfonate units, pyridinium salts, etc.) are introduced on the ligand skeleton,[4] although these additional charges could sometimes interfere with the anion recognition process. To circumvent this problem, uncharged hydrophilic and biocompatible units, such as carbohydrates, could be introduced. Metal-salophen complexes are widely known as anion receptors in organic solvents.[5] Recently, a glucose appended Zn^{2+}-salophen complex has been synthesized from diacetone-D-glucose (1,2:5,6-di-*O*-isopropylidene-α-D-glucose), 5-chloromethyl-salicylaldehyde and 1,2-diaminobenzene as organic starting materials (Figure 15.1).[6] It is a highly selective aminoacid receptor in water through carboxylate-Zn^{2+} coordination and ammonium-glucose hydrogen bond interaction. Unexpectedly, the association constants of the complexes were strongly dependent on the aminoacid structure.

This result prompts modulation of the structures of both carbohydrate and salophen skeleton of the ligand in order to obtain a library of complexes with finely

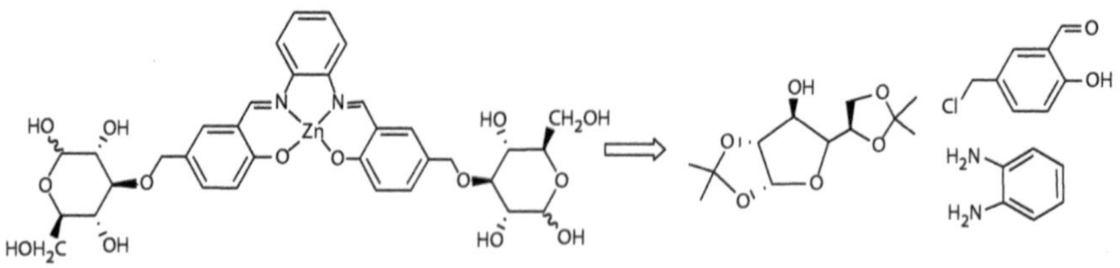

FIGURE 15.1   Glucose-appended Zn^{2+}-salophen complex.

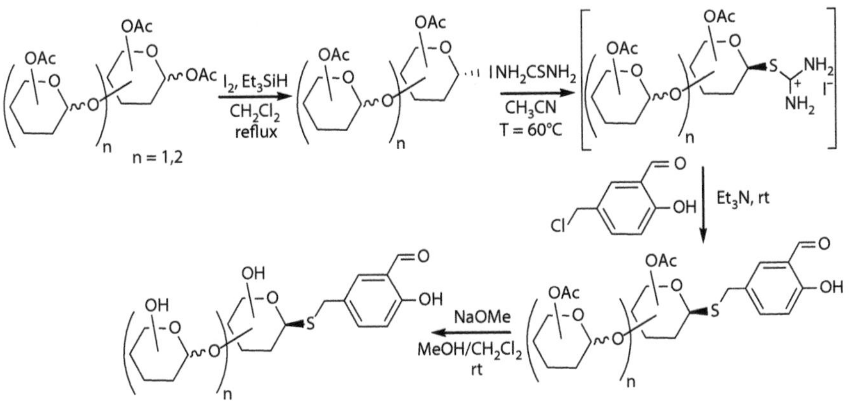

FIGURE 15.2 General scheme for the obtainment of oligosaccharide-salicylaldehyde thio-glycoside derivatives.

tuned selectivity for aminoacid recognition in water. In particular, to guarantee high water solubility of the ligand having variously modified salophen skeleton, oligosaccharides could be selected as appendages. We present here a general method for oligosaccharide-salicylaldehyde conjugation, exploiting the S-alkylation with 5-chloromethyl-salicylaldehyde[7] of glycosyl isothiouronium salts, which are readily available from per-O-acetylated oligosaccharides through glycosyl iodides (Figure 15.2).[8] The procedure is rather fast because three of the four reactions take not more than 1 hour to complete, and only a single chromatographic purification is needed. The conjugation proceeds with high stereoselectivity: in all cases 1,2-*trans* thioglycosides largely predominate, and traces of the minor anomer can be easily separated by chromatography before the de-O-acetylation step. It is worth noting that, unlike most thioglycosylation protocols, this conjugation procedure avoids the use of malodorous thiols or harsh reagents that could be incompatible with the O-glycosidic bonds in oligosaccharides. Indeed, α and β as well as 1→4 and 1→6 linkages were all stable to reaction conditions.

Oligosaccharide-salicylaldehyde conjugates **5-8** (Table 15.1) are ready for one-step metal-salophen hydrosoluble complex formation by metal-templated reductive amination with 1,2-diaminobenzene.

EXPERIMENTAL

GENERAL METHODS

[1]H and ^{13}C nuclear magnetic resonance (NMR) spectra were recorded on Varian INOVA 500 ([1]H: 500 MHz, [13]C: 125 MHz), Bruker DRX-400 ([1]H: 400 MHz, [13]C: 100 MHz), and Varian XL-200 ([1]H: 200 MHz, [13]C: 50 MHz) instruments in D_2O (acetone as internal standard, [1]H: $(CH_3)_2CO$ at δ 2.22; [13]C: $(CH_3)_2CO$ at δ 30.9) or $CDCl_3$ ([1]H: $CHCl_3$ at δ 7.26; [13]C: $CDCl_3$ at δ 77.0). [1]H spectra in D_2O were measured by presaturating the HDO signal. [1]H chemical shift assignments were made from decoupling experiments. Positive matrix-assisted laser desorption ionization

TABLE 15.1
Synthesis of Di- and Trisaccharide-salicylaldehyde Thioglycoside Derivatives

entry[a]	per-O-acetyl oligosaccharide	thioglycoside (% yield)
1	cellobiose	**1** R=Ac (64%) **5** R=H (99%)
2	lactose	**2** R=Ac (65%) **6** R=H (98%)
3	gentiobiose	**3** R=Ac (53%) **7** R=H (97%)
4[b]	maltotriose	**4** R=Ac (56%) **8** R=H (99%)

[a] Reaction conducted on 0.66 mmol, unless otherwise specified
[b] Reaction conducted on 0.22 mmol

mass spectrometry (MALDI-MS) spectra were recorded on an Applied Biosystem Voyager DE-PRO MALDI-TOF mass spectrometer in the positive mode: compounds were dissolved in CH_3CN at a concentration of 1 mg/mL, and 1 µL of these solutions was mixed with 1 µL of a 20 mg/mL solution of 2,5-dihydroxybenzoic acid in 7:3 v/v CH_3CN/water. Optical rotations were measured on a JASCO P-1010 polarimeter. Elemental analyses were performed on a Carlo Erba 1108 instrument. Flash chromatography was performed on Merck Kieselgel 60 (40–63 mesh).

THIOGLYCOSYLATION PROCEDURE

In a round-bottom flask, per-O-acetylated oligosaccharide (660 μmol) was dissolved in anhydrous CH_2Cl_2 (3.2 mL) and treated with I_2 (226 mg, 890 μmol) and then Et_3SiH (142 μL, 890 μmol).* The solution was stirred at reflux with a water-cooled condenser fitted with a drying tube (anhydrous $CaCl_2$). After 30 min, the solution was cooled to rt, diluted with CH_2Cl_2 (80 mL), and washed with 1:1 v/v 1M $NaHCO_3$/10% $Na_2S_2O_3$ mixture (80 mL). The organic layer was dried over anhydrous Na_2SO_4, filtered, and concentrated. The residue was dissolved in acetonitrile (3.2 mL) and treated with thiourea (55.2 mg, 725 μmol). The flask was closed with an air condenser, and the mixture was stirred at 60°C. After a few minutes the mixture turned to a clear yellowish solution, and stirring was continued for 1 hour. After cooling to rt, 5-chloromethyl-salicylaldehyde (191 mg, 1.12 mmol) was added, followed by triethylamine (367 μL, 2.64 mmol). The stirring was continued while a white precipitate was gradually formed. After 1 hour the mixture was concentrated, and the residue was chromatographed (4:1→2:3 v/v hexane–ethyl acetate).

DEPROTECTION PROCEDURE

Thioglycoside (100 μmol) was dissolved in 4:1 MeOH–CH_2Cl_2 (7.5 mL),† cooled to 0°C, and treated with freshly prepared 0.15M methanolic NaOMe (1.3 mL). The yellow solution was stirred at rt in a round-bottom flask closed with a glass stopper. After 5 hours the solution was treated with Amberlyst® 15 (H^+ form) until it turned colorless. It was then filtered and concentrated.

m-FORMYL-*p*-HYDROXYBENZYL 2,3,4,6-TETRA-*O*-ACETYL-β-D-GLUCOPYRANOSYL-(1→4)-2,3,6-TRI-*O*-ACETYL-1-THIO-β-D-GLUCOPYRANOSIDE (1)

White powder; R_f (1:2 v/v hexane-ethyl acetate) 0.50; $[\alpha]_D$ −36 (c 0.5, CH_2Cl_2); ^1H NMR (500 MHz, $CDCl_3$): δ 10.95 (bs, 1H, OH), 9.85 (s, 1H, CHO), 7.48 (d, 1H, J_{meta} 1.5 Hz, H-Ar), 7.44 (dd, 1H, J_{ortho} 8.5 Hz, J_{meta} 1.5 Hz, H-Ar), 6.92 (d, 1H, J_{ortho} 8.5 Hz, H-Ar), 5.13 (t, 1H, $J_{3,4}=J_{3,2}$ 9.5 Hz, H-3^I), 5.12 (t, 1H, $J_{3,4}=J_{3,2}$ 9.5 Hz, H-3^{II}), 5.04 (t, 1H, $J_{4,3}=J_{4,5}$ 9.5 Hz, H-4^{II}), 4.97 (t, 1H, $J_{2,3}=J_{2,1}$ 9.5 Hz, H-2^I), 4.89 (t, 1H, $J_{2,3}=J_{2,1}$ 9.5 Hz, H-2^{II}), 4.51 (m, 2H, H-1^{II}, H-$6a^I$), 4.34 (dd, 1H, J_{gem} 12.5 Hz, $J_{6a,5}$ 4.0 Hz, H-$6a^{II}$), 4.31 (d, 1H, $J_{1,2}$ 9.5 Hz, H-1^I), 4.06 (dd, 1H, J_{gem} 12.0 Hz, $J_{6b,5}$ 5.0 Hz, H-$6b^I$), 4.02 (dd, 1H, J_{gem} 12.5 Hz, $J_{6b,5}$ 1.5 Hz, H-$6b^{II}$), 3.87 (d, 1H, J_{gem} 13.0 Hz, SC*H*HAr), 3.79 (d, 1H, J_{gem} 13.0 Hz, SCH*H*Ar), 3.75 (t, 1H, $J_{4,3}=J_{4,5}$ 9.5 Hz, H-4^I), 3.64 (m, 1H, H-5^{II}), 3.54 (m, 1H, H-5^I), 2.11 (s, 3H, CH_3CO), 2.05 (s, 3H, CH_3CO), 2.02 (s, 3H, CH_3CO), 2.00 (s, 3H, CH_3CO), 1.99 (s, 3H, CH_3CO), 1.98 (s, 3H, CH_3CO), 1.96 (s, 3H, CH_3CO). ^{13}C NMR (125 MHz, $CDCl_3$): δ 196.2 (CHO), 170.3–168.9 (COCH_3), 160.8 (C-OH Ar), 137.6, 133.7, 128.3, 120.4, 117.9 (C-Ar), 100.6 (C-1^{II}), 81.8 (C-1^I), 77.2, 76.2, 73.2, 72.8, 71.9, 71.5, 69.9, 67.7, 61.9, 61.4 (C-2^I, C-2^{II}, C-3^I, C-3^{II}, C-4^I,

---

* *Caution*: exothermic reaction.
† Deprotection of trisaccharide **4** (Table 15.1, entry 4) required, after ~4 hours, addition of more MeOH (5.0 mL) to dissolve the precipitate (partially deprotected species) and allow the completion of the global deacetylation.

C-4^{II}, C-5^{I}, C-5^{II}, C-6^{I}, C-6^{II}), 32.5 (SCH_2Ar), 20.7–20.4 (CH_3CO). TOF-MS: [M + H]$^+$ calcd. for $C_{34}H_{42}O_{19}S$ (m/z), 786.20; found, 809.12 ([M + Na]$^+$). Anal. Calcd for $C_{34}H_{42}O_{19}S$: C, 51.91; H, 5.38; S, 4.08. Found: C, 51.80; H, 5.45; S, 4.00.

m-FORMYL-p-HYDROXYBENZYL 2,3,4,6-TETRA-O-ACETYL-β-D-GALACTOPYRANOSYL-(1→4)-2,3,6-TRI-O-ACETYL-1-THIO-β-D-GLUCOPYRANOSIDE (2)

White powder; R_f (1:2 v/v hexane-ethyl acetate) 0.55; [α]$_D$ –39.4 (c 2.0, CH_2Cl_2); ^{1}H NMR (500 MHz, CDCl$_3$): δ 10.98 (s, 1H, OH), 9.88 (s, 1H, CHO), 7.50 (d, 1H, J_{meta} 2.0 Hz, H-Ar), 7.47 (dd, 1H, J_{ortho} 8.5 Hz, J_{meta} 2.0 Hz, H-Ar), 6.95 (d, 1H, J_{ortho} 8.5 Hz, H-Ar), 5.34 (d, 1H, $J_{4,3}$ 3.0 Hz, H-4^{II}), 5.17 (t, 1H, $J_{3,4}=J_{3,2}$ 9.5 Hz, H-3^{I}), 5.10 (dd, 1H, $J_{2,1}$ 10.5 Hz, $J_{2,3}$ 8.0 Hz, H-2^{II}), 4.99 (t, 1H, $J_{2,1}=J_{2,3}$ 10.0 Hz, H-2^{I}), 4.96 (dd, 1H, $J_{3,2}$ 10.5 Hz, $J_{3,4}$ 3.5 Hz, H-3^{II}), 4.51 (dd, 1H, J_{gem} 12.5 Hz, $J_{6a,5}$ 2.0 Hz, H-6a^{I}), 4.48 (d, 1H, $J_{1,2}$ 8.0 Hz, H-1^{II}), 4.33 (d, 1H, $J_{1,2}$ 10.5 Hz, H-1^{I}), 4.09 (m, 3H, H-6a^{II}, H-6b^{I}, H-6b^{II}), 3.88 (m, 2H, H-5^{II}, SCHHAr), 3.81 (d, 1H, J_{gem} 12.5 Hz, SCHHAr), 3.78 (t, 1H, $J_{4,3}=J_{4,5}$ 9.5 Hz, H-4^{I}), 3.56 (m, 1H, H-5^{I}), 2.15 (s, 3H, CH_3CO), 2.13 (s, 3H, CH_3CO), 2.05 (s, 3H, CH_3CO), 2.04 (s, 6H, 2 CH_3CO), 2.03 (s, 3H, CH_3CO), 1.96 (s, 3H, CH_3CO). ^{13}C NMR (100 MHz, CDCl$_3$): δ 196.2 (CHO), 170.3–169.0 (COCH$_3$), 160.9 (C-OH Ar), 137.7, 133.7, 128.3, 120.4, 118.0 (C-Ar), 101.0 (C-1^{II}), 81.8 (C-1^{I}), 76.8, 76.1, 73.6, 70.9, 70.7, 70.1, 69.1, 66.5, 62.1, 60.7 (C-2^{I}, C-2^{II}, C-3^{I}, C-3^{II}, C-4^{I}, C-4^{II}, C-5^{I}, C-5^{II}, C-6^{I}, C-6^{II}), 32.6 (SCH_2Ar), 20.8–20.5 (CH_3CO). TOF-MS: [M + H]$^+$ calcd. for $C_{34}H_{42}O_{19}S$ (m/z), 786.20; found, 809.09 ([M + Na]$^+$). Anal. Calcd for $C_{34}H_{42}O_{19}S$: C, 51.91; H, 5.38; S, 4.08. Found: C, 51.74; H, 5.50; S, 3.98.

m-FORMYL-p-HYDROXYBENZYL 2,3,4,6-TETRA-O-ACETYL-β-D-GLUCOPYRANOSYL-(1→6)-2,3,4-TRI-O-ACETYL-1-THIO-β-D-GLUCOPYRANOSIDE (3)

White powder; R_f (1:2 v/v hexane-ethyl acetate) 0.50; [α]$_D$ –29.3 (c 1.0, CH_2Cl_2); ^{1}H NMR (500 MHz, CDCl$_3$): δ 9.90 (s, 1H, CHO), 7.53 (d, 1H, J_{meta} 2.0 Hz, H-Ar), 7.49 (dd, 1H, J_{ortho} 8.5 Hz, J_{meta} 2.0 Hz, H-Ar), 6.98 (d, 1H, J_{ortho} 8.5 Hz, H-Ar), 5.21 (t, 1H, $J_{3,4}=J_{3,2}$ 9.5 Hz, H-3^{II}), 5.13 (t, 1H, $J_{3,4}=J_{3,2}$ 9.5 Hz, H-3^{I}), 5.09 (t, 1H, $J_{4,3}=J_{4,5}$ 9.5 Hz, H-4^{II}), 5.04 (m, 2H, H-2^{I}, H-2^{II}), 4.89 (t, 1H, $J_{4,3}=J_{4,5}$ 9.5 Hz, H-4^{I}), 4.54 (d, 1H, $J_{1,2}$ 7.5 Hz, H-1^{II}), 4.30 (dd, 1H, J_{gem} 12.0 Hz, $J_{6a,5}$ 4.5 Hz, H-6a^{II}), 4.22 (d, 1H, $J_{1,2}$ 10.0 Hz, H-1^{I}), 4.13 (dd, 1H, J_{gem} 12.0 Hz, $J_{6b,5}$ 2.0 Hz, H-6b^{II}), 3.95 (d, 1H, J_{gem} 13.0 Hz, SCHHAr), 3.89 (d, 1H, J_{gem} 8.5 Hz, H-6a^{I}), 3.78 (d, 1H, J_{gem} 13.0 Hz, SCHHAr), 3.70 (m, 1H, H-5^{II}), 3.59 (m, 2H, H-5^{I}, H-6b^{I}), 2.11 (s, 3H, CH_3CO), 2.03 (s, 9H, 3 CH_3CO), 2.01 (s, 3H, CH_3CO), 1.99 (s, 3H, CH_3CO), 1.98 (s, 3H, CH_3CO). ^{13}C NMR (125 MHz, CDCl$_3$): δ 196.0 (CHO), 170.6–169.4 (COCH$_3$), 160.9 (C-OH Ar), 137.9, 134.1, 128.1, 120.4, 118.2 (C-Ar), 100.7 (C-1^{II}), 81.2 (C-1^{I}), 73.6, 72.7, 72.0, 71.1, 71.0, 69.7, 68.9, 68.5, 68.3, 61.7 (C-2^{I}, C-2^{II}, C-3^{I}, C-3^{II}, C-4^{I}, C-4^{II}, C-5^{I}, C-5^{II}, C-6^{I}, C-6^{II}), 32.2 (SCH_2Ar), 20.8–20.5 (CH_3CO). TOF-MS: [M + H]$^+$ calcd. for $C_{34}H_{42}O_{19}S$ (m/z), 786.20; found, 809.00 ([M + Na]$^+$). Anal. Calcd for $C_{34}H_{44}O_{19}S$: C, 51.91; H, 5.38; S, 4.08. Found: C, 51.81; H, 5.45; S, 3.99.

m-FORMYL-*p*-HYDROXYBENZYL 2,3,4,6-TETRA-*O*-ACETYL-α-D-GLUCOPYRANOSYL-(1→4)-2,3,6-TRI-*O*-ACETYL-α-D-GLUCOPYRANOSYL-(1→4)-2,3,6-TRI-*O*-ACETYL-1-THIO-β-D-GLUCOPYRANOSIDE (4)

White powder; R_f (1:1 v/v hexane-ethyl acetate) 0.20; $[\alpha]_D$ +53 (*c* 0.7, CH$_2$Cl$_2$); ^{1}H NMR (500 MHz, CDCl$_3$): δ 11.03 (s, 1H, OH), 9.92 (s, 1H, CHO), 7.53 (bs, 1H, H-Ar), 7.49 (dd, 1H, J_{ortho} 8.5 Hz, J_{meta} 1.5 Hz, H-Ar), 6.98 (d, 1H, J_{ortho} 8.5 Hz, H-Ar), 5.38 (m, 3H, H-1^{II}, H-3^{II}, H-3^{III}), 5.27 (d, 1H, $J_{1,2}$ 4.0 Hz, H-1^{III}), 5.23 (t, 1H, $J_{3,4} = J_{3,2}$ 9.5 Hz, H-3^I), 5.07 (t, 1H, $J_{4,3} = J_{4,5}$ 10.0 Hz, H-4^{III}), 4.90 (t, 1H, $J_{2,3} = J_{2,1}$ 9.5 Hz, H-2^I), 4.86 (dd, 1H, $J_{2,3}$ 10.0 Hz, $J_{2,1}$ 4.0 Hz, H-2^{III}), 4.72 (dd, 1H, $J_{2,3}$ 10.0 Hz, $J_{2,1}$ 3.5 Hz H-2^{II}), 4.46 (m, 2H, H-6a^I, H-6a^{II}), 4.34 (m, 2H, H-1^I, H-6b^I), 4.25 (dd, 1H, J_{gem} 12.5 Hz, $J_{6a,5}$ 3.5 Hz, H-6a^{III}), 4.19 (dd, 1H, J_{gem} 12.0 Hz, $J_{6b,5}$ 3.5 Hz, H-6b^{II}), 4.06 (dd, 1H, J_{gem} 12.5 Hz, $J_{6b,5}$ 1.5 Hz, H-6b^{III}), 4.05–3.92 (m, 5H, H-4^I, H-4^{II}, H-5^{II}, H-5^{III}, SC*H*HAr), 3.80 (d, 1H, J_{gem} 13.0 Hz, SCH*H*Ar), 3.64 (m, 1H, H-5^I), 2.20 (s, 3H, CH$_3$CO), 2.16 (s, 3H, CH$_3$CO), 2.10 (s, 3H, CH$_3$CO), 2.05 (s, 6H, 2 CH$_3$CO), 2.04 (s, 3H, CH$_3$CO), 2.00 (s, 9H, 3 CH$_3$CO), 1.98 (s, 3H, CH$_3$CO). ^{13}C NMR (125 MHz, CDCl$_3$): δ 196.0 (CHO), 170.1–169.0 (*C*OCH$_3$), 160.5 (*C*-OH Ar), 137.3, 133.5, 128.0, 120.1, 117.7 (C-Ar), 95.4, 95.2 (C-1^{II}, C-1^{III}), 80.8 (C-1^I), 75.8, 73.5, 72.1, 71.2, 70.3, 70.0, 69.8, 68.9, 68.6, 68.1, 67.5, 62.8, 62.0, 61.0, 59.9 (C-2^I, C-2^{II}, C-2^{III}, C-3^I, C-3^{II}, C-3^{III}, C-4^I, C-4^{II}, C-4^{III}, C-5^I, C-5^{II}, C-5^{III}, C-6^I, C-6^{II}, C-6^{III}), 32.1 (S*C*H$_2$Ar), 20.6–20.1 (*C*H$_3$CO). TOF-MS: [M + H]$^+$ calcd. for C$_{46}$H$_{58}$O$_{27}$S (*m/z*), 1074.29; found, 1096.99 ([M + Na]$^+$). Anal. Calcd for C$_{46}$H$_{58}$O$_{27}$S: C, 51.39; H, 5.44; S, 2.98. Found: C, 51.51; H, 5.55; S, 3.06.

m-FORMYL-*p*-HYDROXYBENZYL β-D-GLUCOPYRANOSYL-(1→4)-1-THIO-β-D-GLUCOPYRANOSIDE (5)

White powder; R_f (8:2 v/v CH$_2$Cl$_2$-MeOH) 0.30; $[\alpha]_D$ -50 (*c* 0.5, H$_2$O); ^{1}H NMR (500 MHz, D$_2$O): δ 9.91 (s, 1H, CHO), 7.72 (d, 1H, J_{meta} 2.0 Hz, H-Ar), 7.63 (dd, 1H, J_{ortho} 9.0 Hz, J_{meta} 2.0 Hz, H-Ar), 7.00 (d, 1H, J_{ortho} 9.0 Hz, H-Ar), 4.48 (d, 1H, $J_{1,2}$ 8.0 Hz, H-1^{II}), 4.31 (d, 1H, $J_{1,2}$ 10.0 Hz, H-1^I), 4.03–3.27 (m, 14H). ^{13}C NMR (50 MHz, CDCl$_3$): δ 197.7 (CHO), 159.5 (*C*-OH Ar), 138.7, 134.1, 130.6, 121.4, 118.1 (C-Ar), 103.1 (C-1^{II}), 84.6 (C-1^I), 79.2, 76.6, 76.3, 76.2, 73.8, 72.6, 70.1, 70.0, 61.2, 60.8 (C-2^I, C-2^{II}, C-3^I, C-3^{II}, C-4^I, C-4^{II}, C-5^I, C-5^{II}, C-6^I, C-6^{II}), 33.4 (S*C*H$_2$Ar). TOF-MS: [M + H]$^+$ calcd for C$_{20}$H$_{28}$O$_{12}$S (*m/z*), 492.13; found, 515.21 ([M + Na]$^+$). Anal. Calcd for C$_{20}$H$_{28}$O$_{12}$S: C, 48.77; H, 5.73; S, 6.51. Found: C, 48.54; H, 5.83; S, 6.40.

m-FORMYL-*p*-HYDROXYBENZYL β-D-GALACTOPYRANOSYL-(1→4)-1-THIO-β-D-GLUCOPYRANOSIDE (6)

White powder; R_f (8:2 v/v CH$_2$Cl$_2$-MeOH) 0.30; $[\alpha]_D$ -54 (*c* 0.6, H$_2$O); ^{1}H NMR (500 MHz, D$_2$O): δ 9.92 (s, 1H, CHO), 7.71 (d, 1H, J_{meta} 2.0 Hz, H-Ar), 7.62 (dd, 1H, J_{ortho} 9.0 Hz, J_{meta} 2.0 Hz, H-Ar), 7.00 (d, 1H, J_{ortho} 9.0 Hz, H-Ar), 4.42 (d, 1H, $J_{1,2}$ 7.0 Hz, H-1^{II}), 4.31 (d, 1H, $J_{1,2}$ 10.0 Hz, H-1^I), 4.02 (d, 1H, J_{gem} 14.0 Hz, SC*H*HAr), 3.94–3.36 (m, 13H). ^{13}C NMR (50 MHz, CDCl$_3$): δ 198.1 (CHO), 159.7 (*C*-OH Ar), 138.9, 134.4,

130.8, 121.5, 118.2 (C-Ar), 103.5 (C-1^{II}), 84.6 (C-1^{I}), 79.3, 78.8, 76.4, 76.0, 73.2, 72.5, 71.6, 69.2, 61.7, 60.8 (C-2^{I}, C-2^{II}, C-3^{I}, C-3^{II}, C-4^{I}, C-4^{II}, C-5^{I}, C-5^{II}, C-6^{I}, C-6^{II}), 33.5 (SCH_2Ar). TOF-MS: [M + H]$^+$ calcd for $C_{20}H_{28}O_{12}S$ (m/z), 492.13; found, 515.08 ([M + Na]$^+$). Anal. Calcd for $C_{20}H_{28}O_{12}S$: C, 48.77; H, 5.73; S, 6.51. Found: C, 48.49; H, 5.81; S, 6.33.

m-FORMYL-p-HYDROXYBENZYL β-D-GLUCOPYRANOSYL-(1→6)-1-THIO-β-D-GLUCOPYRANOSIDE (7)

White powder; R_f (8:2 v/v CH_2Cl_2-MeOH) 0.20; [α]$_D$ −62 (c 0.5, H_2O); ^1H NMR (500 MHz, D_2O): δ 9.94 (s, 1H, CHO), 7.75 (d, 1H, J_{meta} 2.0 Hz, H-Ar), 7.64 (dd, 1H, J_{ortho} 8.5 Hz, J_{meta} 2.0 Hz, H-Ar), 7.01 (d, 1H, J_{ortho} 8.5 Hz, H-Ar), 4.45 (d, 1H, $J_{1,2}$ 8.0 Hz, H-1^{II}), 4.32 (d, 1H, $J_{1,2}$ 10.0 Hz, H-1^{I}), 4.11-3.29 (m, 14H). ^{13}C NMR (50 MHz, D_2O): δ 197.8 (CHO), 159.6 (C-OH Ar), 138.9, 134.1, 130.7, 121.4, 118.1 (C-Ar), 103.4 (C-1^{II}), 85.1 (C-1^{I}), 79.3, 77.8, 76.6, 76.4, 73.8, 72.6, 70.3, 69.9, 69.2, 61.4 (C-2^{I}, C-2^{II}, C-3^{I}, C-3^{II}, C-4^{I}, C-4^{II}, C-5^{I}, C-5^{II}, C-6^{I}, C-6^{II}), 33.5 (SCH_2Ar). TOF-MS: [M + H]$^+$ calcd for $C_{20}H_{28}O_{12}S$ (m/z), 492.13; found, 515.07 ([M + Na]$^+$). Anal. Calcd for $C_{20}H_{28}O_{12}S$: C, 48.77; H, 5.73; S, 6.51. Found: C, 48.48; H, 5.85; S, 6.35.

m-FORMYL-p-HYDROXYBENZYL α-D-GLUCOPYRANOSYL-(1→4)-α-D-GLUCOPYRANOSYL-(1→4)-1-THIO-β-D-GLUCOPYRANOSIDE (8)

White powder; R_f (7:3 v/v CH_2Cl_2-MeOH) 0.45; [α]$_D$ −4.8 (c 0.4, H_2O); ^1H NMR (500 MHz, D_2O): δ 9.92 (s, 1H, CHO), 7.71 (d, 1H, J_{meta} 2.0 Hz, H-Ar), 7.62 (dd, 1H, J_{ortho} 8.5 Hz, J_{meta} 2.0 Hz, H-Ar), 6.99 (d, 1H, J_{ortho} 8.5 Hz, H-Ar), 5.39 (d, 1H, $J_{1,2}$ 3.5 Hz), 5.36 (d, 1H, $J_{1,2}$ 3.5 Hz), 4.31 (d, 1H, $J_{1,2}$ 10.0 Hz, H-1^{I}), 4.02 (d, 1H, J_{gem} 13.5 Hz, SCHHAr), 3.94–3.35 (m, 19H). ^{13}C NMR (100 MHz, D_2O): δ 198.0 (CHO), 159.7 (C-OH Ar), 138.9, 134.4, 130.8, 121.5, 118.1 (C-Ar), 100.4, 100.1 (C-1^{II}, C-1^{III}), 84.6 (C-1^{I}), 79.0, 78.3, 77.4, 77.3, 74.0, 73.5, 73.4, 72.6, 72.4, 72.1, 71.8, 70.0, 61.4, 61.1, 61.0 (C-2^{I}, C-2^{II}, C-2^{III}, C-3^{I}, C-3^{II}, C-3^{III}, C-4^{I}, C-4^{II}, C-4^{III}, C-5^{I}, C-5^{II}, C-5^{III}, C-6^{I}, C-6^{II}, C-6^{III}), 33.4 (SCH_2Ar). TOF-MS: [M + H]$^+$ calcd for $C_{26}H_{38}O_{17}S$ (m/z), 654.18; found, 677.00 ([M + Na]$^+$). Anal. Calcd for $C_{26}H_{38}O_{17}S$: C, 47.70; H, 5.85; S, 4.90. Found: C, 47.45; H, 5.99; S, 4.70.

ACKNOWLEDGMENTS

NMR and MS facilities of CIMCF (Centro di Metodologie Chimico-Fisiche) of Università di Napoli "Federico II" are gratefully acknowledged.

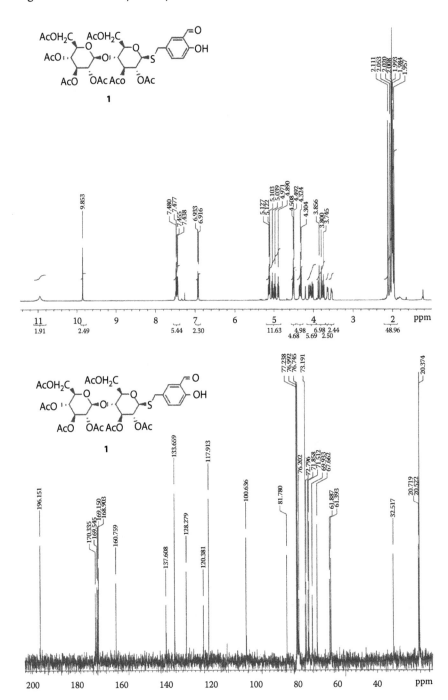

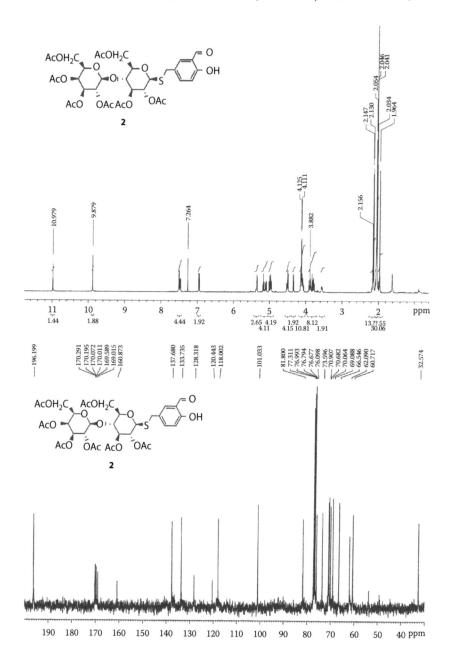

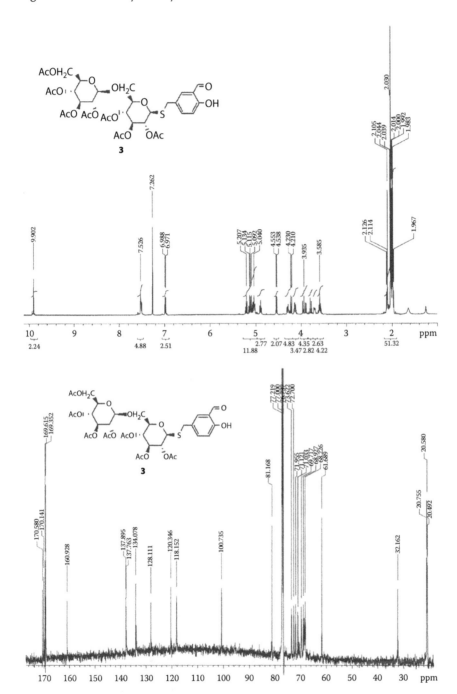

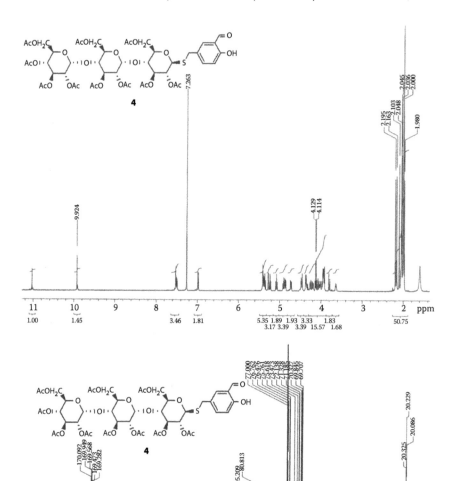

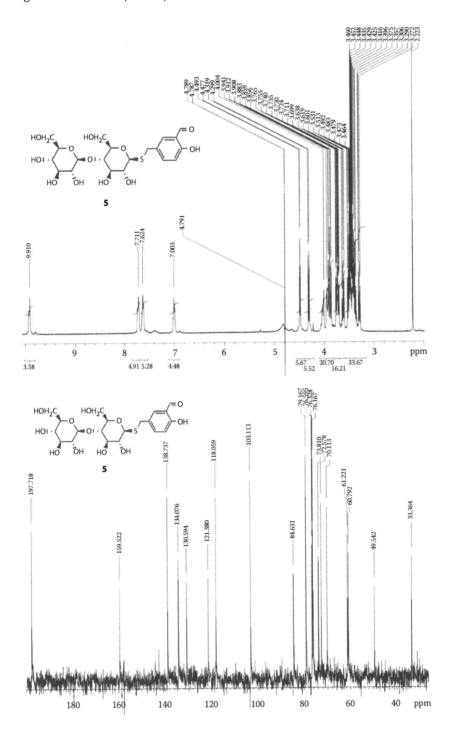

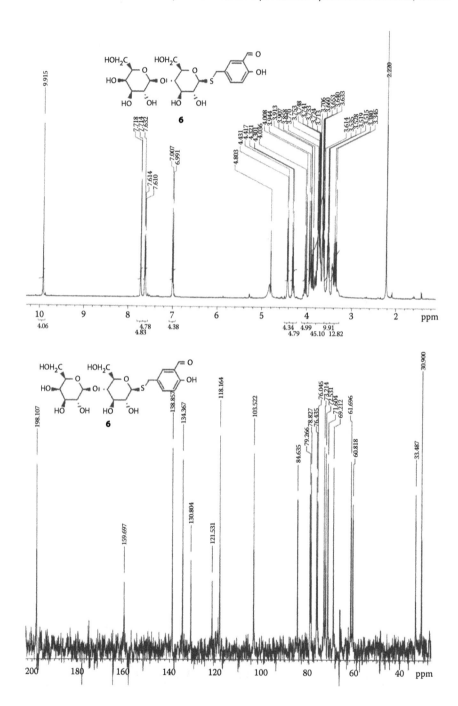

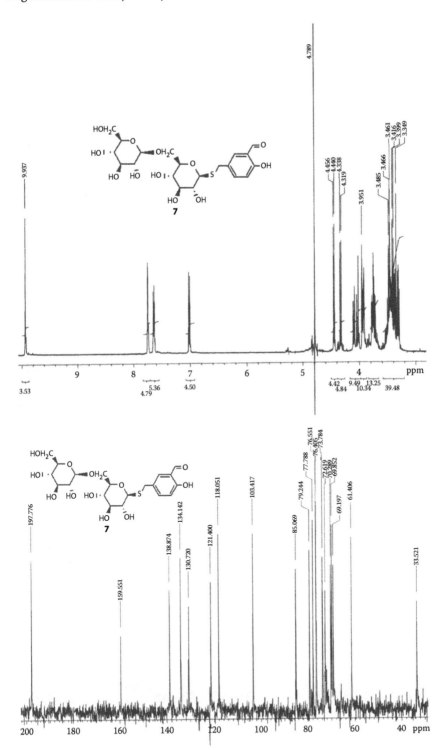

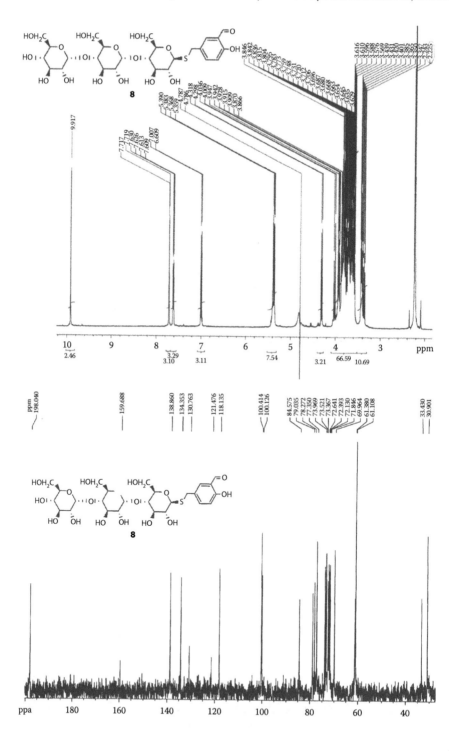

REFERENCES

1. (a) Beer, P. D.; Gale, P. A. *Angew. Chem. Int. Ed.*, 2001, *40,* 486–516; (b) Suksai, C.; Tuntulani, T. *Chem. Soc. Rev.* 2003, *32*, 192–202; (c) Ilioudis, C. A.; Tocher, D. A.; Steel, J. W. *J. Am. Chem. Soc.* 2004, *126*, 12395–12402; (d) Katayev, E. A.; Ustynyuk, Y. A.; Sessler, J. L. *Coord. Chem. Rev.* 2006, *250*, 3004–3037; (e) Kang, S. O.; Begum, R. A.; Bowamn-James, K. *Angew. Chem. Int. Ed.* 2006, *45*, 7882–7894.
2. (a) Schmuck, C.; Bickert, V. *Org. Lett.* 2003, *5*, 4579–4581; (b) Imai, H.; Munakata, H.; Uemori, Y.; Sakura, N. *Inorg. Chem.* 2004, *43*, 1211–1213; (c) Arena, G.; Casnati, A.; Contino, A.; Magri, A.; Sansone, F.; Sciotto, F.; Ungaro, R. *Org. Biomol. Chem.* 2006, *4*, 243–249.
3. Kubik, S. *Chem. Soc. Rev.*, 2010, *39,* 3648–3663.
4. Oshovsky, G. V.; Reinhoudt, D. N.; Verboom, W. *Angew. Chem. Int. Ed.*, 2007, *46,* 2366–2393.
5. Dalla Cort, A.; De Bernardin, P.; Forte, G.; Yafteh Mihan, F. *Chem. Soc. Rev.* 2010, *39,* 3863–3874.
6. Dalla Cort, A.; De Bernardin, P.; Schiaffino, L. *Chirality* 2009, *21*, 104–109.
7. Dalla Cort, A.; Mandolini, L.; Pasquini, C.; Schiaffino, L. *Org. Biomol. Chem.* 2006, *4*, 4543–4546.
8. Valerio, S.; Iadonisi, A.; Adinolfi, M.; Ravidà, A. *J. Org. Chem.* 2007, *72*, 6097–6106.

16 Allyl 3,4,6-Tri-*O*-Benzyl-α-D-Mannopyranoside

Chakree Wattanasiri
Somsak Ruchirawat
Mélanie Platon[*]
Siwarutt Boonyarattanakalin[†]

CONTENTS

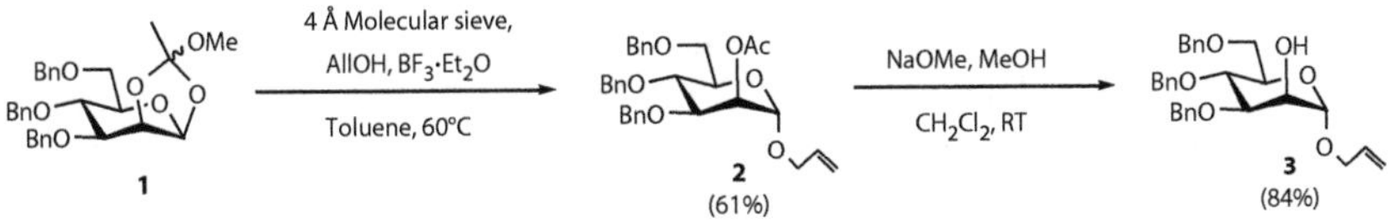

SCHEME 16.1

Oligosaccharides play a critical role in many important biological processes. However, due to minute amounts found in nature, isolation from natural sources in amounts sufficient for characterization and biological studies is not always feasible.

Mannose-rich oligosaccharides are found on the surface of many pathogens. One of the most common glycosidic bonds found in the mannose-rich oligosaccharides is the α (1→2)-glycosidic bond. Many building blocks suitable for construction of this linkage, including compound 3,[1,2] have been designed.[3] In this work, we report a short synthesis of 3 from mannosyl orthoester 1.[4,5] The bicyclic orthoester 1 was opened with allyl alcohol using $BF_3 \cdot Et_2O$ as a catalyst to give fully protected compound 2,[1,2] whose acetyl group was removed to provide the desired glycosyl acceptor 3.

---

[*] Checker under supervision of X. Liu: xinyuliu@pitt.edu.
[†] Corresponding author: siwarutt@siit.tu.ac.th.

EXPERIMENTAL

GENERAL METHODS

Nuclear magnetic resonance (NMR) spectra were recorded with a Bruker AVANCE III (300 MHz) spectrometer for solutions in $CDCl_3$ with chemical shifts referenced to $CDCl_3$ (7.26 ppm for ^1H and 77.0 ppm for ^{13}C). Splitting patterns are indicated as s, singlet; br s, broad singlet; d, doublet; dd, doublet of doublet; m, multiplet. High-resolution mass spectral (HRMS) analyses were performed by the MS-service at Chulabhorn Research Institute (CRI). Peaks are reported as m/z. Optical rotations were measured at 26°C using a JASCO P-1020 polarimeter. Analytical thin-layer chromatography (TLC) was performed on Merck Silica Gel 60 F254 plates (0.25 mm). Compounds were visualized by staining with cerium sulfate-ammonium molybdate (CAM) solution or phosphomolybdic acid (PMA) solution. Pulverized crystalline metal aluminosilicate molecular sieves 4A were activated by heating at 300°C under vacuum for 5 minutes. Allyl alcohol was distilled over magnesium and kept over potassium carbonate (K_2CO_3). Flash column chromatography was performed on Fluka Kieselgel 60 (230–400 mesh). All solvents (e.g., hexane and ethyl acetate) for flash column chromatography were distilled prior to use.

ALLYL 2-*O*-ACETYL-3,4,6-TRI-*O*-BENZYL-α-D-MANNOPYRANOSIDE (2)

The mannosyl orthoester $1^{4,5}$ (614.9 mg, 1.214 mmol) was co-evaporated with toluene three times. Activated molecular sieves (900 mg) were added, followed by allyl alcohol (1.65 mL, 24.276 mmol) and toluene (5 mL), and the mixture was stirred for 30 minutes at 60°C under argon. $BF_3 \cdot Et_2O$ (60 μL, 0.486 mmol) was added in a single portion via syringe, and the mixture was stirred at 60°C for 1 hour or until the reaction was complete. Triethylamine (5 drops) was added, and the mixture was filtered through a Celite pad, which was washed with CH_2Cl_2. After concentration, the crude product was chromatographed to yield 2 as a colorless syrup (395.8 mg, 61%). $R_f = 0.5$ (3:10 ethyl acetate–hexane); $[\alpha]_D^{27}$ +33.3 (c 1, $CHCl_3$); (lit.[2] $[\alpha]_D$ +34.5 (c 3.4, $CHCl_3$) , ^1H NMR (300 MHz, $CDCl_3$) δ 7.05-7.57 (m, 15H, arom.), 5.74–6.03 (m, 1H, –$CH_2CH=CH_2$), 5.45–5.51 (s, 1H, H–C(1)), 5.34–5.45 (s, 1H, H–C(2)), 5.07–5.34 (m, 2H, –$CH_2CH=CH_2$), 4.78–5.04 (m, 2H, OCHHPh), 4.60–4.78 (m, 2H, OCHHPh), 4.34–4.60 (m, 4H, OCH_2Ph, –CH_2CHCH_2), 3.52–4.22 (m, 5H, H–C(3), H–C(4). H–C(5), –H_bC(6), –H_aC(6)), 2.01–2.25 (s, 3H, COCH_3); ^{13}C NMR (300 MHz, $CDCl_3$) δ 170.49(COCH_3), 138.29, 138.15, 137.90, 133.36, 128.37, 128.29, 128.06, 127.87, 127.77, 127.71, 127.62, 127.57, 117.75, 96.85(C(1)), 78.18, 75.18, 74.25, 73.39, 71.77, 71.35, 68.75, 68.70, 68.09, 21.12(COCH_3);. ESI-HRMS (m/z): [M + Na$^+$] calculated for $C_{32}H_{36}O_7Na^+$, 555.23593; found: 555.3687.

ALLYL 3,4,6-TRI-*O*-BENZYL-α-D-MANNOPYRANOSIDE (3)

Intermediate 2 (395.8 mg, 0.743 mmol) was dissolved in methanol (3.5 mL) and CH_2Cl_2 (8 mL). Sodium methoxide (40 mg, 0.743 mmol) was added to the solution and the mixture was stirred for 12 hours, diluted with water (30 mL), and extracted with CH_2Cl_2 (3 × 30 mL). The combined organic layers were dried with Na_2SO_4 and concentrated in vacuo. The crude product was purified by flash column chromatography to give **3** (306.4 mg, 84%) as a colorless syrup. R_f = 0.26 (ethyl acetate/hexane 3:10); $[\alpha]_D^{27}$ +58.2 (*c* 1, $CHCl_3$); (lit.[2] $[\alpha]_D$ +60.2 (c 1.2, $CHCl_3$); ^{1}H NMR (300 MHz, $CDCl_3$) δ 7.08–7.45 (m,15H, arom), 5.77–5.99 (m, 1H, $-CH_2CH=CH_2$), 5.10–5.38 (m, 2H, $-CH_2CH=CH_2$), 4.89–5.02 (s, 1H, *H*-C(1)), 4.59–4.89 (m, 6H, 3 × OCH_2Ph), 4.41–4.59 (m, 2H, $-CH_2CHCH_2$), 3.62–4.25 (m, 6H, *H*–C(2), *H*–C(3), *H*-C(4), *H*-C(5), $-H_bC(6)$, $-H_aC(6)$), 2.38–2.63 (s, 1H, $-OH$); ^{13}C NMR (300 MHz, $CDCl_3$) δ 138.25, 138.19, 137.90, 133.66, 128.49, 128.31, 128.29, 127.91, 127.88, 127.81, 127.79, 127.64, 127.53, 117.45, 98.34(C(1)), 80.20, 75.11, 74.26, 73.40, 71.95, 71.67, 68.86, 68.32, 67.90; ESI-HRMS (*m/z*): [M + Na+] calculated for $C_{30}H_{34}O_6Na^+$, 513.2253; found: 513.5696.

ACKNOWLEDGMENTS

This research was supported by the National Research University Project of Thailand, Office of Higher Education Commission, and the Thailand Research Fund (RSA5580059). We thank the Center of Excellence on Environmental Health, Toxicology and Management of Chemicals (ETM) for the scholarship for CW.

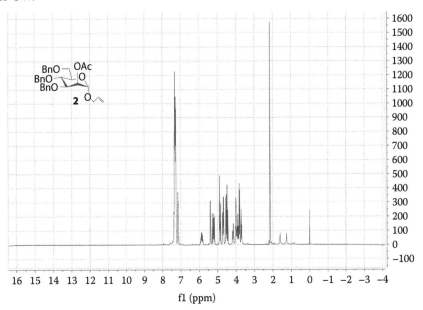

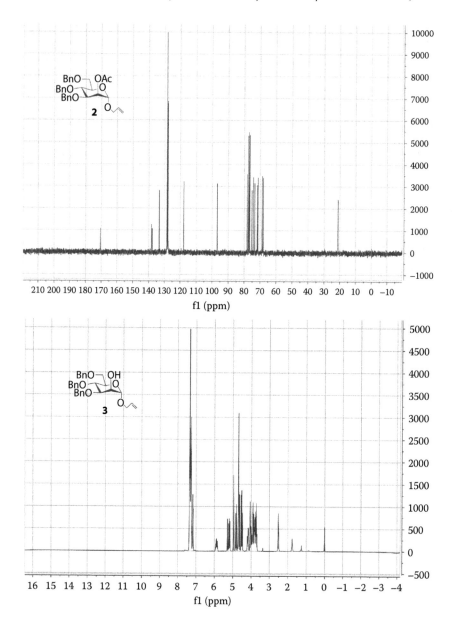

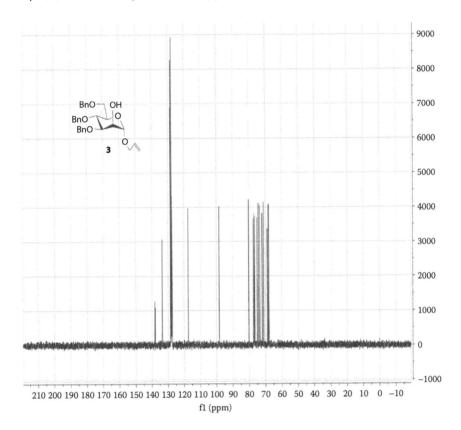

REFERENCES

1. Ogawa, T.; Nukada, T. *Carbohydr. Res.* 1985, *136*, 135.
2. Goddat, J.; Grey, A. A.; Hricovini, M.; Grushcow, J.; Carver, J. P.; Shah, R. N. *Carbohydr. Res.* 1994, *252*, 159.
3. Cao, B.; Williams, S. J. *Nat. Prod. Rep.* 2010, *27*, 919.
4. Boonyarattanakalin, S.; Liu, X.; Michieletti, M.; Lepenies, B.; Seeberger, P. H. *J. Am. Chem. Soc.* 2008, *130*, 16791.
5. Ravida, A.; Liu, X.; Kovacs, L.; Seeberger, P. H. *Org. Lett.* 2006, *8*, 1815.

17 Synthesis of a Spacer-Armed 6′-Sialyl-N-Acetyllactosamine

G. V. Pazynina

A. A. Chinarev

A. B. Tuzikov

V. V. Nasonov

N. N. Malysheva[*]

N. V. Bovin[†]

CONTENTS

REAGENTS AND CONDITIONS

Recently we have described an efficient and simple method for the synthesis of (α2-6) sialooligosaccharides by the classical Koenigs-Knorr reaction, in which the simplest sialyl donor (i.e., the glycosyl chloride of the peracetylated methyl ester of sialic acid) is used for glycosylation of galacto-4,6-diols (mono- and disaccharides). The reaction is high yielding, regio- and stereoselective, and takes place at room temperature in the presence of Ag_2CO_3 or $Ag_2CO_3/AgOTf$ as promoters. The approach is applicable to various forms of sialic acid, namely, Neu5Ac, Neu5Gc, 9-deoxy-9-N-Ac-Neu5Ac, and (α2-8) oligosaccharides of Neu5Ac.[1,2] Here we apply the cited method for the synthesis of the spacer-armed 6′-sialyl-N-acetyllactosamine **5-α**, α-Neu5Ac-(2 → 6)-β-D-Gal-(1 → 4)-β-D-GlcNAc-O(CH₂)₃NH₂.

---

[*] Corresponding author: bovin@carb.ibch.ru.
[†] Checker: nelly.malyschewa@gmail.com.

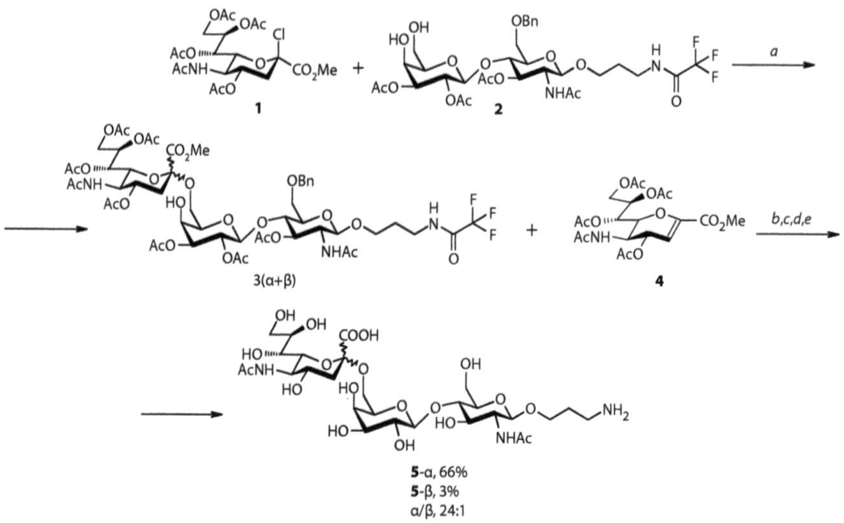

SCHEME 17.1 (a) Ag$_2$CO$_3$, MS 4Å, CH$_2$Cl$_2$, 4 days, r. t. (b) H$_2$, Pd/C, MeOH, 2 h, r. t. (c) 2M MeONa, dry MeOH, 30 min, r. t. (d) 0.1 M aqueous NaOH, 12 h, r. t. (e) Separation of the reaction products. The yield of **5** is given based on diol **2**.

The salient points involved in the sialylation described below are as follows. When the reaction of 4′,6′-diol **2**,[*] glycosyl chloride **1**,[†] and promoter, Ag$_2$CO$_3$,[‡,§] is complete, the mixture is filtered and concentrated, and the combined products are deprotected by sequential catalytic hydrogenolysis, deacetylation, and hydrolysis of the methyl ester and *N*-trifluoroacetyl groups. The resulting mixture of products, the target sialooligosaccharides, the Neu5Ac glycal, and the deprotected acceptor is resolved by cation-exchange chromatography on Dowex H$^+$ to give a mixture of **5**-α/β. Finally, the individual anomers can be separated by reverse-phase high-performance liquid chromatography (HPLC) (see Figure 17.1).[¶] The structure of the target compounds is confirmed by ^{1}H and ^{13}C NMR data (Figures 17.2 and 17.3); the spectra of **5** are identical with those described before.[7]

[*] It should be pointed out that the use of the acceptor with the only free hydroxyl group at C-6 instead of the less hindered 4,6-diol **2** significantly worsens the results of sialylation.[3] The detailed synthesis of **2** was described earlier.[4]

[†] Neu5Ac glycosyl choride **1** is readily prepared by treatment of solution of the corresponding glycosyl acetate in chloroform with HCl (generated in situ from MeOH and AcCl) for 2 to 3 days, and used after simple workup.[5]

[‡] Silver carbonate is prepared by scrupulously following the protocol by McCloskey and Coleman[6]; it is advised to strictly follow the described procedures and conditions of handling Ag$_2$CO$_3$.[6]

[§] No silver triflate is added to the reaction mixture; the yield of **3**-β and conversion rate of chloride **1** into glycal **4** are noticeably increased in the presence of AgOTf.

[¶] As **5**-β is present in the anomeric mixture in a very minor amount, its separation is often not required, and the mixture of anomers can be used after the removal of impurities by gel-permeation chromatography on Sephadex LH-20, elution with 1:1 MeCN-H$_2$O. Alternatively, the anomers of **5** could be separated by ion-exchange chromatography on DEAE Sepharose A-25, elution with 0.01 M aqueous pyridine acetate.

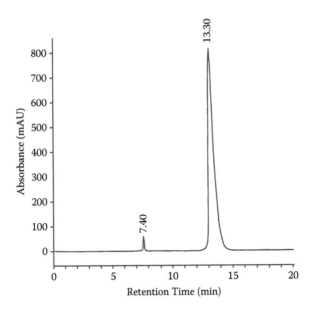

FIGURE 17.1   Analytical HPLC chromatogram (Phenomenex C18 Luna, 5 µm, 4.6 × 250 mm; eluent: water, 0.5 mL/min, 30°C; UV detection at 195 nm) of the **5** α + β mixture obtained by separation of the sialylation products on ion-exchange resin. Peak assignment: **5**-β (7.40 min) and **5**-α (13.30 min).

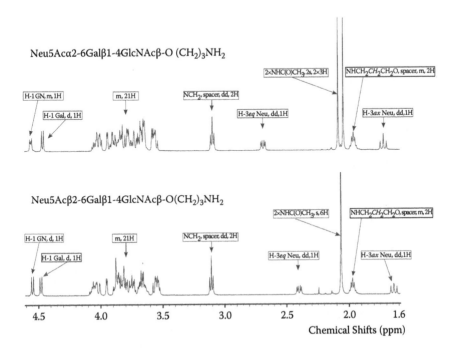

FIGURE 17.2   ¹H NMR spectra (500 MHz, D₂O, 30°C) of **5**-α and **5**-β.

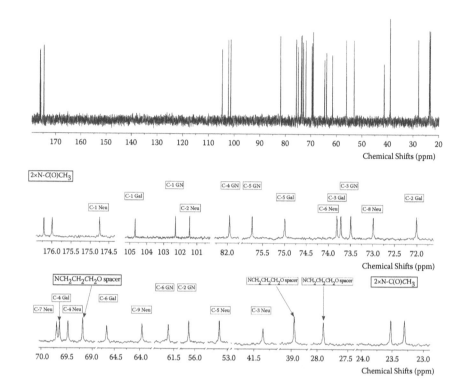

FIGURE 17.3 ^{13}C NMR spectrum (125 MHz, D$_2$O, 30°C) of **5-α** and its expanded fragments.

The presented synthesis is superior to other sialylation methods, which tend to give lower yields and stereoselectivity, and require more complex modifications of sialyl donors and reaction conditions.[8] In addition, the separation of the target product described here is faster and less laborious than chromatography on normal phase silica gel. The approach can be used for the multi-gram preparation of **5-α**.

EXPERIMENTAL

GENERAL METHODS

Reactions were performed using commercial reagents (Aldrich, Merck, and Fluka), and solvents were purified according to standard procedures. Glycosyl chloride **1**[5] diol **2**[4], and Ag$_2$CO$_3$[6] were synthesized as described. Thin-layer chromatography (TLC) was carried out on silica gel 60 F$_{254}$-coated aluminum foil (Merck). Spots were visualized by dipping the plates in 10% aqueous H$_3$PO$_4$, or in a solution of ninhydrin (2.5 g in a mixture of 96:4:1 Me$_2$CO–H$_2$O–AcOH, 500 mL), and heating. ^{1}H and ^{13}C NMR spectra were recorded at 30°C for solutions in D$_2$O with a Bruker WM-500 instrument at 500 MHz and 125 MHz, respectively. The ^{1}H chemical shifts are referenced to the signal of the residual HDO (δ_H 4.75); the ^{13}C chemical shifts are referenced to the signal of external 1,4-dioxane (δ_C 67.4). Matrix-assisted laser desorption ionization time-of-flight mass spectra (MALDI-TOF MS) were recorded with a Vision-2000 MALDI-TOF spectrometer. HPLC was performed with a HP Agilent 1100 chromatograph equipped with a ultraviolet (UV) detector. Specific optical rotation ($[\alpha]_D$) was measured with a PerkinElmer 341LC polarimeter.

3-AMINOPROPYL [(5-ACETAMIDO-3,5-DIDEOXY-D-*GLYCERO*-α-D-*GALACTO*-NON-2-ULOPYRANOSYL)ONIC ACID]-(2 → 6)-(β-D-GALACTOPYRANOSYL)-(1 → 4)-2-ACETAMIDO-2-DEOXY-β-D-GLUCOPYRANOSIDE (5-α)

A solution of chloride **1** (305 mg, 0.6 mmol) in dry CH_2Cl_2 (3 mL) was added to a mixture of diol **2** (150 mg, 0.2 mmol), Ag_2CO_3 (500 mg, 1.8 mmol), freshly activated molecular sieves 4Å (1.0 g), and dry CH_2Cl_2 (7 mL), which had been stirred for 10 min at room temperature, and the formed suspension was vigorously stirred in the dark at room temperature in a tightly closed flask for 4 d. Progress of the reaction was monitored by TLC: R_f 0.44 (**4**), R_f 0.29 (**3-β**), R_f 0.21 (**3-α**), R_f 0.18 (**2**), 9:1 $CHCl_3$–MeOH. $CHCl_3$–Py (100:1, 10 mL) was added, the mixture was filtered, the solids were washed with 1:1 $CHCl_3$–MeOH (5 × 20 mL), the combined filtrates were concentrated with co-evaporation of toluene and dried in vacuo, to give 450 mg of a yellow solid. The latter was dissolved in MeOH (20 mL), 10% Pd/C (600 mg) was added, and the mixture was stirred under H_2 at ambient temperature and pressure for 16 h. The catalyst was filtered off, washed with MeOH (3 × 10 mL), and the combined filtrates were concentrated to dryness with co-evaporation of toluene. The colorless residue (390 mg) was dissolved in dry MeOH (6 mL), 2M MeONa in MeOH (0.3 mL) was added, and the mixture was kept for 30 min at room temperature and concentrated without heating. A solution of the residue in H_2O (6 mL) was kept for 12 h at room temperature and applied on a column of Dowex 50X4-400 ion-exchange resin (H^+-form in H_2O, 5 mL). The resin was washed sequentially with H_2O (20 mL) and 1M aqueous Py (25 mL) to give anomers of **5** (110 mg). Subsequent elution with 1M aqueous NH_3 (25 mL) gave the deprotected acceptor (15 mg; recovery, 17%). The crude mixture of anomers (20–30 mg) was subjected to HPLC on reversed-phased C18 silica gel (Phenomenex Luna, 21.2 × 250 mm, 5 μm, pore size - 100Å) by elution with water (10.0 mL/min, 30°C); retention times –7.5 min (**5-β**), 13.5 min (**5-α**). The fractions containing products were collected, combined, and concentrated to dryness to afford **5-β** (4 mg, 3%) and **5-α** (97 mg, 66%).

  5-α: colorless solid; $[\alpha]_D^{20}$ –20° (*c* 1.0, H_2O, internal salt), lit.: $[\alpha]_D^{20}$ –21° (*c* 0.86, H_2O, Na-salt)[7]; TLC: R_f 0.46 (3:3:2 MeOH–MeCN–H_2O); ^{1}H NMR (500 MHz): δ 1.72 (dd, 1H, $J_{4,3ax}$ ~$J_{3ax,3eq}$ 12.2 Hz, Neu H-3*ax*), 1.90–2.00 (m, 2H, NCH$_2$*CH$_2$*CH$_2$O spacer), 2.05, 2.09 (2s, 2 × 3H, 2 × NC(O)CH$_3$ Neu, GN), 2.69 (dd, 1H, $J_{4,3eq}$ 4.6 Hz, $J_{3ax,3eq}$ 12.5 Hz, Neu H-3*eq*), 3.07 (dd, 2H, J_{hem}~J_{vic} 7.0 Hz, NCH$_2$ spacer), 3.52–4.08 (m, 21H), 4.47 (d, 1H, $J_{1,2}$ 7.8 Hz, H-1 Gal), 4.56 (m, 1H, GN); ^{13}C NMR (175 MHz): δ 23.2, 23.4 (2 × N-C(O)CH$_3$), 27.8 (NCH$_2$CH$_2$CH$_2$O spacer), 39.8 (NCH$_2$CH$_2$CH$_2$O spacer), 41.2 (C-3 Neu), 53.0 (C-5 Neu), 56.0 (C-2 GN), 61.5 (C-6 GN), 63.8 (C-9 Neu), 64.5 (C-6 Gal), 69.1 (NCH$_2$CH$_2$CH$_2$O spacer), 69.4 (C-4 Neu), 69.5 (C-4 Gal), 69.6 (C-7 Neu), 71.9 (C-2 Gal), 72.8 (C-8 Neu), 73.4 (C-3 GN), 73.6 (C-3 Gal), 73.7 (C-6 Neu), 74.9 (C-5

Gal), 75.6 (C-5 GN), 81.9 (C-4 GN), 101.3 (C-2 Neu), 102.2 (C-1 GN), 104.6 (C-1 Gal), 174.6 (C-1, Neu), 175.8, 176.1 (2 × N-C(O)CH$_3$); MALDI-MS: 732 [M]$^+$, 755 [M + Na]$^+$, 771 [M + K]$^+$; Anal. Calcd for C$_{28}$H$_{49}$N$_3$O$_{19}$: C 45.96; H 6.75; N 5.74. Found: C 45.37; H 6.96; N 5.42.^{e}

ACKNOWLEDGMENTS

This work was supported by the RAS Presidium program "Molecular and Cell Biology."

REFERENCES

1. Pazynina, G.; Tuzikov, A.; Chinarev, A.; Obukhova, P.; Bovin, N. *Tetrahedron Lett.* 2002, *43*, 8011–8013.
2. Pazynina, G.; Nasonov, V.; Belaynchikov, I.; Brossmer, R.; Maisel, M.; Tuzikov, A.; Bovin, N. V. *Int. J. Carb. Chem.* 2010, doi:10.1155/2010/59424.
3. Khorlin, A. Y.; Privalova, I. M.; Bystrova, I. V. *Carbohydr. Res.* 1971, *19*, 271–275.
4. Pazynina, G.; Severov, V.; Bovin, N. *Rus. J. Bioorg. Chem. (Engl. Transl.)* 2008, *34*, 625–631.
5. Byramova, N. E.; Tuzikov, A. B.; Bovin, N. V. *Carbohydr. Res.* 1992, *237*, 161–175.
6. McCloskey, C. M.; Coleman, G.H. *Org. Synth.* 1945, *26*, 53–54.
7. Sherman, A.; Yudina, O.; Shashkov, A.; Menshov, V.; Nifant'ev, N. *Carb. Res.* 2001, *330*, 445–458.
8. (a) Boons, G.-J.; Demchenko, A. V. *Chem. Rev.* 2000, *100*, 4539–4565; (b) Halcomb, R. L.; Chappell, M. D. *J. Carb. Chem.* 2002, *21*, 723–768; (c) Liu, Y.; Ruan, X.; Li, X.; Li, Y. *J. Org. Chem.* 2008, *73*, 4287–4290.

18 Synthesis of β-(1→3)-mannobiose

Sebastien Picard
Mikael Thomas
*Anne Geert Volbeda**
Xavier Guinchard
David Crich†

CONTENTS

* Checker under supervision of J. Codée: jcodee@chem.leidenuniv.nl.
† Corresponding author: dcrich@chem.wayne.edu.

α-(1→3)-mannobiose is a dimer of mannose that has been obtained by chemical[1–3] and enzymatic[4] synthesis, or via degradation of natural mannan.[5] Its synthetic and commercial availability has given rise to a number of studies in which α-(1→3)-mannobiose has been used as a ligand for lectins,[6–8] as a substrate for mannosidases,[9] and in the functionalization of multivalent arrays.[10,11] In sharp contrast, β-(1→3)-mannobiose has never been described most likely due to its challenging synthesis despite the fact that the stereocontrolled synthesis of β-mannopyranosides was largely solved.[12] The present protocol describes the preparation of β-(1→3)-mannobiose via the 4,6-O-benzylidene acetal controlled method, thereby making this compound available.

EXPERIMENTAL

GENERAL METHODS

Optical rotations were measured on a JASCO P-1010 polarimeter. ^{1}H NMR-spectra (300 or 500 MHz) and ^{13}C NMR-spectra (75 or 125 MHz) were recorded on Brüker spectrometers. Chemical shifts are given in ppm (δ) and were referenced to the internal solvent signal or to TMS used as an internal standard. Multiplicities are described as follows: s (singlet), br s (broad singlet), d (doublet), t (triplet), q (quadruplet), dd (doublet of doublet), ddd (doublet of doublet of doublet), dt (doublet of triplet), td (triplet of doublet), and m (multiplet). Coupling constants J are given in Hz. Infrared (IR) spectra were recorded on a Perkin-Elmer Fourier transform infrared (FTIR) system using diamond window Dura SamplIR II, and the data are reported in reciprocal centimeters (υ, cm^{-1}). Mass spectra were recorded on Micromass LCT (ESI). Reactions were performed using oven-dried glassware under an atmosphere of dry argon. Silica gel 60 (35–70 μm) was used for flash chromatography. All reactions were monitored by thin-layer chromatography (TLC) (Merck 60 F$_{254}$ aluminum sheets), the spots being visualized with ultraviolet light and/or by charring with vanillin (1%) + sulfuric acid (5%) in EtOH. Reagents were from commercial sources and were used as received, unless stated otherwise. Trifluoromethanesulfonic anhydride (Tf_2O) was distilled under argon over P_2O_5 before use. Dichloromethane was distilled under argon over CaH_2. Toluene and dimethylformamide were obtained from Acros Organics (Extra Dry over Molecular Sieve, Acroseal®) and were used as received. Solutions in organic solvents were dried with anhydrous $MgSO_4$ and concentrated under reduced pressure.

(4-METHYLPHENYL) 2-O-BENZYL-4,6-O-BENZYLIDENE-3-O-(2-NAPHTHALENYLMETHYL)-1-THIO-α-D-MANNOPYRANOSIDE (2)

A mixture of (4-methylphenyl) 4,6-O-benzylidene-1-thio-α-D-mannopyranoside (**1**)[13] (900 mg, 2.4 mmol) and Bu_2SnO (722 mg, 2.9 mmol), in dry toluene (25 mL), was heated under reflux overnight. The clear solution was cooled to room temperature and concentrated under reduced pressure. The residue was dissolved in dry DMF (10 mL), and CsF (730 mg, 4.8 mmol) was added, followed by solid 2-naphthylmethyl bromide (NAPBr, 780 mg, 3.6 mmol). The mixture was stirred for 6 h at room temperature, diluted with CH_2Cl_2, and filtered through Celite®. The filtrate was concentrated, the residue was dissolved in dry DMF (10 mL), cooled to 0°C, and NaH (231 mg of a 50% dispersion in mineral oil, 4.8 mmol) was added. After 10 min of stirring at that temperature, benzyl bromide (570 μL, 4.8 mmol) was added. The mixture was stirred overnight at room temperature, the reaction was quenched with saturated aqueous $NaHCO_3$, and the mixture was extracted several times with CH_2Cl_2. The combined organic layers were washed with brine, dried, and concentrated. The crude product was chromatographed (10% t-BuOMe in heptane) to give (4-methylphenyl) 2-O-benzyl-4,6-O-benzylidene-3-O-(2-naphthalenylmethyl)-1-thio-α-D-mannopyranoside (1.32 g, 2.18 mmol, 91%) as a clear viscous oil. [α]$_D^{22}$ + 102.2 (c 1.0, CHCl$_3$). ^{1}H NMR (500 MHz, CDCl$_3$) δ (ppm): 2.33 (s, 3H, CH$_3$), 3.90 (t, J = 10.0 Hz, 1H, H-6b), 4.03–4.08 (m, 2H, H-2, H-3), 4.24 (dd, J = 10.0 Hz, J = 4.5 Hz, 1H, H-6a), 4.27–4.38 (m, 2H, H-4, H-5), 4.71–4.77 (m, 2H, CH$_2$Ph), 4.82 (d, J = 12.5 Hz, 1H, CHaHbNap), 4.95 (d, J = 12.5 Hz, 1H, CHaHbPh), 5.45 (s, 1H, H-1), 5.68 (s, 1H, benzylidene-H), 7.09 (d, J = 7.5 Hz, 2H, aromatic), 7.23–7.50 (m, 13H, aromatic), 7.52–7.55 (m, 2H, aromatic), 7.71–7.75 (m, 1H, aromatic), 7.79–7.83 (m, 3H, aromatic). ^{13}C NMR (75 MHz, CDCl$_3$) δ (ppm): 21.3 (CH$_3$), 65.6 (C-5), 68.7 (C-6), 73.1 (CH$_2$Nap), 73.2 (CH$_2$Ph), 76.4 (C-3), 78.1 (C-2), 79.3 (C-4), 87.6 (C-1), 101.8 (benzylidene-C), 125.8, 126.0, 126.2, 126.4, 126.5, 127.8, 128.0, 128.1, 128.2, 128.3, 128.4, 128.6, 129.1, 130.0, 130.1, 132.5, 133.2, 133.5, 136.0, 137.8, 137.9, 138.1. IR (cm^{-1}): 3032, 2861, 1492, 1453, 1372, 1348, 1087, 1015, 904, 810, 734, 696. ESIHRMS: $C_{38}H_{40}NO_5S$ Calcd for [M + NH$_4$]$^+$: 622.2627. Found: 622.2641.

BENZYL 2-O-BENZYL-4,6-O-BENZYLIDENE-3-O-(2-NAPHTHALENYLMETHYL)-β-D-MANNOPYRANOSIDE 3

To a cooled (–78°C) solution of **2** (480 mg, 0.8 mmol), 2,4,6-tri-*tert*-butylpyrimidine (TTBP) (995 mg, 4 mmol), and 1-benzenesulfinyl piperidine (BSP) (201 mg, 0.96 mmol) in dry CH_2Cl_2 (16 mL), was added Tf_2O (160 μL, 0.96 mmol). The orange reaction mixture was stirred 1 h at $-78°C$ before neat benzyl alcohol (125 μL, 1.2 mmol) was added. The reaction mixture was stirred 20 min at $-78°C$, warmed to room temperature and stirred 2 h at room temperature, quenched with saturated aqueous $NaHCO_3$, and then extracted several times with CH_2Cl_2. The combined organic layers were washed with brine, dried over $MgSO_4$, and concentrated under reduced pressure. The crude product was purified on silica gel (20% EtOAc in heptane) to give benzyl 2-O-benzyl-4,6-O-benzylidene-3-O-(2-naphthalenylmethyl)-α-D-mannopyranoside (398 mg, 0.67 mmol, 84%) as a clear viscous oil. $[\alpha]_D^{22}$ -66.0 (c 1.0, $CHCl_3$). ^{1}H NMR (500 MHz, $CDCl_3$) δ (ppm): 3.33 (td, $J = 9.5$ Hz, $J = 5.0$ Hz, 1H), 3.61 (dd, $J = 10.0$ Hz, $J = 3.5$ Hz, 1H), 3.95–4.02 (m, 2H), 4.34 (t, $J = 9.5$ Hz, 1H), 4.41 (dd, $J = 10.5$ Hz, $J = 4.5$ Hz, 1H), 4.55 (s, 1H), 4.62 (d, $J = 12.0$ Hz, 1H), 4.79 (d, $J = 12.5$ Hz, 1H), 4.85 (d, $J = 12.5$ Hz, 1H), 4.98 (d, $J = 12.0$ Hz, 1H), 5.03 (d, $J = 12.0$ Hz, 1H), 5.09 (d, $J = 12.0$ Hz, 1H), 5.71 (s, 1H), 7.34–7.60 (m, 18H), 7.72–7.74 (m, 1H), 7.78–7.82 (m, 2H), 7.84–7.87 (m, 1H). ^{13}C NMR (75 MHz, $CDCl_3$) δ (ppm): 67.7, 68.7, 71.2, 72.3, 74.8, 75.8, 77.9, 78.7, 101.0 ($^1J_{CH} = 153.9$ Hz), 101.6, 125.6, 125.8, 126.1, 126.2, 126.3, 127.6, 127.7, 127.9, 128.1, 128.2, 128.3, 128.5, 128.7, 128.5, 128.7, 128.9, 133.0, 133.3, 135.8, 137.2, 137.7, 137.7, 138.5. IR (cm^{-1}): 3031, 2867, 2247, 1453, 1086, 1044, 907, 814, 727, 696. ESIHRMS: $C_{38}H_{36}O_6Na$ Calcd for $[M + Na]^+$: 611.2410, found 611.216. Anal. Calcd for $C_{38}H_{36}O_5S$: C, 75.47; H, 6.00; O, 13.23. Found: C, 75.10; H, 6.14; O, 13.65.

BENZYL 2-O-BENZYL-4,6-O-BENZYLIDENE-β-D-MANNOPYRANOSIDE (4)

Compound **3** (0.61 g, 1.04 mmol) was dissolved in a mixture of 20:1 CH_2Cl_2–H_2O (15 mL) and DDQ (0.38 g, 1.66 mmol) was added. The mixture was stirred for 90 min at room temperature, quenched with saturated aqueous $NaHCO_3$ and the mixture was extracted several times with CH_2Cl_2. The combined organic layers were washed with brine, dried, concentrated and the crude product was chromatographed (hexanes:EtOAc 6:1 to 3:1) to give benzyl 2-O-benzyl-4,6-O-benzylidene-β-D-mannopyranoside (0.406 g, 0.90 mmol, 87%) as a clear viscous oil. $[\alpha]_D^{26}$ -109.2 (c 1.0, $CHCl_3$). ^{1}H NMR (500 MHz, $CDCl_3$) δ (ppm): 2.37 (d, $J = 9.0$ Hz, 1H, OH), 3.35 (td, $J = 9.5$ Hz, $J = 5.0$ Hz, 1H, H-5), 3.76 (td, $J = 9.0$ Hz, $J = 3.5$ Hz, 1H, H-3), 3.86 (t, $J = 9.5$ Hz, 1H, H-4), 3.91 (t, $J = 10.0$ Hz, 1H, H-6b), 3.95 (d, $J = 4.0$ Hz, 1H, H-2), 4.36 (dd, $J = 10.5$ Hz, $J = 5.0$ Hz, 1H, H-6a), 4.64 (d, $J = 12.0$ Hz, 1H, CHaHbPh), 4.66 (s, 1H, H-1), 4.70 (d, $J = 11.5$ Hz, 1H, CHaHbPh), 5.01 (d, $J = 12.0$ Hz, 1H, CHaHbPh), 5.12 (d, $J = 11.5$ Hz, 1H, CHaHbPh), 5.56 (s, 1H, benzylidene-H), 7.28–7.40 (m, 13H, aromatic), 7.46–7.49 (m, 2H, aromatic). ^{13}C NMR (75 MHz, $CDCl_3$) δ (ppm): 67.4 (C-5), 68.8 (C-6), 71.0 (C-3), 71.4 (CH_2Ph), 75.9 (CH_2Ph), 78.7 (C-2), 79.5 (C-4), 101.1 (C-1), 102.2 (benzylidene-C), 126.5, 128.0, 128.1, 128.4, 128.5, 128.7, 129.3, 137.3, 137.4, 138.3. IR (cm^{-1}): 3482, 3031, 2872, 2249, 1497, 1454, 1384, 1355, 1209, 1088, 1025, 909, 730, 696. ESIHRMS: $C_{27}H_{32}NO_6$ Calcd for $[M + NH_4]^+$: 466.2230, found 466.2040. Anal. Calcd for $C_{27}H_{28}O_6$: C, 72.30; H, 6.29; O, 21.40. Found: C, 72.24; H, 6.49; O, 20.98.

Benzyl 2-*O*-benzyl-4,6-*O*-benzylidene-3-*O*-(2-*O*-benzyl-4,6-*O*-benzylidene-α-D-manno-pyranosyl)-(1→3)-β-D-mannopyranoside (5α) and benzyl 2-*O*-benzyl-4,6-*O*-benzylidene-3-*O*-(2-*O*-benzyl-4,6-*O*-benzylidene-β-D-mannopyranosyl)-(1→3)-β-D-mannopyranoside (5β)

To a cooled (–78°C) solution of **2** (1.21 g, 2.0 mmol), TTBP (1.10 g, 4.4 mmol), and BSP (500 mg, 2.4 mmol) in dry CH_2Cl_2 (40 mL), was added Tf_2O (400 μL, 2.4 mmol). The orange reaction mixture was stirred for 1 h at –78°C before a solution of **4** (1.0 g, 2.2 mmol) in dry CH_2Cl_2 (10 mL) was added. The reaction mixture was stirred for 2 h at –78°C, the reaction was quenched with saturated aqueous $NaHCO_3$, and extracted twice with CH_2Cl_2. The combined organic layers were washed with brine, dried, concentrated, and the crude product was chromatographed (20% EtOAc in heptane) to furnish an inseparable mixture of diastereomeric disaccharides (1.98 g). The foregoing material was dissolved in 10:1 CH_2Cl_2–water (24 mL) and DDQ (580 mg, 2.56 mmol) was added. The mixture was stirred at room temperature for 6 h, saturated aqueous $NaHCO_3$ was added and the mixture was extracted three times with CH_2Cl_2. The combined organic layers were washed with brine, dried, and concentrated under vacuum. Repeated chromatographic purification on silica gel (performed with a Combiflash companion with RediSep prepacked column, 230–400 mesh, solid application, eluent 20% EtOAc in heptane) yielded compound **5β** (1.28 g, 1.62 mmol, 76%, colorless oil). $[\alpha]_D^{24}$ –130.3 (*c* 1.0, $CHCl_3$). ^1H NMR (500 MHz, $CDCl_3$) δ (ppm): 2.45 (d, *J* = 9.5 Hz, 1H, OH), 3.00 (td, *J* = 9.5 Hz, *J* = 4.5 Hz, 1H, H-5′), 3.42 (td, *J* = 9.5 Hz, *J* = 5.0 Hz, 1H, H-5), 3.46–3.54 (m, 1H, H-3′), 3.59 (d, *J* = 3.5 Hz, 1H, H-2′), 3.67–3.76 (m, 2H, H-3, H-6′), 3.95–4.07 (m, 3H, H-2, H-4, H-6′), 4.12–4.17 (m, 3H, H-1′, H-4′, H-6), 4.38 (dd, *J* = 10.5 Hz, *J* = 4.5 Hz, 1H, H-6), 4.54 (d, *J* = 11.0 Hz, 1H, C*H*aHbPh), 4.63 (s, 1H, H-1), 4.64 (d, *J* = 12.0 Hz, 1H, C*H*aHbPh), 4.82 (d, *J* = 12.5 Hz, 1H, C*H*aHbPh), 4.99–5.05 (m, 3H, CHa*H*bPh ×3), 5.30 (s, 1H, benzylidene-H), 5.60 (s, 1H, benzylidene-H), 7.16–7.46 (m, 25H, aromatic). ^{13}C NMR (75 MHz, $CDCl_3$) δ (ppm): 67.1 (C-5′), 68.0 (C-5), 68.6 (C-6′), 68.8 (C-6), 70.3 (C-3′), 71.5 (*C*H$_2$Ph), 73.2, 73.7, 74.6 (*C*H$_2$Ph), 75.1 (*C*H$_2$Ph), 76.8, 77.6, 79.7, 97.3 ($^1J_{CH}$ = 159.4 Hz, C-1′), 101.4 ($^1J_{CH}$ = 155.5 Hz, C-1), 101.9 (benzylidene-C), 102.0 (benzylidene-C), 126.4, 127.9, 128.0, 128.2, 128.3 (×3), 128.4, 128.5, 128.6, 128.7, 129.0, 129.1 (×2), 137.2, 137.4, 137.5, 138.2, 138.3. IR (cm^{-1}): 3549, 2870, 1454, 1384, 1265, 1211, 1184, 1086, 1024, 915, 730, 695. ESIHRMS: $C_{47}H_{48}O_{11}Na$ Calcd for [M + Na]$^+$: 811.3094, found 811.3085. Anal. Calcd for $C_{47}H_{48}O_{11}$: C, 71.56; H, 6.13; O, 22.31. Found: C, 71.32; H, 6.14; O, 22.20.

Compound **5α** was isolated as a viscous oil in various amounts depending on the chromatographic run; $[\alpha]_D^{24}$ –11.8 (*c* 1.0, $CHCl_3$). ^1H NMR (500 MHz, $CDCl_3$) δ (ppm): 2.28 (d, *J* = 9.0 Hz, 1H, OH), 3.40 (td, *J* = 10.0 Hz, *J* = 5.0 Hz, 1H, H-5), 3.64 (td, *J* = 9.5 Hz, *J* = 4.5 Hz, 1H, H-3′), 3.73 (t, *J* = 10.0 Hz, 1H, H-4′), 3.82 (t, *J* = 10.0 Hz, 1H, H-6′), 3.86–4.00 (m, 5H, H-2, H-2′, H-3, H-5′, H-6), 4.14 (dd, *J* = 10.5 Hz, *J* = 5.0 Hz, 1H, H-6′), 4.19 (d, *J* = 11.5 Hz, 1H, C*H*aHbPh), 4.23 (t, *J* = 9.5 Hz, 1H, H-4), 4.33 (d, *J* = 11.5 Hz, 1H, CHa*H*bPh), 4.37 (dd, *J* = 10.5 Hz, *J* = 4.5 Hz, 1H, H-6), 4.61–4.65 (m, 2H, H-1, C*H*aHbPh), 4.88 (d, *J* = 12.0 Hz, 1H, CHa*H*bPh), 4.99 (d, *J* = 12.0 Hz, 1H, CHa*H*bPh), 5.00 (d, *J* = 11.5 Hz, 1H, CHa*H*bPh), 5.34

(s, 1H, H-1′), 5.53 (s, 1H, benzylidene-H), 5.60 (s, 1H, benzylidene-H), 7.02–7.06 (m, 2H, aromatic), 7.22–7.50 (m, 23H, aromatic). ^{13}C NMR (75 MHz, CDCl$_3$) δ (ppm): 63.9 (C-3′), 67.3 (C-5), 68.1 (C-4′), 68.6 (CH_2Ph, C-6′), 71.2 (CH_2Ph), 72.8(C-6), 75.0 (CH_2Ph), 75.9 (CH), 77.3 (CH), 78.3 (CH), 79.0 (CH), 79.4 (CH), 98.9 ($^1J_{CH}$ = 172.5 Hz, C-1′), 100.9 ($^1J_{CH}$ = 157.0 Hz, C-1), 102.1 (benzylidene-C, ×2), 126.1, 126.3, 127.6, 127.8, 127.9, 128.0, 128.2, 128.3, 128.4, 128.4, 128.5, 129.1, 129.3, 137.0, 137.4, 137.8. IR (cm^{-1}): 3031, 2923, 2869, 1496, 1453, 1382, 1087, 1025, 749, 696. ESIHRMS: $C_{47}H_{48}O_{11}$Na Calcd for [M + Na]$^+$: 811.3094. Found: 811.3117.

3-O-β-D-MANNOPYRANOSYL-D-MANNOSE (6β)

Benzyl 2-O-benzyl-4,6-O-benzylidene-3-O-(2-O-benzyl-4,6-O-benzylidene-D-manno-pyranosyl)-(1→3)-β-D-mannopyranoside (5β, 550 mg, 0.7 mmol) was dissolved in dry methanol (30 mL) and 10% Pd/C (125 mg) was added. The flask was placed in a hydrogenation apparatus at 40°C/2 bar. After 4 h of shaking, the mixture was filtered through a pad of Celite and evaporated to dryness to give 3-O-β-D-mannopyranosyl-D-mannose (240 mg, 0.7 mmol, 100%) as a white solid. ^{1}H NMR (500 MHz, CD$_3$OD) δ (ppm), major diastereomer: 3.24–3.30 (ddd, J = 2.4, 6.3 and 9.1 Hz, 1H, H-5′), 3.51 (dd, J = 3.1 and 9.5 Hz, 1H, H-3′), 3.58 (t, J = 9.5 Hz, 1H, H-4′), 3.70 (dd, J = 6.4 and 11.9 Hz, 1H, H-6′), 3.73–3.87 (m, 4H, 2H-6, H-4 and H-5), 3.89 (dd, J = 2.1 and 11.9 Hz, 1H, H-6′), 3.95-3.97 (m, 1H, H-2′), 3.97–3.99 (m, 1H, H-2), 3.98–4.05 (m, 1H, H-3), 4.72 (s, 1H, H-1′β), 5.12 (s, 1H, H-1α). Partial signals for minor diastereoisomer: 4.74 (s, 1H, H-1′β), 4.77 (s, 1H, H-1β). ^{13}C NMR (75 MHz, CD$_3$OD) δ (ppm), major diastereomer: 63.0 (*2, C-6 and C-6′), 67.0 (C-4 or C-5), 68.7 (C-4′), 70.1 (C-2′), 72.9 (C-2), 73.8 (C-4 or C-5), 75.3 (C-3′), 78.6 (C-5′), 79.7 (C-3), 95.8 (C-1α), 99.0 (C-1′β). Partial signals for minor diastereoisomer: 66.6, 68.6, 72.6, 77.7, 81.4, 95.5 (C-1β), 98.4 (C-1′β). IR (cm^{-1}): 3330, 2932, 2500, 1649, 1563, 1414, 1240, 1129, 1088, 1055, 976, 820. ESIHRMS: $C_{12}H_{22}O_{11}$Na Calcd for [M + Na]$^+$: 365.1060. Found: 365.1062.

  3-O-α-D-Mannopyranosyl-D-mannose (6α) can also be obtained from **5α**. Benzyl 2-O-benzyl-4,6-O-benzylidene-3-O-(2-O-benzyl-4,6-O-benzylidene-α-D-mannopyranosyl)-(1→3)-β-D-mannopyranoside (5α, 110 mg, 0.14 mmol) was dissolved in methanol (6 mL), 10% Pd/C (50 mg) was added, and the mixture was shaken under hydrogen at 40°C/2 bar. After 4 h the mixture was filtered through Celite and concentrated to dryness, to give 3-O-α-D-mannopyranosyl-D-mannose (47 mg, 0.14 mmol, 100%) as a white solid. ^{1}H NMR (500 MHz, CD$_3$OD) δ (ppm), major diastereomer: 3.61 (t, J = 9.5 Hz, 1H, H-4′), 3.66–3.76 (m, 2H, H-6 or H-6′), 3.76–3.87 (m, 6H, H-4, H-5, H-5′, H-3′, H-6 or H-6′), 3.87–3.91 (m, 1H, H-3), 3.96–4.00 (m, 1H, H-2′), 4.00–4.02 (m, 1H, H-2), 5.01 (d, J = 1.8 Hz, 1H, H-1α), 5.05 (d, J = 1.5 Hz, 1H, H-1′α). Partial signals for minor diastereoisomer: 3.23–3.28 (m, 1H), 4.03 (d, J = 2.7 Hz, H-2), 4.76 (s, 1H, H-1β), 5.10–5.12 (m, 1H, H-1′α). ^{13}C NMR (75 MHz, CD$_3$OD) δ (ppm): 63.0 (×2, C-6 and C-6′), 67.9, 68.9 (C-4′), 72.2 (C-2 or C-2′ or C-3′), 72.4 (C-2 or C-2′ or C-3′), 72.6 (C-2 or C-2′ or C-3′), 74.3, 75.0, 80.3 (C-3), 96.0 (C-1α), 104.0 (C-1′α). Partial signals for minor diastereoisomer: 63.1 (×2, C-6 and C-6′), 67.5, 69.0, 72.2, 75.1, 78.2, 82.7, 95.6 (C-1β), 103.4

(C-1′α). IR (cm⁻¹): 3330, 2932, 2500, 1649, 1563, 1414, 1240, 1129, 1088, 1055, 976, 820. ESIHRMS: $C_{12}H_{22}O_{11}Na$ Calcd for [M + Na]⁺: 365.1060. Found: 365.1051.

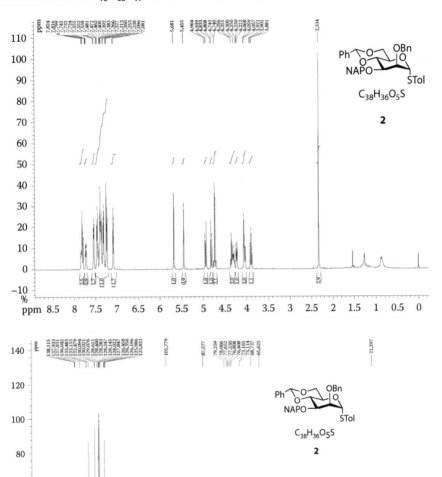

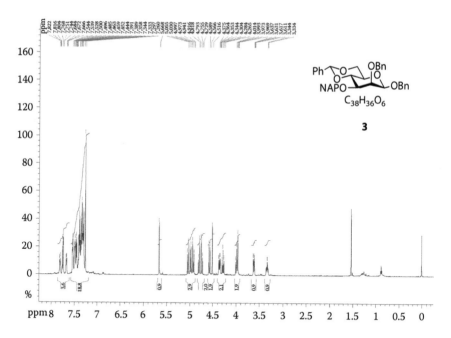

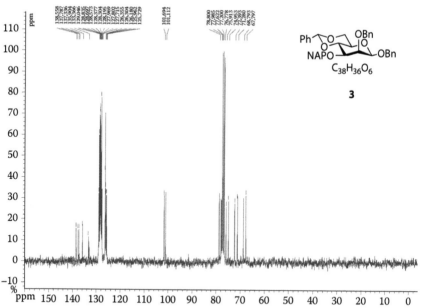

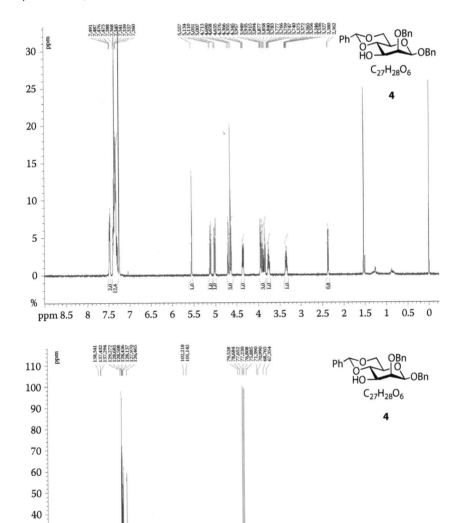

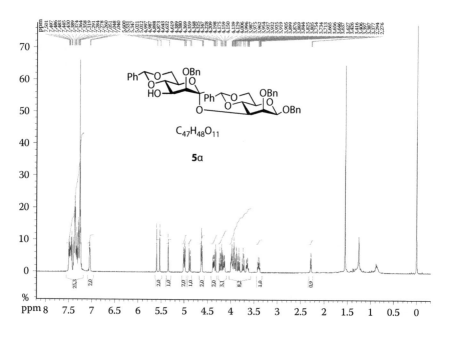

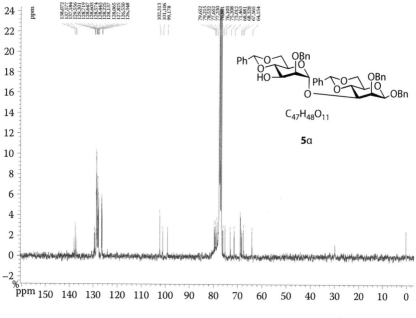

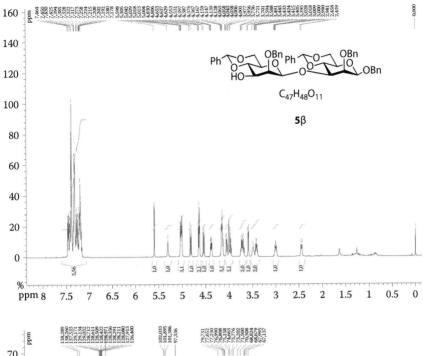

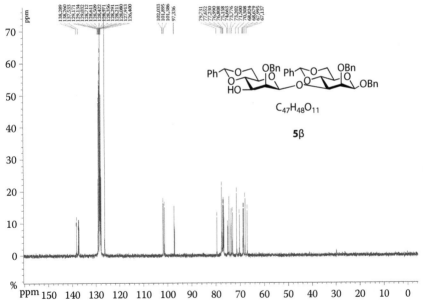

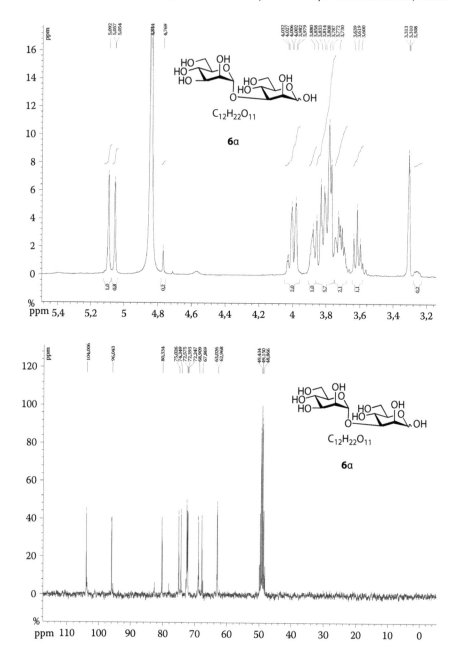

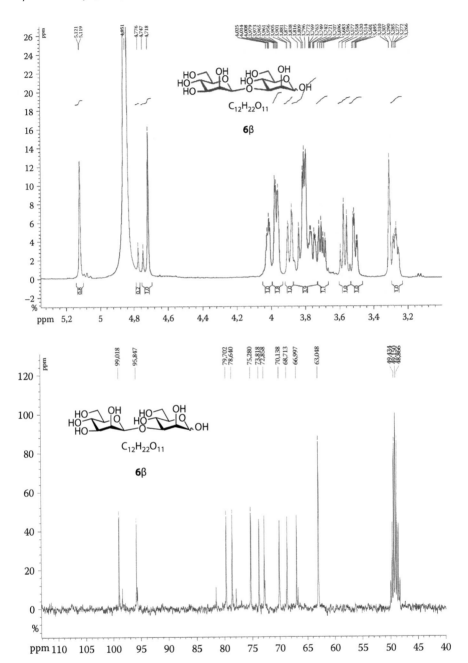

REFERENCES

1. Ponpipom, M. M. *Carbohydr. Res.* 1977, *59*, 311.
2. Awad, L. F.; El Ashry, E. S. H.; Schuerch, C. *Carbohydr. Res.* 1983, *122*, 69.
3. Ogawa, T.; Yamamoto, H. *Agric. Biol. Chem.* 1985, *49*, 475.
4. Ajisaka, K.; Matsuo, I.; Isomura, M.; Fujimoto, H.; Shirakabe, M.; Okawa, M. *Carbohydr. Res.* 1995, *270*, 123.
5. Ikeda, R.; Maeda, T. *Carbohydr. Res.* 2004, *339*, 503.
6. Nomura, K.; Takahashi, N.; Hirose, M.; Nakamura, S.; Yagi, F. *Carbohydr. Res.* 2005, *340*, 2004.
7. Adhya, M.; Singha, B.; Chatterjee, B. P. *Carbohydr. Res.* 2009, *344*, 2489.
8. Ishii, M.; Matsumura, S.; Toshima, K. *Angew. Chem. Int. Ed.* 2007, *46*, 8396.
9. Athanasopoulos, V. I.; Niranjan, K.; Rastall, R. A. *Carbohydr. Res.* 2005, *340*, 609.
10. Hayes, W.; Osborn, H. M. I.; Osborne, S. D.; Rastall, R. A.; Romagnoli, B. *Tetrahedron* 2003, *59*, 7983.
11. Hayes, W.; Osborn, H. M. I.; Osborne, S. D.; Rastall, R. A.; Romagnoli, B. *Tetrahedron Lett.* 2002, *43*, 7683.
12. Crich, D. *Acc. Chem. Res.* 2010, *43*, 1144.
13. Cumpstey, I.; Chayajarus, K.; Fairbanks, A. J.; Redgrave, A. J.; Seward, C. M. P. *Tetrahedron: Asymmetry* 2004, *15*, 3207.

19 Phenyl 4,6-O-Benzylidene-1-thio-α-D-mannopyranoside

Min Huang
*Huu-Anh Tran**
Luis Bohé†
David Crich

CONTENTS

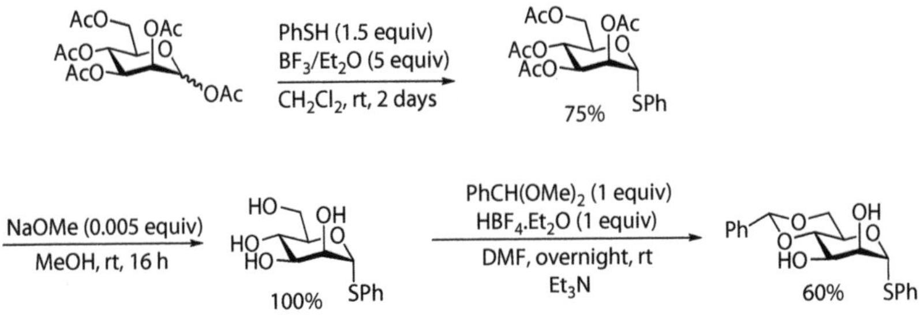

SCHEME 19.1

The classical problem in carbohydrate chemistry of the stereocontrolled synthesis of β-mannopyranosides was largely solved by the introduction of the 4,6-O-benzylidene acetal controlled method in the mid-1990s.[1] The main drawback to this method,

---

* Checker under supervision of D. Bundle: dave.bundle@ualberta.ca.
† Corresponding author : bohe@icsn.cnrs-gif.fr.

which has been widely applied in the synthesis of a variety of oligosaccharides, gly-coconjugates, and natural products,[2] is the need for careful control in the introduction of the 4,6-*O*-benzylidene acetal so as to limit the formation of the corresponding over-reaction product—the 2,3:4,6-di-*O*-benzylidene acetal. Chromatographic puri-fication has typically been necessary to obtain the pure monoacetal thereby lim-iting reaction scale. This situation was remedied in 2005 by the group of Mallet, who described a protocol for 4,6-*O*-benzylidenation employing fluoroboric acid as catalyst that enabled purification of the product by simple crystallization.[3] The present method describes the use of the Mallet conditions, along with the prepara-tion of the starting tetraol, for the facile production of multigram quantities of phenyl 4,6-*O*-benzylidene-1-thio-α-D-mannopyranoside thereby overcoming the main obsta-cle to the use of the benzylidene acetal-directed β-mannosylation on a larger scale.

EXPERIMENTAL

GENERAL METHODS

Melting points were recorded with a Büchi Melting Point B-450 apparatus. Proton nuclear magnetic resonance (^{1}H) spectra were recorded with a BrukerAvance 500 (500 MHz) spectrometer or a Bruker 800 US^2 spectrometer (800 MHz); multiplici-ties are given as singlet (s), doublet (d), doublet of doublets (dd), triplet (t), apparent triplet (at), or multiplet (m). Carbon nuclear magnetic resonance (^{13}C) spectra were recorded with BrukerAvance 500 (125 MHz) Bruker 800 US^2 spectrometer (200 MHz) spectrometers. Residual solvent signals were used as an internal reference. High-resolution (HRMS) electrospray ionization time-of-flight (ESI-TOF) mass spectra were recorded using a Micromass LCT instrument.

PHENYL 2,3,4,6-TETRA-*O*-ACETYL-1-THIO-α-D-MANNOPYRANOSIDE

To a solution of peracetyl α,β-D-mannopyranose[4] (α:β = 4:1, 49.87 g, 127.8 mmol) in dichloromethane (140 mL), were added successively, under argon with stirring at room temperature thiophenol (19.5 mL, d = 1.078, 190.7 mmol) and boron trifluoride diethyl etherate (79.4 mL, d = 1.15, 894 mmol). The yellow-orange solution formed was stirred at room temperature for 2 days, during which it turned deep purple. The mixture was washed successively with saturated sodium hydrogen carbonate solution (3 × 100 mL), 5% aqueous sodium hydroxide solution (5 × 100 mL), and brine, dried over MgSO$_4$, filtered, and concentrated under reduced pressure. The residue, a pale yellow oil, was crystallized from diethyl ether (200 mL)/cyclohexane (300 mL) to give the title compound as a white powder (42.13 g, 75%), mp 83–84°C (from diethyl ether/cyclohexane), lit:[5] 87°C (from diethyl ether/light petroleum); [α]$_D^{26}$ +106.2 (*c* 1, CHCl$_3$), lit:[5] [α]$_D^{27}$ +107.2 (*c* 1, CHCl$_3$); R_f = 0.46 (AcOEt/Heptane, 1:1).

^{1}H NMR (500 MHz, CDCl$_3$) δ 7.31–7.49 (m, 5 H, aromatic), 5.49 (m, 2H, H-1, H-3), 5.33 (m, 2H, H-2, H-4), 4.54 (m, 1H, H-5), 4.30 (dd, 1 H, $J_{5,6a}$ 5.8 Hz, $J_{6a,6b}$ 12.4 Hz, H-6a), 4.10 (dd, 1H, $J_{5,6b}$ 1.9 Hz, $J_{6a,6b}$ 12.4 Hz, H-6b), 2.15 (s, 3H), 2.07 (s, 3H), 2.05 (s, 3H), 2.02 (s, 3H); ^{13}C NMR (125 MHz, CDCl$_3$) δ 170.67, 170.05, 169.95, 169.88, 132.79, 132.21 (2C), 129.35 (2C), 128.27, 85.85 (C-1), 71.07 (C-3), 69.69 (C-5),

69.54 (C-2), 66.55 (C-4), 62.61 (C-6), 21.02, 20.85 (2C), 20.78. HRMS (ESI): [M + NH$_4$]$^+$ calcd for C$_{20}$H$_{28}$NO$_9$S, 458.1485; found, 458.1491. Anal. Calcd for C$_{20}$H$_{24}$O$_9$S: C, 54.54; H, 5.49. Found: C, 54.73; H, 5.42.

PHENYL 1-THIO-α-D-MANNOPYRANOSIDE

To a stirred solution of phenyl 2,3,4,6-tetra-*O*-acetyl-1-thio-α-D-mannopyranoside (42.1 g, 95.6 mmol) in anhydrous methanol (240 mL), 100 μL of NaOMe (25% in MeOH) was added at room temperature under argon. The mixture was stirred overnight (16 h), then neutralized with ion exchange Amberlite® 120 (H$^+$) resin (about 1.0 g, 5 min), filtered, and concentrated under reduced pressure. The residue was dried over P$_2$O$_5$ overnight giving 26.4 g of the crude product in the form of a white foam, which was used in the next step without further purification. An additional trituration of the crude product with diethyl ether gave the analytical sample, mp 127°C (from diethyl ether); lit:[5] 128–129°C (from diethyl ether/ethanol) [α]$_D$^{25} +242.1 (*c* 3, absolute EtOH), lit:[5] [α]$_D$^{27} +253.2 (*c* 3, EtOH); R_f = 0.12 (MeOH/DCM, 1:9). ^{1}H NMR (500 MHz, DMSO-d$_6$) δ 7.27–7.50 (m, 5 H, aromatic), 5.34 (s, 1H, H-1), 5.11 (d, 1H, $J_{2,OH}$ 4.3 Hz), 4.86 (d, 1H, $J_{4,OH}$ 5.2 Hz), 4.77 (d, 1H, $J_{3,OH}$ 5.3Hz), 4.49 (t, 1H, $J_{6,OH}$ 6.0Hz), 3.88 (broad s, 1 H, H-2), 3.77 (at, 1H, *J* 7.0 Hz, H-5), 3.66 (dd, 1 H, $J_{5,6a}$ 5.4 Hz, $J_{6a,6b}$ 11.5 Hz, H-6a), 3.49 (m, 3H, H-6b, H-4, H-3); ^{13}C NMR (125 MHz, DMSO-d$_6$) δ 134.9, 131.0 (2C), 129.0 (2C), 127.0, 88.8 (C-1), 75.4 (C-5), 71.9 (C-2), 71.5 (C-3), 67.0 (C-4), 60.9 (C-6). HRMS (ESI): [M + NH$_4$]$^+$ calcd for C$_{12}$H$_{20}$NO$_5$S, 290.1062; found, 290.1062. Anal. Calcd for C$_{12}$H$_{16}$O$_5$S: C, 52.93; H, 5.92; S, 11.77. Found: C, 52.86; H, 6.12, S, 11.75.

PHENYL 4,6-*O*-BENZYLIDENE-1-THIO-α-D-MANNOPYRANOSIDE

To a solution of phenyl 1-thio-α-D-mannopyranoside (10.69 g, 39.3 mmol, previously dried over P$_2$O$_5$) in anhydrous DMF (80 mL) were added, successively with stirring under argon at 0°C, α,α-dimethoxytoluene (6 mL, d = 1.014, 40 mmol) and tetrafluoroboric acid diethyl ether complex (5.3 mL, d = 1.19, 38.9 mmol). The mixture was stirred at room temperature overnight,* neutralized with Et$_3$N (20 mL), and concentrated under reduced pressure. The residue, a yellow-orange solid, was crystallized from absolute ethanol (575 mL) to give the 4,6-*O*-benzylidene acetal as a white powder (8.46 g, 60%), mp 207°C (from absolute EtOH), lit:[5] 200°C (from EtOH); [α]$_D$^{26} +229 (*c* 0.5, DMSO), lit:[6] [α]$_D$^{20} +292 (*c* 0.5, MeOH); R_f = 0.29 (AcOEt/Heptane,1:1). ^{1}H NMR (800 MHz, DMSO-d$_6$) δ 7.31–7.48 (m, 10 H, aromatic), 5.63 (s, 1H, benzylidene-H), 5.54 (d, 1H, $J_{2,OH}$ 4.1 Hz), 5.46 (s, 1H, H-1), 5.22 (d, 1H, $J_{3,OH}$ 6.08 Hz), 4.06 (m, 2H, H-5, H-6a), 4.00 (m, 1 H, H-2), 3.95 (at, 1H, *J* 9.4 Hz, H-4), 3.77 (at, 1H, *J* 11.8 Hz, H-6b), 3.75 (m, 1H, H-3); ^{13}C NMR (200 MHz, DMSO-d$_6$) δ 137.8, 133.6, 131.3 (2C), 129.3 (2C), 128.8, 128.0 (2C), 127.4, 126.4 (2C), 101.2 (benzylidene), 89.2 (C-1), 78.4 (C-4), 72.4 (C-2), 68.0 (C-3), 67.6 (C-6), 65.3 (C-5), HRMS (ESI):[M + H]$^+$ calcd for C$_{19}$H$_{21}$O$_5$S, 361.1110; found,

---

* Thin-layer chromatography (TLC) frequently shows a considerable amount of the starting material at this stage, resulting from hydrolysis of the benzylidene group on the plate.

361.1107. Anal. Calcd for $C_{19}H_{20}O_5S$: C, 63.32; H, 5.59; S, 8.90. Found: C, 63.20; H, 5.63; S, 8.72.

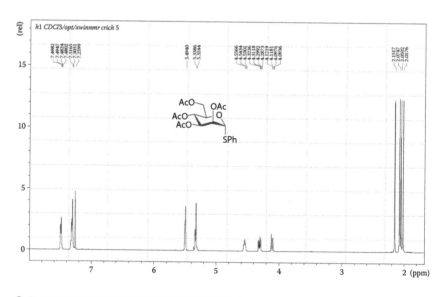

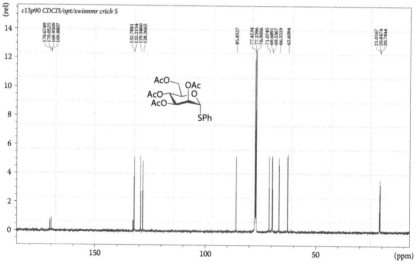

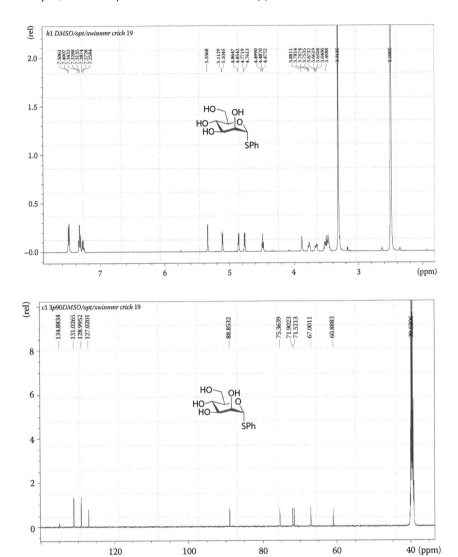

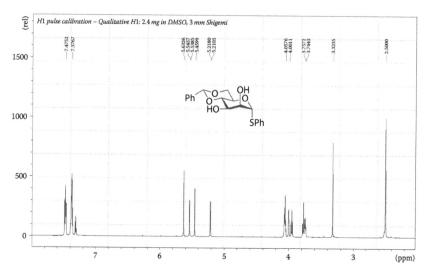

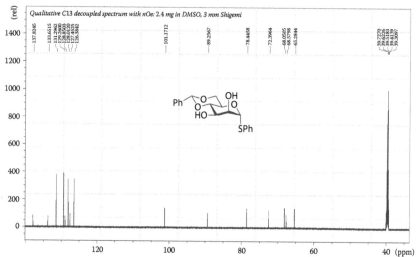

REFERENCES

1. (a) Crich, D. *Acc. Chem. Res.* 2010, *43*, 1144–1153. (b) Aubry, S.; Sasaki, K.; Sharma, I.; Crich, D. *Top. Curr. Chem.* 2011, *301*, 141–188.
2. (a) Cai, F.; Wu, B.; Crich, D. In *Adv. Carbohydr. Chem. Biochem.*; Horton, D., Ed.; Elsevier: 2009; Vol. 62, pp. 251–309. (b) Picard, S.; Crich, D. *Chimia* 2011, *65*, 59–64. (c) Guinchard, X.; Picard, S.; Crich, D. In *Modern Tools for the Synthesis of Complex Bioactive Molecules*; Arseniyadis, S., Cossy, J., Eds.; Wiley: Hoboken, NJ, 2012, 395–432. (d) Matsui, R.; Seto, K.; Sato, Y.; Suzuki, T.; Nakazaki, A.; Kobayashi, S. *Angew. Chem. Int. Ed.* 2011, *50*, 680–683.

3. Chevalier, R.; Esnault, J.; Vandewalle, P.; Sendid, B.; Colombel, J.-F.; Poulain, D.; Mallet, J.-M. *Tetrahedron* 2005, *61*, 7669–7677.
4. Watt, J. A.; Williams, S. J. *Org. Biomol. Chem.* 2005, *3*, 1982–1992.
5. Maity, S. K.; Dutta, S. K.; Banerjee, A. K.; Achari, B.; Singh, M. *Tetrahedron* 1994, *50*, 6965–6974.
6. Cherif, S.; Clavel, J.-M.; Monneret, C. J. *J. Carbohydr. Chem.* 1998, *17*, 1203–1218.

20 Synthesis of 1,3,4,6-Tetra-O-acetyl-α-D-glucopyranose Revisited

*Swati S. Nigudkar**
*Scott J. Hasty**
Monica Varese†
Papapida Pornsuriyasak
Alexei V. Demchenko‡

CONTENTS

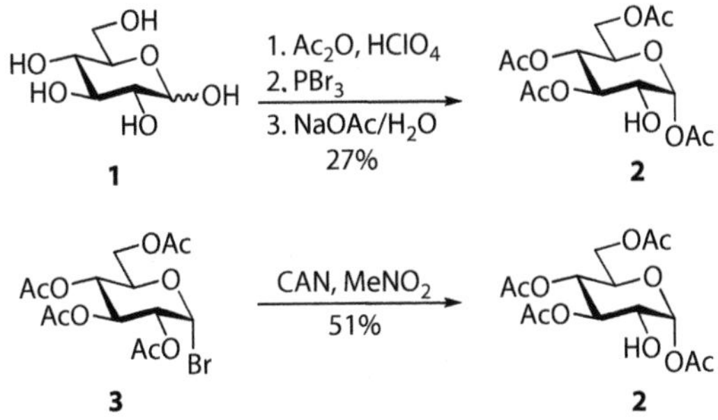

SCHEME 20.1

* These authors contributed equally to this project.
† Checker under supervision of L. Panza: luigi.panza@pharm.unipmn.it.
‡ Corresponding author: demchenkoa@msx.umsl.edu.

Regioselective, partial protection is an important transformation in carbohydrate chemistry as it simplifies preparation of differentially substituted synthetic interme- diates. One such procedure is the synthesis of tetraacetyl glucose **2**. The method for obtaining **2** was originally presented by Helferich and Zirner[1] and involved a multi- step reaction that eventually led to a mixture of hemiacetal (1-OH derivative) and **2**, from which the latter could be isolated in pure form by crystallization in a modest yield of 27%. Chromatographic purification of the material in the mother liquor is impractical due to the propensity of **2** to rapidly isomerize into its more stable hemi- acetal counterpart. Since **2** is an important building block, the original protocol is still in frequent use.

Herein we present an optimized version of the original procedure[1] involving opti- mized reaction temperature, which should be meticulously followed. Nevertheless, the problem of low yield of **2** remains unsolved and hard to explain, particularly in the context of a reliable and high-yielding procedure that is available for the synthe- sis of its 2-hydroxyl galactose counterpart.[2] We also present an alternative protocol by which a notably higher yield of 51% can be achieved from acetobromoglucose **3**[3] and CAN without chromatography.

EXPERIMENTAL

GENERAL METHODS

The reactions were performed using commercial reagents (Aldrich and Acros). CH_3NO_2 was distilled from CaH_2. Ammonium cerium (IV) nitrate (CAN) was dried for 4 h under reduced pressure. Reactions were monitored by thin-layer chromatog- raphy on Kieselgel 60 F_{254} (EM Science). The compounds were detected by charring with 10% sulfuric acid in methanol. Solvents were removed under reduced pres- sure at <40°C. Optical rotations were measured with a Jasco P-1020 polarimeter. ^{1}H NMR (nuclear magnetic resonance) spectra were recorded at 300 MHz, and ^{13}C NMR spectra were recorded at 75 MHz (Bruker Avance) for solutions in $CDCl_3$. The ^{1}H chemical shifts are referenced to the signal of the residual $CHCl_3$ ($\delta_H = 7.27$ ppm). The ^{13}C chemical shifts are referenced to the central signal of $CDCl_3$ ($\delta_C = 77.23$ ppm). High-resolution mass spectra (HRMS) were recorded with a JEOL MStation (JMS-700) Mass Spectrometer.

1,3,4,6-TETRA-*O*-ACETYL-α-D-GLUCOPYRANOSE (2)

Method A, from D-glucose (1)

Acetic anhydride (100 mL, 1.06 mol) was added to a three-necked flask equipped with a stirrer and thermometer. While stirring, D-glucose (**1**, 400 mg, 2.22 mmol) and 70% aq. $HClO_4$ (0.4 mL) were added, followed by addition, with stirring, of more **1** (26.0 g, 144.3 mmol) in small portions, maintaining the internal temperature within

40–45°C. When the addition was complete (~45 min), the reaction was stirred for 1 h at room temperature. The mixture was cooled to 15°C, PBr$_3$ (17.2 mL, 0.18 mol) was added dropwise at 25–30°C, and the mixture was stirred for an additional 30 min. Water (9.2 mL, 0.51 mol) was added dropwise, maintaining an internal temperature at 40–45°C, and the mixture was stirred for 1.5 h at 10°C, followed by the addition of 20 mL of aq. NaOAc·3H$_2$O (obtained by dissolving 80 g, 0.59 mol, of NaOAc·3H$_2$O in 100 mL of H$_2$O, **~170 mL solution**). Upon this, the internal temperature rose to 60°C, and external cooling was needed to reduce the temperature to 40°C. The remaining NaOAc·3H$_2$O solution (~150 mL) was added dropwise, maintaining the internal temperature at 45–50°C. The cooling was removed, the mixture was stirred for 45 min at room temperature, transferred into a 2 L separatory funnel containing a mixture of ice and water (~500 mL total) and extracted with CH$_2$Cl$_2$ (3 × 500 mL). The combined organic layer (~1.5 L) was washed successively with ice-cold water (400 mL), chilled (10°C) sat. aq. NaHCO$_3$ (2 × 400 mL), and ice-cold water (3 × 400 mL). The organic layer was separated, dried over MgSO$_4$, and concentrated in vacuo. The residue was crystallized from dry diethyl ether (~150 mL) to afford the title compound (14 g, 27% yield) as white crystals. R_f = 0.39 (7:3 EtOAc–hexanes); m.p. 98–99°C (diethyl ether); $[\alpha]_D^{23}$ +120 (*c* 1, CHCl$_3$); Lit. data:[1] m.p. 98–100°C, [α] D +141 (*c* 3.2, CHCl$_3$). ^{1}H-NMR: δ, 2.03, 2.06, 2.08, 2.18 (4s, 12H, 4 × COCH$_3$), 2.66 (d, 1H, $J_{2,OH}$ = 8.2 Hz, OH), 3.88 (ddd, 1H, $J_{2,3}$ = 9.8 Hz, H-2), 3.98–4.04 (m, 1H, $J_{5,6a}$ = 2.5 Hz, $J_{5,6b}$ = 4.3 Hz, H-5), 4.05 (dd, 1H, $J_{6a,6b}$ = 12.6 Hz, H-6a), 4.25 (dd, 1H, H-6b), 5.07 (dd, 1H, $J_{4,5}$ = 9.8 Hz, H-4), 5.25 (dd, 1H, $J_{3,4}$ = 9.8 Hz, H-3), 6.22 (d, 1H, $J_{1,2}$ = 3.8 Hz, H-1) ppm; ^{13}C-n.m.r.: δ, 20.7, 20.8, 20.9, 21.1 (4 × CH$_3$), 61.7 (C-6), 67.6 (C-4), 69.7 (C-5), 69.8 (C-2), 73.2 (C-3), 91.4 (C-1), 169.4, 169.7, 170.9, 171.5 (4 × C = O) ppm; HR-FAB MS [M + Na]$^+$ calcd for C$_{14}$H$_{20}$O$_{10}$Na 371.0949, found 371.0959.

Method B, from 2,3,4,6-tetra-*O*-acetyl-α-D-glucopyranosyl bromide (3)

CAN (25.6 g, 46.7 mmol) was added to a stirred solution of **3**[3] (12 g, 29.3 mmol) in dry CH$_3$NO$_2$ (75 mL), and the reaction mixture was stirred for 18 h at room temperature. After that, the mixture was transferred into a separatory funnel containing ice-cold water (300 mL) and extracted with CH$_2$Cl$_2$ (3 × 250 mL). The combined organic extract (~0.8 L) was washed with sat. aq. NaHCO$_3$ (2 × 300 mL) and water (3 × 300 mL). The organic layer was separated, dried over MgSO$_4$, and concentrated in vacuo.* The residue was crystallized from dry diethyl ether (~100 mL) to afford the title compound (5.2 g, 51% yield) as white crystals.

ACKNOWLEDGMENTS

This work was supported by an award from the National Science Foundation (CHE-1058112).

---

* After the organic layer was separated, dried over MgSO$_4$, and concentrated in vacuo, the residue was co-evaporated with MeOH (2 × 20 mL) to aid the removal of CH$_3$NO$_2$.

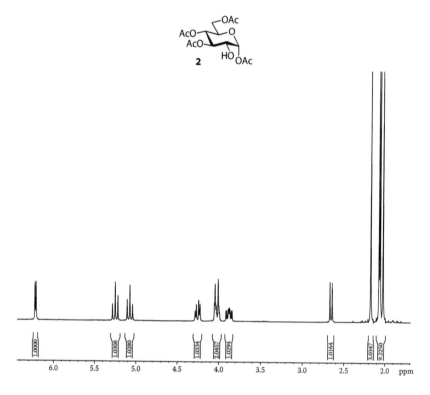

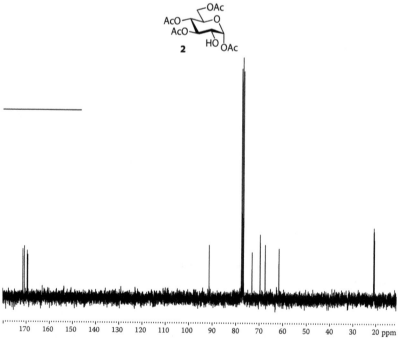

REFERENCES

1. Helferich, B.; Zirner, J. *Chem. Ber.* **1962**, *95*, 2604–2611.
2. Chittenden, G. J. F. *Carbohydr. Res.* **1988**, *183*, 140–143.
3. Lemieux, R. U. In *Methods in Carbohydrate Chemistry*; Whistler, R. L., Wolform, M. L., Eds.; Academic Press: New York, 1963; Vol. 2, pp. 221–222.

21 Synthesis of 4,6-O-Benzylidene Acetals of Methyl α-D-Glucopyranoside, Ethyl 1-Thio-β-D-gluco- and Galactopyranoside

Papapida Pornsuriyasak
Jagodige P. Yasomanee
*Edward I. Balmond**
Cristina De Meo
Alexei V. Demchenko†

CONTENTS

* Checker under supervision of M. C. Galan: M.C.Galan@bristol.ac.uk.
† Corresponding author: demchenkoa@msx.umsl.edu.

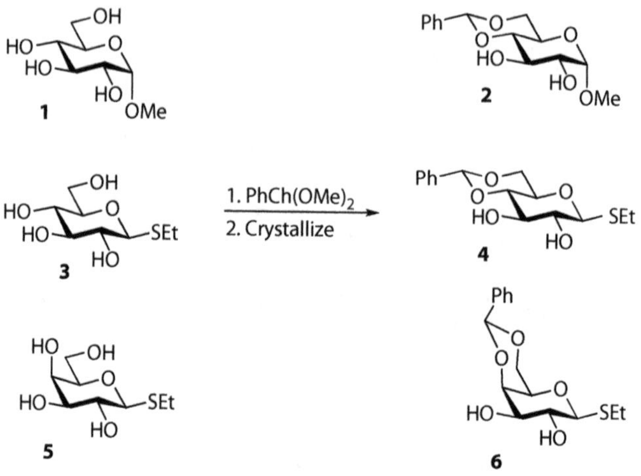

SCHEME 21.1

Benzylidene acetals have found broad application in synthetic carbohydrate chemistry. A new chiral center (PhC*H<) is formed during their formation, wherein the bulky phenyl substituent adopts predominantly the thermodynamically more stable equatorial orientation. Our laboratory has previously developed a simple protocol for the preparation of methyl 4,6-O-benzylidene-α-D-glucopyranoside (2) from commercially available methyl α-D-glucopyranoside (1). The purification of the product was achieved by simple precipitation without the need for chromatographic separation.[1] Herein, we describe the application of this procedure to the synthesis of benzylidene acetals of ethyl thioglycosides 3 and 5. All product purification can be achieved by crystallization/precipitation only. Additional quantities and hence higher yields can be achieved by subsequent chromatography of the mother liquor.

EXPERIMENTAL

GENERAL METHODS

Reactions were performed using commercial reagents (Aldrich and Acros) and solvents were purified according to standard procedures. Thin-layer chromatography (TLC) was performed on Kieselgel 60 F_{254} (EM Science). Compounds were detected by ultraviolet (UV) light and by charring with 10% sulfuric acid in methanol. Solvents were removed under reduced pressure at <40°C. CH_2Cl_2 and CH_3CN were distilled from CaH_2. Optical rotations were measured with a Jasco P-1020 polarimeter. ^{1}H NMR (nuclear magnetic resonance) spectra were recorded at 300 MHz or 500 MHz. ^{13}C NMR spectra were recorded at 75 MHz (Bruker Avance). The ^{1}H chemical shifts are referenced to the signal of the residual $CHCl_3$ (δ_H = 7.27 ppm, δ_C = 77.23 ppm). High-resolution mass spectra (HRMS) were determined with a JEOL MStation (JMS-700) Mass Spectrometer.

METHYL 4,6-O-BENZYLIDENE-α-D-GLUCOPYRANOSIDE (2)

Benzaldehyde dimethyl acetal (0.70 mL, 4.64 mmol) and 10-camphorsulfonic acid (30 mg, 0.13 mmol) were added to a vigorously stirred suspension of methyl α-D-glucopyranoside (**1**, 500 mg, 2.58 mmol) in CH_3CN (10 mL), and the mixture was heated under reflux for 15–20 min.[*] The mixture was allowed to cool to room temperature, neutralized with Et_3N (~0.2 mL), diluted with ethyl acetate (~50 mL), and washed with water (3 × 15 mL).[†] The organic layer was dried over $MgSO_4$, concentrated in vacuo, and hexane (70–80 mL) was added with stirring to a solution of the residue[‡] in CH_2Cl_2 (10–15 mL) to give the title compound as white crystals (590 mg, 81%). m.p. 166–167°C (CH_2Cl_2 – hexanes), lit. data:[2] m.p. 165–166°C; $[\alpha]_D^{23}$ +103.9 (c 2, $CHCl_3$), lit. data[3]: $[\alpha]_D^{23}$ +89 (c 1, $CHCl_3$); R_f = 0.5 (methanol – CH_2Cl_2, 1/9, v/v); ¹H NMR (300 MHz): δ 2.41, 2.91 (2 br. s, 2H, 2-OH, 3-OH), 3.46 (s, 3H, OCH_3), 3.49 (dd, 1H, $J_{4,5}$ = 9.2 Hz, H-4), 3.62 (dd, 1H, $J_{2,3}$ = 9.2 Hz, H-2), 3.74 (dd, 1H, $J_{6a,6b}$ = 9.2 Hz, H-6a), 3.80 (dd, 1H, $J_{5,6a}$ = 10.3 Hz, $J_{5,6b}$ = 3.8 Hz, H-5), 3.93 (dd, 1H, $J_{3,4}$ = 9.2 Hz, H-3), 4.28 (dd, 1H, H-6b), 4.79 (d, 1H, $J_{1,2}$ = 3.9 Hz, H-1), 5.53 (s, 1H, CHPh), 7.36–7.52 (m, 5H, aromatic), ppm; ¹³C NMR data agreed with those published,[3a] and minor differences noted were due to differences in conditions of measurement; HR FAB-MS: $[M + H]^+$ calcd for $C_{14}H_{19}O_{13}$, 283.1182; found, 283.1181.

ETHYL 4,6-O-BENZYLIDENE-1-THIO-β-D-GLUCOPYRANOSIDE (4)

Benzaldehyde dimethyl acetal (0.70 mL, 4.64 mmol) and 10-camphorsulfonic acid (30 mg, 0.13 mmol) were added to a vigorously stirred suspension of ethyl 1-thio-β-D-glucopyranoside (**3**,[4] 558 mg, 2.58 mmol) in CH_3CN (10 mL), the mixture was refluxed[§] for 15–20 min and worked-up as described for the synthesis of **2** to give **4** as white crystals (645 mg, 80%).[¶] m.p. 131.5–132.0°C (CH_2Cl_2 – hexanes), lit. data[5]: m.p.; 129–130°C; $[\alpha]_D^{24}$ –70.3 (c 1, $CHCl_3$), lit. data[6]: $[\alpha]_D^{23}$ –65 (c 1, $CHCl_3$); R_f = 0.49 (methanol – CH_2Cl_2, 1/9, v/v); ¹H NMR (300 MHz): δ 1.33 (t, 3H, J = 7.4 Hz, SCH_2CH_3), 2.72–2.80 (m, 3H, 2-OH, SCH_2CH_3), 3.01 (d, 1H, J = 2.1 Hz, 3-OH), 3.46–3.54 (m, 2H, $J_{2,3}$ = 10.9 Hz, H-2, H-6a), 3.57 (dd, 1H, $J_{4,5}$ = 9.2 Hz, H-4), 3.76 (dd, 1H, $J_{5,6a}$ = 10.3 Hz, $J_{5,6b}$ = 4.8 Hz, H-5), 3.82 (dd, 1H, $J_{3,4}$ = 8.8 Hz, H-3), 4.35 (dd, 1H, $J_{6a,6b}$ = 10.7 Hz, H-6b), 4.46 (d, 1H, $J_{1,2}$ = 9.8 Hz, H-1), 5.54 (s, 1H, CHPh), 7.35–7.54 (m, 5H, aromatic) ppm; ¹³C NMR: δ, 15.4, 24.8, 68.7, 70.6, 73.5, 74.7, 80.5, 86.6, 102.0, 126.5 (×2), 128.5 (×2), 129.4, 137.2 ppm; FAB-MS: $[M + Na]^+$ calcd for $C_{15}H_{20}NaO_5S$, 335.0929; found, 335.0925.

---

[*] As the reaction progresses, the solid **1** dissolves. After 15-20 min at reflux, the reaction should be complete, as indicated by TLC.

[†] Additional quantities of the product could be obtained by extracting the combined water washing with ethyl acetate.

[‡] Alternatively, the residue could be purified by column chromatography on silica gel (methanol - CH_2Cl_2 gradient elution) to give **2** (658 mg, 91%).

[§] As the reaction progresses, the solid starting material dissolves. After 15-20 min at reflux, the reaction should be complete, as indicated by TLC.

[¶] Alternatively, the residue could be chromatographed on silica gel (methanol – CH_2Cl_2 gradient elution) to give **4** (694 mg, 86%).

ETHYL 4,6-*O*-BENZYLIDENE-1-THIO-β-D-GALACTOPYRANOSIDE (6)

Benzaldehyde dimethyl acetal (0.63 mL, 4.17 mmol) and 10-camphorsulfonic acid (27 mg, 0.12 mmol) were added to a vigorously stirred suspension of ethyl 1-thio-β-D-galactopyranoside (**5**,[4] 500 mg, 2.32 mmol) in CH_3CN (10 mL), and the mixture was refluxed for 15–20 min. After work-up, as described above, crystallization gave **6** as white crystals (542 mg, 75%).[*] m.p. 157.5–158.5°C (CH_2Cl_2 – hexanes), $[\alpha]_D^{23}$ –87.2 (*c* 1, $CHCl_3$); lit. data:[4] m. p. 157.6–158.6°C, $[\alpha]_D^{27}$ –60.6 (*c* 0.7, $CHCl_3$); R_f= 0.5 (methanol – CH_2Cl_2, 1/9, v/v); ^{1}H NMR (500 MHz): δ 1.33 (t, 3H, *J* = 7.4 Hz, SC*H$_2$*C*H$_3$*), 2.67–2.83 (m, 4H, 2-OH, 3-OH, SC*H$_2$*C*H$_3$*), 3.44 (d, 1H, $J_{5,6a}$ = 1.7 Hz, $J_{5,6b}$ = 1.3 Hz, H-5), 3.63 (dd, 1H, $J_{3,4}$ = 3.7 Hz, H-3), 3.77 (dd, 1H, $J_{2,3}$ = 9.4 Hz, H-2), 3.97 (dd, 1H, $J_{6a,6b}$ = 12.5 Hz, H-6a), 4.17 (dd, 1H, $J_{4,5}$ = 0.9 Hz, H-4), 4.29 (dd, 1H, H-6b), 4.30 (d, 1H, $J_{1,2}$ = 9.4 Hz, H-1), 5.49 (s, 1H, C*H*Ph), 7.33–7.47 (m, 5H, aromatic) ppm; ^{13}C NMR: δ 15.4, 23.6, 69.5, 69.9, 70.3, 74.0, 75.8, 85.5, 101.6, 126.6 (×2), 128.5 (×2), 129.5, 137.8 ppm; fast atom bombardment mass spectrometry (FABMS): $[M + Na]^+$ calcd for $C_{15}H_{20}NaO_5S$, 335.0929; found, 335.0925.

ACKNOWLEDGMENTS

This work was supported by an award from the National Science Foundation (CHE-1058112).

---

[*] Alternatively, the residue could be chromatographed on silica gel (methanol – CH_2Cl_2 gradient elution) to give **6** (631 mg, 87%).

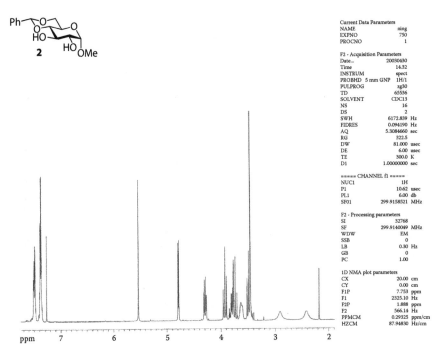

Current Data Parameters
NAME   ning
EXPNO   750
PROCNO   1

F2 - Acquisition Parameters
Date...   20030430
Time   14.32
INSTRUM   spect
PROBHD 5 mm GNP   1H/1
PULPROG   zg30
TD   65536
SOLVENT   CDCl3
NS   16
DS   2
SWH   6172.839 Hz
FIDRES   0.094190 Hz
AQ   5.3084660 sec
RG   322.5
DW   81.000 usec
DE   6.00 usec
TE   300.0 K
D1   1.00000000 sec

===== CHANNEL f1 =====
NUC1   1H
P1   10.62 usec
PL1   6.00 db
SF01   299.9158521 MHz

F2 - Processing parameters
SI   32768
SF   299.9140049 MHz
WDW   EM
SSB   0
LB   0.30 Hz
GB   0
PC   1.00

1D NMA plot parameters
CX   20.00 cm
CY   0.00 cm
F1P   7.753 ppm
F1   2325.10 Hz
F2P   1.888 ppm
F2   566.14 Hz
PPMCM   0.29325 ppm/cm
HZCM   87.94830 Hz/cm

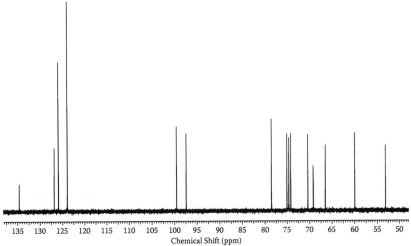

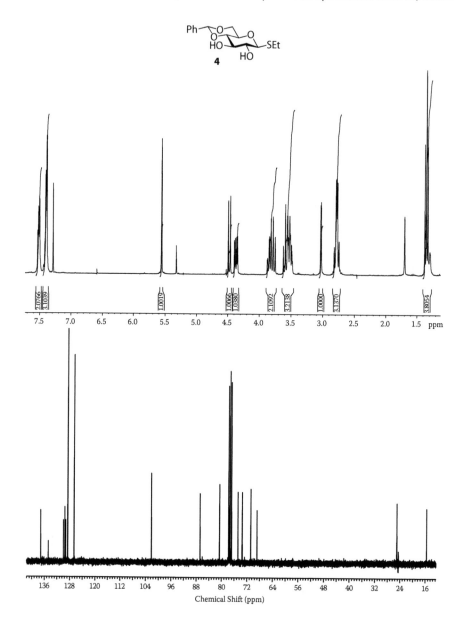

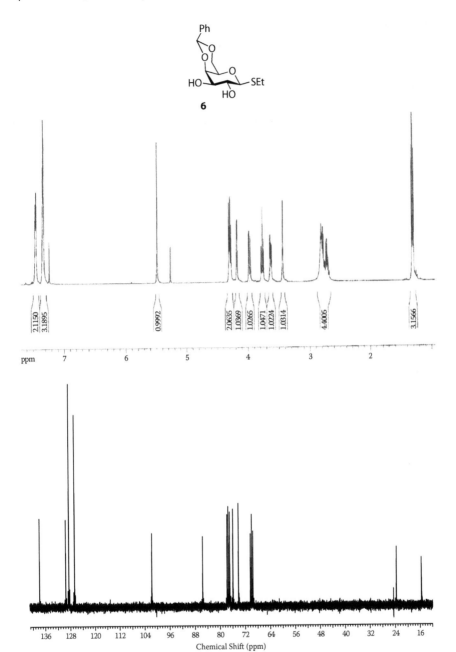

REFERENCES

1. Demchenko, A. V.; Pornsuriyasak, P.; De Meo, C. *J. Chem. Educ.* 2006, *83*, 782–784.
2. Horton, D.; Lauterback, J. H. *J. Org. Chem.* 1969, *34*, 86–92.
3. Soderberg, E.; Westman, J.; Oscarson, S. *J. Carbohydr. Chem.* 2001, *20*, 397–410; (a) Conway, E.; Guthrie, R. D.; Gero, S. D.; Lukacs, G.; Sepulchre, A. M. *J. Chem. Soc. Perkin Trans.,* 1974, *2*, 542–546.
4. Gridley, J. J.; Hacking, A. J.; Osborn, H. M. I.; Spackman, D. G. *Tetrahedron* 1998, *54*, 14925–14946.
5. Nakagawa, T.; Ueno, K.; Kashiwa, M.; Watanabe, J. *Tetrahedron Lett.* 1994, *35*, 1921–1924.
6. Van Steijn, A. M. P.; Kamerling, J. P.; Vliegenthart, J. F. G. *Carbohydr. Res.* 1992, *225*, 229–245.

22 Synthesis of α-*N*-Acetylneuraminic Acid Methyl Glycoside

*Leonid O. Kononov**
Alexander A. Chinarev
Alexander I. Zinin
Chase Gobble†

CONTENTS

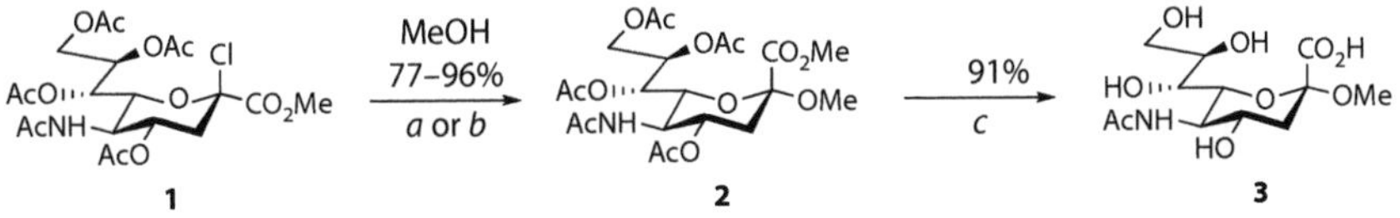

SCHEME 22.1 Reagents and conditions: (a) MeOH, ~20°C. (b) 1. MeOH, ~20°C; 2. Ac₂O, Py. (c) MeONa, MeOH, then H₂O, ~20°C.

N-Acetylneuraminic acid (Neu5Ac) methyl α-glycoside (**3**) and its *O*-acetylated derivative (**2**) are valuable intermediates for preparation of a variety of sialic acid analogues.[1] In addition, methyl sialoside **3** is a useful reference compound in biochemical studies,[2a–c] and also can be used for probing of carbohydrate-binding proteins.[2d,e]

---

* Corresponding author: kononov@ioc.ac.ru, leonid.kononov@gmail.com.
† Checker under supervision of C. DeMeo; e-mail: cdemeo@siue.edu.

Usually, *O*-acetylated methyl ester methyl glycoside **2** is prepared by glycosidation, and conventional saponification of **2** gives methyl sialoside **3**. Although several methods for the preparation of *O*-acetylated methyl ester methyl glycoside **2** have been developed,[1,3,5a] most of them rely on the use of heavy metal salts or other toxic or expensive promoters.

The highly stereoselective synthesis of methyl sialoside **3** and its *O*-acetylated methyl ester derivative **2** described here[4] is experimentally simple, high yielding, and devoid of drawbacks inherent to the known approaches. Our procedure relies on the reaction of Neu5Ac methyl ester glycosyl chloride **1**,[4,5,6] readily accessible in quantitative yield from the corresponding Neu5Ac methyl ester glycosyl acetate,[5,7] with anhydrous methanol in the absence of any specially added promoter[4,8] to give the target protected methyl sialoside **2** in high yield (one-product conversion **1→2**, thin-layer chromatography [TLC]). A few examples of "non-catalyzed" glycosidations with acetobromohexoses, leading to the corresponding de-*O*-acetylated methyl glycosides, have been described.[9] Furthermore, various bromo analogs of **1** were used for sialylation of methanol and benzyl alcohol in the absence of metal salt promoters (but in the presence of collidinium salt) to give glycosides in 70–95% yields, with α/β ratio of 1:1–9:1.[5c]

The product of substantial purity (see Figure 22.1 for a copy of ^{1}H NMR spectrum of crude **2**), suitable for most applications, can be isolated from the reaction mixture by simple evaporation of the solvent and co-concentration with CCl_4 to remove traces of HCl.* Most impurities in the crude **2** originate from de-*O*-acetylation by transesterification under acidic conditions, which was shown to take place under the reaction conditions.[4] If a product of high purity is needed, the crude **2** can be treated by acetic anhydride in pyridine to ensure complete acetylation of all hydroxyl groups and subsequently purified by chromatography or crystallization as described in the Experimental section to give *O*-acetylated methyl sialoside **2** in 77–96% yield. If the unprotected methyl sialoside **3** is the desired product, the crude **2** can be saponified to give the methyl sialoside **3** after purification by ion-exchange chromatography in 91% overall yield with respect to the starting Neu5Ac chloride; the protocol is applicable to large-scale preparation of **3**.

EXPERIMENTAL

GENERAL METHODS

The reactions were performed using commercial reagents (Aldrich and Fluka) and solvents purified according to standard procedures. Fully acetylated Neu5Ac glycosyl chloride **1** was synthesized as described before.[6] TLC was carried out on silica gel 60 F_{254}-coated aluminum foil (Merck). Spots were visualized by heating the plates after immersion in a 1:10 (v/v) mixture of 85% aqueous H_3PO_4 and 95% EtOH. The ^{1}H and ^{13}C NMR spectra were recorded at 30°C for solutions in $CDCl_3$ or D_2O with a Bruker Avance 600 spectrometer (600 MHz and 150 MHz, respectively) or a Bruker Avance 700 spectrometer (700 MHz and 175 MHz, respectively). The ^{1}H chemical shifts are referenced to the signal of the residual $CHCl_3$ (δ_H 7.27) for solutions in

---

* Complete removal of HCl at this step is essential.

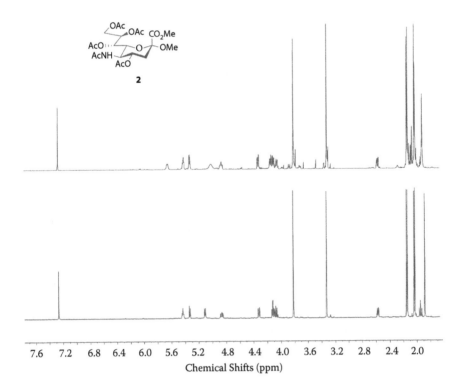

FIGURE 22.1   ¹H NMR spectra (obtained at 700 MHz) of *O*-acetylated αNeu5Ac methyl ester methyl glycoside **2** before (top) and after acetylation step and chromatography or crystallization (bottom).

CDCl₃ or HDO (δ_H 4.75) for solutions in D₂O; the ¹³C chemical shifts are referenced to the central signal of CDCl₃ (δ_C 77.0) for solutions in CDCl₃ or to the signal of external 1,4-dioxane (δ_C 67.4 in D₂O) for solutions in D₂O. Matrix-assisted laser desorption ionization time-of-flight (MALDI-TOF) mass spectra (MALDI-MS) were recorded with a Vision-2000 MALDI-TOF spectrometer. The specific optical rotation ($[\alpha]_D$) of **2** and **3** was measured with a JASCO DIP-360 polarimeter. Analytical high-performance liquid chromatography (HPLC) was performed with a HP Agilent 1100 chromatograph equipped with ultraviolet (UV) detector (195 nm) on a reversed-phase C18 stainless steel column (4.6 × 250 mm, Phenomenex Luna C18, 5 µm particle size, 100 Å pore size) by elution with 50 mM aqueous LiClO₄ (0.5 mL/min, at 30°C).

METHYL (METHYL 5-ACETAMIDO-4,7,8,9-TETRA-*O*-ACETYL-3,5-DIDEOXY-D-*GLYCERO*-α-D-*GALACTO*-NON-2-ULOPYRANOSID)ONATE (2)

Crude 2

Glycosyl chloride **1** (1.02 g, 2 mmol)⁶ was dissolved in anhydrous MeOH (80 mL), and the mixture was kept at room temperature (~20°C) for 50 min (TLC: R_f 0.50 (**1**), R_f

0.40 (**2**), AcOEt). The solvent was evaporated, CH_2Cl_2 (5 mL) and CCl_4 (10 mL) were added to the residue, and the solution was concentrated again (this co-concentration with CH_2Cl_2–CCl_4 mixture was repeated five times).* Then the amorphous residue was dissolved in $CHCl_3$ (20 mL), and the solution was washed with saturated aqueous $NaHCO_3$ (5 mL),† water (5 mL), the organic phase was dried with Na_2SO_4, concentrated, and dried in vacuo to give crude **2** (1.04 g; for purity, see Figure 22.1).

Pure 2

A 1:1 (v/v) Ac_2O–Py mixture (8 mL) was added to the crude **2** (500 mg, 0.99 mmol), the mixture was kept overnight (16 h), cooled in ice-water bath, then MeOH (8 mL) was added. After 30 min the mixture was concentrated and co-concentrated with toluene (5 × 5 mL) to dryness. The residue was dissolved in $CHCl_3$ (20 mL), the chloroform solution was consequently washed with 10% aq. H_2SO_4 (5 mL), saturated aqueous $NaHCO_3$ (5 mL) and H_2O (5 mL), dried over anhydrous Na_2SO_4 (2.0 g), filtered, and then concentrated to dryness to give essentially pure product, which was dissolved in toluene–AcOEt (2:1, 6 mL) at 70°C. Hexane (4 mL) was added, the solution was kept overnight at –20°C, the precipitated colorless crystals were collected and dried in vacuo to give methyl sialoside **2** as colorless crystals (385 mg, 76%;‡ for purity, see Figures 22.1, 22.2, and 22.3), R_f 0.31 ($CHCl_3$–i-PrOH, 15:1), R_f 0.40 (AcOEt).§ M. p. 140–142°C, lit.: 135–137°C.[3a] $[\alpha]_D^{20}$ –19.0 (c 1.0, MeOH), lit.: $[\alpha]_D^{20}$ –18 (c 4.0, MeOH),[5a] $[\alpha]_D^{20}$ –19 (MeOH),[3a] $[\alpha]_D^{20}$ –5.0 (c 4.0, MeOH).[3b] ^{1}H NMR ($CDCl_3$): δ_H 1.88 (s, 3H, NC(O)CH_3), 1.95 (dd, 1H, $J_{4,3ax}$ 12.5 Hz, H-3ax), 2.04, 2.05, 2.15, 2.16 (4 s, 3H each, 4 × OAc), 2.58 (dd, 1H, $J_{4,3eq}$ 4.6 Hz, $J_{3ax,3eq}$ 12.9 Hz, H-3eq), 3.34 (s, 3H, OCH_3), 3.83 (s, 3H, CO_2CH_3), 4.08 (m, 1H, H-5), 4.12 (dd, 1H, $J_{9a,9b}$ 12.4 Hz, $J_{9a,8}$ 5.7 Hz, H-9$_a$), 4.13 (dd, 1H, $J_{7,6}$ 2.2 Hz, $J_{6,5}$ 10.7 Hz, H-6); 4.33 (dd, 1H, $J_{9a,8}$ 2.8 Hz, H-9$_b$), 4.87 (ddd, 1H, $J_{5,4}$ 10.2 Hz, H-4), 5.12 (d, 1H, $J_{NH,5}$ 10.0 Hz, NHAc); 5.34 (dd, 1H, $J_{8,7}$ 8.5 Hz, H-7), 5.44 (ddd, 1H, H-8). ^{13}C NMR ($CDCl_3$): δ_C 20.7, 20.80, 20.83, 21.1 (4 × OC(O)CH_3), 23.1 (NC(O)CH_3), 37.8 (C-3), 49.4 (C-5), 52.4 (OCH_3), 52.7 (CO_2CH_3), 62.4 (C-9), 67.4 (C-7), 68.6 (C-8), 69.1 (C-4), 72.5 (C-6), 99.0 (C-2), 168.2 (C-1, $^3J_{C-1,H-3ax}$ 5.7 Hz[10]), 170.1, 170.2, 170.3, 170.7, 171.0 (5 × CO). MALDI-MS: [M] calcd for $C_{21}H_{31}NO_{13}$, 505.2; [M + Na] calcd for $C_{21}H_{31}NNaO_{13}$, 528.2; [M + K] calcd for $C_{21}H_{31}KNO_{13}$, 544.1; found, m/z: 505, 528, 544. Anal. Calcd for $C_{21}H_{31}NO_{13}$: C 49.9; H 6.18; N 2.77. Found: C 49.5; H 6.23; N 2.70. For previously reported NMR data see the literature.[3b,d,e,f]

---

* One of the checkers noted that addition of CH_2Cl_2 to CCl_4 helps to prevent undesired crystallization of the product, which makes complete removal of HCl difficult.

† This aqueous washing is essential for reproducibility of the procedure. Failure to remove HCl at this step would result in the isolation of the deprotected product **3** contaminated with pyridinium chloride (see footnote below), as the procedure described here makes no effort to remove it (in such a case anion-exchange chromatography may be required for the final purification of **3**).

‡ Concentration of the mother liquor gives more **2** of somewhat lower purity (70 mg, total yield ~90%).

§ Alternatively, methyl glycoside **2** can be purified by chromatography. To this end, the residue obtained after acetylation of crude **2** was dissolved in $CHCl_3$ (5 mL) and applied on a column of silica gel (100 g) packed in $CHCl_3$, and eluted with a gradient i-PrOH–$CHCl_3$ (2:98) → i-PrOH–$CHCl_3$ (10:90) to give **2** (480 mg, 95%) as a colorless solid (for purity, see Figures 22.1 and 22.2).

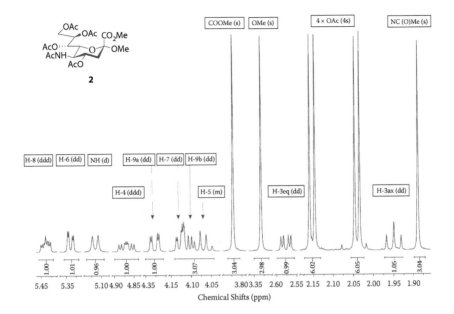

FIGURE 22.2 Expanded fragments of ¹H NMR spectrum (obtained at 700 MHz) of O-acetylated αNeu5Ac methyl ester methyl glycoside **2**.

METHYL 5-ACETAMIDO-3,5-DIDEOXY-D-*GLYCERO*-α-D-*GALACTO*-NON-2-ULOPYRANOSIDE (3)

Crude **2** (500 mg, 0.99 mmol) was dissolved in MeOH (2 × 5 mL) and concentrated (to remove traces of CHCl₃), again dissolved in MeOH (7 mL), then 2 M MeONa in MeOH (3 mL) was added to the solution, the mixture was kept at room temperature (~20°C) for 2 h, diluted with H₂O (10 mL), and left to stand overnight (16 h) at room temperature (~20°C). The mixture was neutralized with AcOH (350 μL) and concentrated to dryness. The residue was dissolved in H₂O (2 mL) and chromatographed on a column packed with Dowex 50 × 4 (200–400 mesh) (pyridinium form, 15 mL, H₂O) using H₂O as the eluent to give after concentration of the respective fractions **3** as a colorless solid (290 mg, 91%,* for purity, see Figures 22.4, 22.5, and 22.6), R_f 0.26 (MeOH–AcOEt–H₂O–AcOH, 2:5:1:0.1). Analytical HPLC: one peak, 5.67 min. $[\alpha]_D^{20}$ −9.0 (*c* 1.0, H₂O), lit.: $[\alpha]_D^{20}$ −9.5 (*c* 1.0, H₂O),[5a] $[\alpha]_D^{20}$ +3.6 (*c* 1.0, H₂O) for the Na⁺-salt.[5c] ¹H NMR (D₂O):† δ_H 1.65 (dd, 1H, $J_{4,3ax}$ 12.0 Hz, H-3*ax*), 2.05 (s, 3H,

---

* Similar results were obtained when pure **2** was subjected to saponification.
† One of the checkers noted minor deviations from the NMR data reported here (e.g., δ(H-3*eq*) 2.63 ppm). These could have resulted from differences in pH or the presence of different counterions or salts in the NMR samples. For example, the product with different NMR data may be contaminated with pyridinium chloride due to incomplete removal of HCl from crude **2** by co-concentration with CCl₄, when this step is omitted.

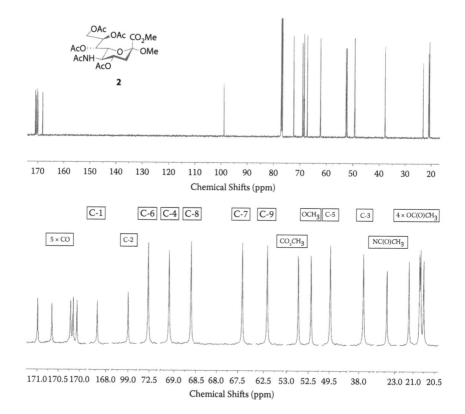

FIGURE 22.3 ^{13}C NMR spectrum (obtained at 150 MHz) of *O*-acetylated αNeu5Ac methyl ester methyl glycoside **2** (top) and its expanded fragments (bottom).

NC(O)CH$_3$), 2.74 (dd, 1H, $J_{4,3eq}$ 4.6 Hz, $J_{3ax,3eq}$ 12.5 Hz, H-3*eq*), 3.36 (s, 3H, OCH$_3$), 3.61 (dd, 1H, $J_{8,7}$ 9.0 Hz, $J_{7,6}$ 1.4 Hz, H-7), 3.66 (dd, 1H, $J_{9a,9b}$ 12.3 Hz, $J_{9a,8}$ 6.5 Hz, H-9$_a$), 3.70 (ddd, 1H, $J_{5,4}$ 8.5 Hz, H-4), 3.73 (dd, 1H, $J_{6,5}$ 10.5 Hz, H-6), 3.83 (ddd, 1H, H-5), 3.90 (dd, 1H, $J_{9b,8}$ 2.4 Hz, H-9$_b$), 3.91 (ddd, 1H, H-8). ^{13}C NMR (D$_2$O): δ_C 23.3 (NC(O)CH$_3$), 40.2 (C-3), 51.6, 52.0 (C-5, OCH$_3$), 62.7 (C-9), 68.30, 68.35 (C-4, C-7), 71.8 (C-8), 72.7 (C-6), 99.0 (C-2), 173.5 (C-1), 175.2 (NC(O)CH$_3$). MALDI-MS: [M] calcd for C$_{12}$H$_{21}$NO$_9$, 323.1; [M + Na] calcd for C$_{12}$H$_{21}$NNaO$_9$, 346.1; [M + K] calcd for C$_{12}$H$_{21}$KNaO$_9$, 362.1; found, *m/z*: 323, 346, 362. Anal. Calcd for C$_{12}$H$_{21}$NO$_9$: C, 44.58; H, 6.55; N, 4.33. Found: C 44.43; H 6.72; N 4.19. For previously reported NMR data see the literature.[3d,f,11a,b]

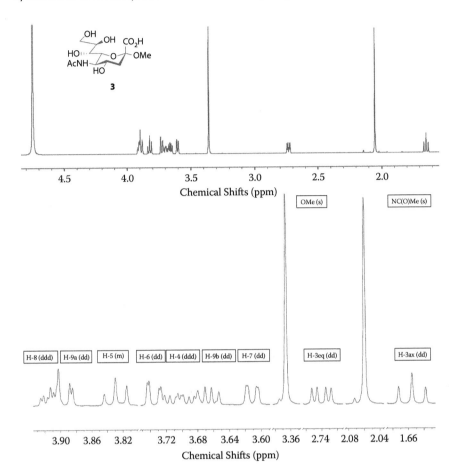

FIGURE 22.4 ¹H NMR spectrum (obtained at 700 MHz) of deprotected αNeu5Ac methyl glycoside **3** (top) and its expanded fragments (bottom).

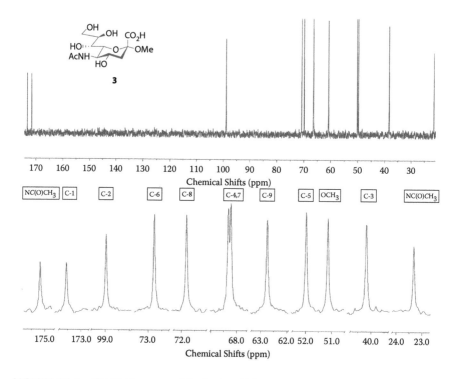

FIGURE 22.5 ^{13}C NMR spectra (obtained at 175 MHz) of deprotected αNeu5Ac methyl glycoside **3** (top) and its expanded fragments (bottom).

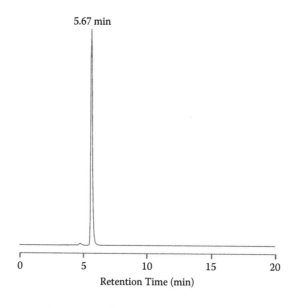

FIGURE 22.6   Reversed-phase HPLC chromatogram of **3** after purification on ion-exchange resin (for details see the General Methods section).

REFERENCES

1. (a) Boons, G.-J.; Demchenko, A. V. *Chem. Rev.*, 2000, *100*, 4539–4565; (b) Okamoto, K.; Goto, T. *Tetrahedron*, 1990, *6*, 5835–5857; (c) Ress, D. K.; Linhardt, R. J. *Curr. Org. Synth.*, 2004, *1*, 31–46; (d) von Itzstein, M.; Thomson, R. J. in: *Glycoscience Synthesis of Oligosaccharides and Glycoconjugates, Top. Curr. Chem.*, 1997, Vol. 186/1997, pp. 119–170, Eds.: Driguez, H.; Thiem, J., Springer, Berlin/Heidelberg, DOI: 10.1007/BFb0119222.

2. (a) Sonnenburg, J. L.; Halbeeek, H.; Varki, A. *J. Biol. Chem.*, 2002, *277*, 17502–17510; (b) Padler-Karavani, V.; Yu, H.; Cao, H.; Chokhawala, H.; Karp, F.; Varki, N.; Chen, X.; Varki, A. *Glycobiology*, 2008, *18*, 818–830; (c) Gamian, A.; Kenne, L. *J. Bacteriol.*, 1993, *175*, 1508–1513; (d) Sparks, M. A.; Williams, K. W.; Lukacs, C.; Screll, A.; Priebe, G.; Spaltenstein, A.; Whitesides, G. M. *Tetrahedron Lett.*, 1993, *49*, 1–12; (e) Tiralongo, J.; Wohlschlager, T.; Tiralongo, E.; Kiefel, M. J. *Microbiology*, 2009, *155*, 3100–3109.

3. (a) Eschenfelder, V.; Brossmer, R. *Carbohydr. Res.*, 1980, *78*, 190–194; (b) Ogura, H.; Furuhata, K.; Itoh, M.; Shitori, Y. *Carbohydr. Res.*, 1986, *158*, 37–52; (c) Isecke, R.; Brossmer, R. *Tetrahedron*, 1994, *50*, 7445–7460; (d) van der Vleugel, D. J. M.; van Heeswijk, W. A. R.; Vliegenthart, J. F. G. *Carbohydr. Res.*, 1982, *102*, 121–130; (e) Gregar, T. Q.; Gervay-Hague, J. *J. Org. Chem.*, 2004, *69*, 1001–1009; (f) Bandgar, B. P.; Zbiral, E. *Monatsh. Chem.*, 1991, *122*, 1075–1088.

4. Kononov, L. O.; Magnusson, G. *Acta. Chem. Scand.*, 1998, *52*, 141–144.

5. (a) Kuhn, R.; Lutz, P.; MacDonald, D. L. *Chem. Ber.*, 1966, *99*, 611–617; (b) Sharma, M. N.; Eby, R. *Carbohydr. Res.*, 1984, *127*, 201–210; (c) Byramova, N. E.; Tuzikov, A. B.; Bovin, N. V. *Carbohydr. Res.*, 1992, *237*, 161–175; (d) Shpirt, A. M.; Kononov, L. O.; Torgov, V. I.; Shibaev, V. N. *Russ. Chem. Bull.*, 2004, *53*, 717–719.

6. Kulikova, N. Y.; Shpirt, A. M.; Chinarev, A.; Kononov, L. O. In: *Carbohydrate Chemistry: Proven Synthetic Methods;* Vol. 1., Kovač, P., (ed.); CRC Press/Taylor & Francis, Boca Raton, FL, 2011, Chapter 27, pp. 245–250.

7. Marra, A.; Sinaÿ, P. *Carbohydr. Res.*, 1989, *190*, 317–322.

8. Orlova, A. V.; Shpirt, A. M.; Kulikova, N. Y.; Kononov, L. O. *Carbohydr. Res.*, 2010, *345*, 721–730.

9. (a) Honig, H.; Weidmann, H. *Synthesis*, 1975, 804; (b) Ovchinnikov, M. V.; Byramova, N. E.; Backinowsky, L. V.; Kochetkov, N. K. *Bioorg. Khim. (Rus.)*, 1983, *9*, 401–406; (c) Byramova, N. E.; Backinowsky, L. V.; Kochetkov, N. K. *Izv. Akad. Nauk SSSR, Ser. Khim. (Rus.)*, 1985, 1140–1145. Engl. transl. - *Bull. Acad. Sci. USSR, Div. Chem. Sci.*, 1985, *34*, 1041–1046; DOI: 10.1007/BF01142798

10. Hori, H.; Nakajima, T.; Nishida, Y.; Ohrui, H.; Meguro, H. *Tetrahedron Lett.*, 1988, *29*, 6317–6320.

11. (a) Sabesan, S.; Bock, K.; Lemieux, R. U. *Can. J. Chem.*, 1984, *62*, 1034–1045; (b) Daman, M. E.; Dill, K. *Carbohydr. Res.*, 1982, *102*, 47–58.

23 Synthesis of Ammonium 3-Deoxy-D-*manno*-oct-2-ulopyranosylonate (Ammonium Kdo)

*Hannes Mikula**
Markus Blaukopf
Georg Sixta
Christian Stanetty
Paul Kosma†

CONTENTS

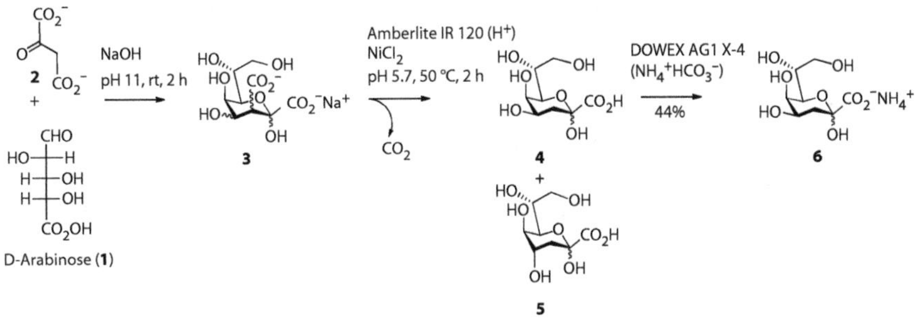

SCHEME 23.1

* Checker under supervision of J. Fröhlich: frohlich@ioc.tuwien.ac.at.
† Corresponding author: paul.kosma@boku.ac.at.

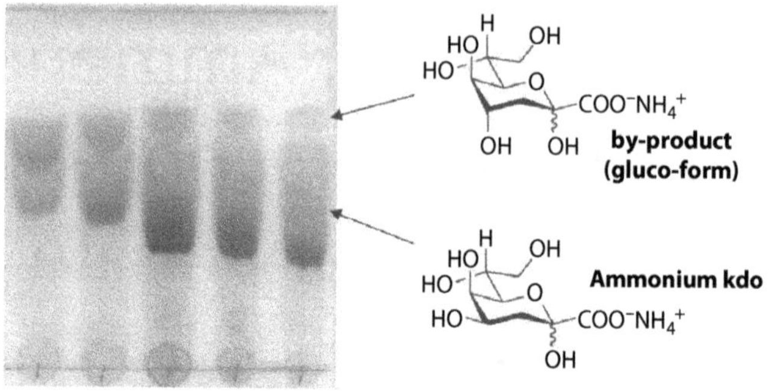

FIGURE 23.1 TLC analysis of selected pooled AEC fractions (CHCl$_3$/MeOH/H$_2$O = 10/10/3, v/v/v).

3-Deoxy-D-*manno*-oct-2-ulosonic acid (Kdo) constitutes an important higher-carbon sugar occurring in capsular polysaccharides and lipopolysaccharides of Gram-negative bacteria as well as in plant rhamnogalacturonans.[1] Considerable progress has been reported in the past exploiting a broad synthetic scope in producing this octulosonic acid (reviewed in references 2 and 3) as well as in elaborating the enzymatic preparation via the condensation of D-arabinose with pyruvate catalyzed by Kdo aldolases.[4] The well-known Cornforth reaction, however, may still be considered as a straightforward, reliable, and scalable approach to give multi-gram amounts of Kdo in a single transformation.[5,6] It is based on an aldol condensation of an excess of D-arabinose **1** (2–3 equivalents) with oxaloacetic acid **2** under alkaline conditions, which leads to the preferential formation of the *manno*-configured adduct. In particular, the detailed description of the synthesis of the crystalline ammonium salt of Kdo by Unger[7] could later be improved by the Ni^{2+}-catalyzed decarboxylation of the intermediate diacid addition product **3** under slightly acidic conditions (pH 5.7), leading to substantially increased overall yields of **6**.[8] Purification of the product mixture and isolation of ammonium Kdo are achieved by immobilizing the epimeric ulosonic acids **4** and **5** on an anion-exchange resin followed by removal of the excess of D-arabinose **1** by exhaustive elution with water.* Since residual amounts of **5** interfere with the crystallization of **6**, removal of the minor amount of the epimeric *gluco*-configured Kdo **5** is achieved by a continuation of anion-exchange chromatography using a gradient of aqueous ammonium hydrogen carbonate as eluent. Careful preparation and equilibration of the HCO$_3^-$ form of the resin and pooling of appropriate fractions are necessary to obtain good product yields (Figure 23.1). Lyophilization of the Kdo-containing pools removes major amounts of the remaining salt, and the product **6** readily crystallizes from a solution in aqueous ethanol. The synthesis has routinely been used in the author's laboratory on a 20–30 g scale.

The ^{1}H NMR (nuclear magnetic resonance) and ^{13}C NMR spectra of **6** are complex due to the presence of anomeric forms. In addition to the predominant presence of the α-pyranose hemiketal, two anomeric furanose forms and a minor proportion of the β-pyranose anomer are observed. The relative proportions of the four

---

* The contaminated D-arabinose present in the effluent must not be reused for the preparation of Kdo since this will inhibit crystallization of the desired product.

anomeric forms may be derived from the intensities of the 3-deoxy proton signals in the range of 2.55–1.83 ppm and these are in agreement with published data.[9,10] An unambiguous assignment of the anomeric configuration of Kdo furanose anomers has not been achieved thus far.[11,12] Based on recent MD-calculations and NMR data of a GlcpA-(1→4)-Kdo disaccharide, the homonuclear coupling constants of the diastereotopic H-3 signals of **6** indicate that the H-3 signal at lowest field should correspond to the β-furanose form.[13]

EXPERIMENTAL

GENERAL METHODS

D-Arabinose and oxaloacetic acid were purchased from SIGMA-Aldrich. Thin-layer chromatography (TLC) was carried out on silica 60 plates (Merck, Darmstadt, Germany). Spots on TLC plates were visualized by ceric ammonium molybdate stain. Preparative ion exchange chromatography was performed using a Sepacore™ Flash System (Büchi, Flawil, Switzerland). NMR spectra were measured on a Bruker Avance 600 instrument. All chemical shifts are given in ppm relative to DSS (δ 0) for ^{1}H and external 1,4-dioxane (δ 67.4) for ^{13}C. Lyophilization was performed on a Labconco FreeZone system (Kansas City, MO), and a MCP500 polarimeter (Anton Paar, Graz, Austria) was used for optical rotation measurements.

AMMONIUM 3-DEOXY-D-*MANNO*-OCT-2-ULOPYRANOSYLONATE (6)

To a solution of $NaHCO_3$ (0.1 g, 1.2 mmol) in distilled water (30 mL), 10 M aqueous NaOH was added at 0°C until pH 10 (as monitored by a pH meter). Oxaloacetic acid (3.17 g, 24.0 mmol) and 10 M aq. NaOH were added simultaneously in small portions over a period of 60 minutes, keeping the pH as close to 10.0 as possible (~9.5–10.5). A solution of D-arabinose (10 g, 66.6 mmol) in water (30 mL) was then added at once, and after removal of the ice-water bath the pH was immediately adjusted to 11.0 by further addition of 10 M aq. NaOH. The reaction mixture was stirred for 2 h at room temperature. After heating to 50°C a solution of $NiCl_2 \cdot 6\ H_2O$ (47 mg, 0.2 mmol) in water (0.5 mL) was added. Cation-exchange resin Amberlite IR120 (Aldrich, hydrogen form) was then added in portions until pH 5.7 (gas evolution) and stirring was continued at 50°C for 2 h. During this period additional resin was added from time to time to keep the pH at 5.7 during the decarboxylation step. The suspension was filtered and the slightly yellow filtrate was applied onto an anion exchange column AG 1X4 (2.5 × 20 cm, HCO_3^-, Bio Rad). The column was washed with water (500 mL) to elute unreacted D-arabinose.

On the following day the column was eluted with water/NH_4HCO_3 (900 mL, linear gradient, 0→0.25 M NH_4HCO_3, 20 mL/min). Fractions (15 mL) were collected (see Figure 23.1) and analyzed by TLC (MeOH/$CHCl_3$/H_2O = 10/10/3, v/v/v). Pure fractions of the desired product (lower spot) were pooled and lyophilized. Fractions being eluted at higher salt concentration (~0.20–0.25M NH_4HCO_3) contained the *gluco*-configured Kdo epimer (**5**) shown as a spot running faster on TLC and these should be discarded or rechromatographed (AEC), since this material interferes with the crystallization of ammonium Kdo. After two AEC separations and subsequent lyophilization a crude material (5.1 g) was obtained, which was dissolved in a minimum amount of water (~10 mL). Ethanol was then added until incipient crystallization (~30 mL) occurred, and the flask was stored at 4°C for 2 days with gradual addition of EtOH (up to 90 mL). The crystals were filtered, washed with dry ethanol (20 mL), and dried in vacuo (20°C, 0.01 mbar) to afford 2.71 g (10.6 mmol, 44%) of the title compound as colorless prisms. mp 120–122°C (with decomposition); lit.[6]: 121–123°C; lit.[7]: 121–124°C; lit.14: 123–125°C $[\alpha]_D^{20}$ +42 (c 1.0, H_2O); lit.[5]: +42.3 (c 1.7, H_2O); lit.[6]: +40.3 (c 1.93, H_2O); ^1H NMR (D_2O) for major α-pyranose form: δ 3.98 (ddd, 1 H, $J_{4,5}$ 3.1, $J_{4,3e}$ 5.2, $J_{4,3a}$ 12.1 Hz, H-4), 3.94 (br d, 1 H, H-5), 3.81 (ddd, 1 H, $J_{7,6}$ 8.9, $J_{7,8a}$ 2.9, $J_{7,8b}$ 5.9 Hz, H-7), 3.73 (dd, 1 H, $J_{8a,8b}$ 12.1 Hz, H-8a), 3.73 (dd, 1 H, H-6), 3.55 (dd, 1 H, H-8b), 1.91 (t, 1 H, $J_{3a,3e}$ 13.0 Hz, H-3a) and 1.81 (ddd, 1 H, $^4J_{5,3e}$ 1.0 Hz, H-3e); ^{13}C NMR (D_2O) for major tautomer (α-pyranose form): δ 176.78 (CO), 96.45 (C-2), 71.18 (C-6), 69.26 (C-7), 66.65 (C-5), 66.25 (C-4), 63.03 (C-8), and 33.67 (C-3).

ACKNOWLEDGMENTS

The authors thank Dr. Andreas Hofinger for recording the NMR spectra. Financial support of this work by a grant from the Austrian Science Fund FWF (grant P 22909-N17) is gratefully acknowledged.

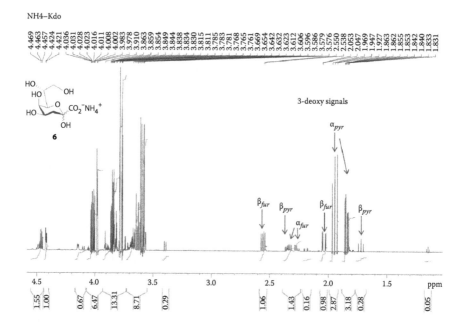

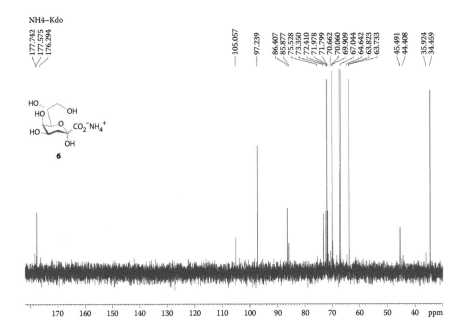

REFERENCES

1. Raetz, C. R.; Whitfield, C. *Annu. Rev. Biochem.* 2002, *71*, 635–700.
2. Li, L. S.; Wu, Y. L. *Curr. Org. Chem.* 2003, *7*, 447–475.
3. Kosma, P.; Zamyatina, A. Chemical synthesis of lipopolysaccharide core. In *Bacterial Lipopolysaccharides*; Knirel, Y. A.; Valvano, M. A.; Eds.; Springer, Wien, New York, 2011, pp. 131–161.
4. Sugai, T.; Shen, G.-J.; Ichikawa, Y.; Wong, C.-H. *J. Am. Chem. Soc.* 1993, *115*, 413–421.
5. Ghalambor, M. A.; Levine, E. M.; Heath, E. C. *J. Biol. Chem.* 1966, *241*, 3207–3215.
6. Hershberger, C.; Davis, M.; Binkley, S. B. *J. Biol. Chem.* 1968, *243*, 1578–1588.
7. Unger, F. M. *Adv. Carbohydr. Chem. Biochem.* 1981, *38*, 323–388.
8. Shirai, R.; Ogura, H. *Tetrahedron Lett.* 1989, *30*, 2263–2264.
9. Brade, H.; Zähringer, U.; Rietschel, E. T.; Christian, R.; Schulz, G.; Unger, F. M. *Carbohydr. Res.* 1984, *134*, 157–166.
10. Cherniak, R.; Jones, R. G.; Gupta, D. S. *Carbohydr. Res.* 1979, *75*, 39–49.
11. Fesik, S. W.; Kohlbrenner, W. E.; Gampe Jr., R. T.; Olejniczak, T. *Carbohydr. Res.* 1986, *153*, 136–140.
12. Kohlbrenner, W. E.; Fesik, S. W. *J. Biol. Chem.* 1985, *260*, 14695–14700.
13. de Córdoba, F. J.; Rodríguez-Carvajal, M. A.; Tejero-Mateo, P.; Gil-Serrano, A. M. *Eur. J. Org. Chem.* 2008, 5815–5822.
14. Birnbaum, G. I.; Roy, R.; Brisson, J-R.; Jennings, H. J. *J. Carbohydr. Chem.* 1987, *6*, 17–39.

24 Synthesis and Characterization of Tetra-*O*-propargyl Pentaerythritol

Peng Xu
*Riccardo Castelli**
Pavol Kováč†

CONTENTS

SCHEME 24.1

The benefit of multivalent interactions over monovalent interactions in various branches of the life sciences has been well established.[1] It is known as the cluster effect or dendritic effect.[2] Pentaerythritol [**1**, tetrakis(hydroxymethyl)methane, 2,2-bis(hydroxymethyl)1,3-propanediol] is a scaffold widely used in the synthesis of glycoclusters and glycodendrimers. It can be derivatized to allow preparation of a

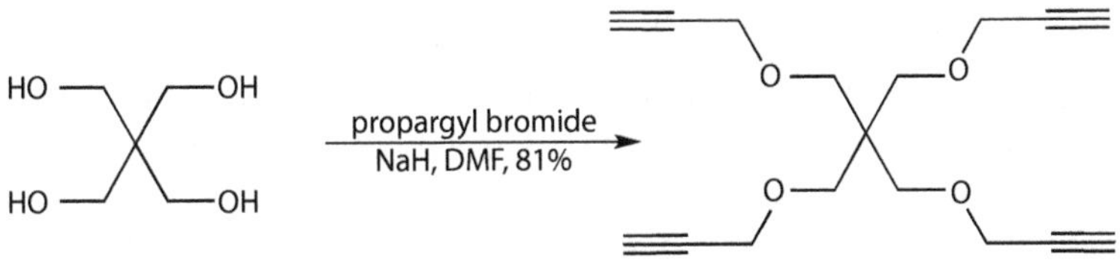

* Checker under supervision of J. Codée: jcodee@chem.leidenuniv.nl.
† Corresponding author: kpn@helix.nih.gov.

large variety of mono-, di-, tri-, or tetra-substituted derivatives.[3,4] Full propargylation of **1** leads to tetrakis(2-propynyloxymethyl)methane (**2**). The latter is an important intermediate in making the first generation of multivalent ligands, for example by Cu[I]-catalyzed azide-alkyne 1,3-dipolar cycloaddition known as "click chemistry."[5,6] Compound **2** has been previously prepared and reported as both an amorphous[7] and a crystalline[8–10] substance. The compound has been fully characterized[9] but we could not reproduce the best reported[9] yield of 87%. According to the protocol described[9] the reaction was conducted in DMSO as solvent without cooling to sub-ambient temperature. According to our experience, in order to obtain high yield of **2** the reaction has to be conducted at temperatures close to 0°C, which the use of DMSO[9] as solvent does not allow. Here we describe a reliable, high-yielding synthesis of **2**.

EXPERIMENTAL

GENERAL METHODS

Pentaerythritol (**1**) and DMF were purchased from Sigma-Aldrich and were used as supplied. The reaction was monitored by thin-layer chromatography (TLC) on silica gel 60-coated glass slides (Analtech Inc.). The TLC spots were visualized by (A) spraying with 0.05% $KMnO_4$ in H_2O or (B) charring with 7% H_2SO_4 in EtOH. Column chromatography was performed using self-packed glass column (ChemGlass) with SiliaFlash F60 silica gel (SILICYCLE, particle size 40–63 μm). Nuclear magnetic resonance (NMR) spectra were measured at 400 MHz (^{1}H) and 100 MHz (^{13}C) with a Bruker Avance spectrometer.

TETRAKIS(2-PROPYNYLOXYMETHYL)METHANE (TETRA-O-PROPARGYL PENTAERYTHRITOL, 2)

Pentaerythritol **1** (10.0 g, 73.4 mmol) is dissolved in anhydrous DMF (500 mL, gentle heating may have to be applied) contained in a 2 L three-neck round bottom flask.[*] The solution is cooled to (-5)–0°C[†] and with mechanical stirring, NaH (60% dispersion in mineral oil, 20.6 g, 514 mmol) was added in four portions during 20 min, and the stirring was continued for another 30 min. Propargyl bromide (45.0 mL, 404 mmol) was then slowly added.[‡,§] When the addition was complete, stirring was continued for another 15 min.[¶] With continued stirring and keeping the

---

[*] The flask was heated from outside by a heat gun while a stream of dry N_2 was allowed to pass through. One side-neck of the flask was equipped with a thermometer; the other side-neck was equipped with an argon balloon.

[†] IpOH–dry ice bath.

[‡] Through a syringe pump or dropwise, during 90 min at this scale. The balloon was removed, and the hydrogen generated during the reaction was allowed to escape through a bubbler.

[§] The color of the reaction mixture changed from gray to light brown to dark brown. During the addition, the stirring should be efficient (~300 rpm) and the inside temperature should be kept at (-3)–0°C by the cooling bath.

[¶] The reaction mixture became thick and dark brown at the end of the 10 min period, with brown foam on the surface.

temperature between (-5)–0°C, 10% acetic acid (~25 mL) was added dropwise, followed by addition of H_2O (500 mL), while the temperature was kept below 25°C. TLC* (1:4 EtOAc–hexane, R_f = 0.6) of the resulting clear, dark brown mixture showed that a major faster-moving product was formed, which could be detected with both reagents A and B. The mixture was extracted with CH_2Cl_2 (2 × 400 mL). The organic layers were combined and washed successively with sat. $NaHCO_3$ aq. and brine, dried (Na_2SO_4), filtered, and the filtrate was concentrated (<45°C) to remove all the DMF. The residue was triturated with EtOAc and filtered through a Celite® pad to remove salts. The filtrate was concentrated to give a dark brown crude product. Chromatography (700 g silica gel, 9:1 hexane–EtOAc) gave **2** (18.9 g, yellow solid). Crystallization from MeOH gave the title compound **2** (18.1 g, 81%) as colorless crystals, m.p. 61.0–61.5°C; Ref. 8: orange solid, 83%, m.p. 52–54°C; Ref. 7: slightly yellow oil, 75%; Ref. 9: 88%, m.p. 57–58°C; Ref. 10: yield 74% (needles), m.p. not reported. NMR data agreed with those reported.[8] electrospray ionization mass spectrometry (ESIMS) (m/z): [M + H]+ calcd for $C_{17}H_{21}O_4$, 289.1440; found: 289.1448; Anal. Calcd for $C_{17}H_{20}O_4$: C 70.81, H 6.99; found: C 70.74, H 7.09.

ACKNOWLEDGMENT

This work was supported by the Intramural Research Program of the NIH, NIDDK.

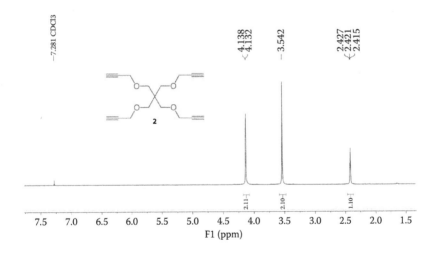

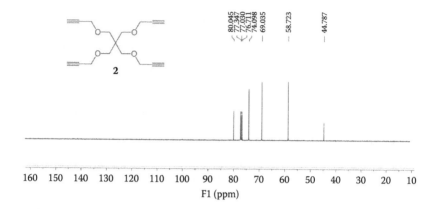

REFERENCES

1. Chabre, Y. M., Roy, R. *Adv. Carbohydr. Chem. Biochem.* 2010, *63*, 165–393.
2. Lee, Y. C.; Lee, R. T. Neoglycoconjugates: Preparation and application, in *Neoglycoconjugates: Preparation and Application*; Academic Press: New York, 1994.
3. Hanessian, S.; Prabhanjan, H.; Qiu, D.; Nambiar, S. *Can. J. Chem.* 1996, *74*, 1731–1737.
4. Hanessian, S.; Qiu, D.; Prabhanjan, H.; Reddy, G. V.; Lou, B. *Can. J. Chem.* 1996, *74*, 1738–1747.
5. Hein, J. E.; Fokin, V. V. *Chem. Soc. Rev.* 2010, *39*, 1302–1315.
6. Bock, V. D.; Hiemstra, H.; van Maarseveen, J. H. *Eur. J. Org. Chem.* 2006, 51–60.
7. Natarajan, A.; Du, W.; Xiong, C.-Y.; DeNardo, G. L.; DeNardo, S. J.; Gervay-Hague, *J. Chem. Commun.* 2007, 695–697.
8. Touaibia, M.; Wellens, A.; Shiao, T. C.; Wang, Q.; Sirois, S.; Bouckaert, J.; Roy, R. *ChemMedChem* 2007, *2*, 1190–1201.
9. Korostova, S. E.; Mikhaleva, A. I.; Shevchenko, S. G.; Sobenina, L. N.; Feldman, V. D.; Shishov, N. I. *Zh. Prikl. Khim.* 1990, *63*, 234–237.
10. Papp, I.; Dernedde, J.; Enders, S.; Haag, R. *Chem. Comm.* 2008, 5851–5853.

25 Improved Synthesis of Ethyl (2,3,4,6-Tetra-O-acetyl-β-D-galactopyranosyl)-(1→3)-4,6-O-benzylidene-2-deoxy-1-thio-2-trichloroacetamido-β-D-glucopyranoside

Sameh E. Soliman[*]
Zoeisha Chinoy[†]
Pavol Kováč[‡]

CONTENTS

[*] On leave of absence from Department of Chemistry, Faculty of Science, Alexandria University, Egypt.
[†] Checker under supervision of Geert-Jan Boons: gjboons@ccrc.uga.edu.
[‡] Corresponding author: kpn@helix.nih.gov.

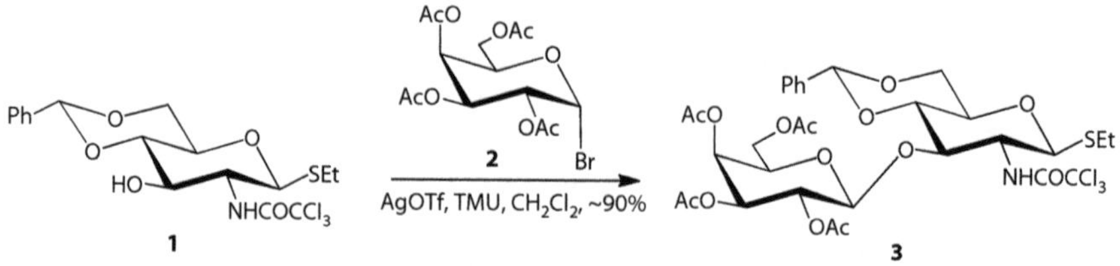

SCHEME 25.1

The 2-deoxy-2-trichloroacetamido group is a powerful stereocontrolling auxiliary in the synthesis of 1,2-*trans*-linked 2-aminosugar-containing oligosaccharides.[1,2] Therefore, the title thioglycoside is a glycosyl donor useful for introduction of the β-D-Gal-(1→3)-D-GlcNAc sequence into many target glycoconjugates important in the life sciences. The compound has been originally prepared by Sherman and coworkers[2] within their study of glycosylation with *N*-trichloroacetyl-protected D-glucosamine derivatives. Side reactions and formation of byproduct were observed, including but not limited to cleavage of the benzylidene group. The highest yield of the desired disaccharide was 60%, when 2,3,4,6-tetra-*O*-acetyl-α-D-galactopyranosyl bromide (acetobromogalactose, **2**)[3] was condensed with ethyl 4,6-*O*-benzylidene-2-deoxy-1-thio-2-trichloroacetamido-β-D-glucopyranoside (**1**) in CH$_2$Cl$_2$ in the presence of molecular sieves and *sym*-collidine (0.5 equiv. relative to **2**). We recently reported that higher yield of the compound (90%)[4] could be obtained by using a larger relative amount of base during the glycosylation step. However, when the same protocol was later repeated in our own laboratory using different batches of molecular sieves, we observed that the yield of **3** was inconsistent. We hypothesized that the basicity of molecular sieves from different batches/suppliers varied, which made it virtually impossible to rationalize the optimum amount of base to be used. To ensure isomerization of the intermediate orthoester,[5,6] the reaction had to be conducted at base-deficient conditions, while the conditions had to be mild to keep the acid-labile benzylidene group intact. The revised protocol (Scheme 25.1) presented here, which we have performed successfully many times, also on multi-gram scale, has consistently afforded disaccharide **3** in ~90% yield. It employs 1,1′,3,3′-tetramethylurea as base[5] and does not require use of molecular sieves. Disaccharide **3** has now been obtained in crystalline form for the first time.

EXPERIMENTAL

GENERAL METHODS

Optical rotation was measured at ambient temperature for solution in CHCl$_3$ with a Perkin-Elmer automatic polarimeter, Model 341. Melting point was measured on a Kofler hot stage. Thin-layer chromatography (TLC) was performed on silica gel 60 coated glass slides. Column chromatography was performed by elution from prepacked (Varian, Inc.) columns of silica gel with the Isolera Flash Chromatograph (Biotage), the latter being connected to the external Evaporative Light Scattering Detector, Model 380-LC (Varian, Inc.). Nuclear magnetic resonance (NMR) spectra were measured at 600 MHz (^{1}H) and 150 MHz (^{13}C) with Bruker Avance spectrometer. Bromide **2** was prepared as described.[3] Solutions in organic solvents were dried with anhydrous Na_2SO_4, and concentrated at 40°C/2 kPa.

ETHYL (2,3,4,6-TETRA-*O*-ACETYL-β-D-GALACTOPYRANOSYL)-(1→3)-4,6-*O*-BENZYLIDENE-2-DEOXY-1-THIO-2-TRICHLOROACETAMIDO-β-D-GLUCOPYRANOSIDE (3)

A solution of acetobromogalactose* (**2**, 307.5 mg, 0.75 mmol) in anhydrous CH_2Cl_2 (1.5 mL) was added, at –30°C in one portion, to a mixture of glycosyl acceptor (**1**,[2] 228.4 mg, 0.5 mmol), 1,1,3,3-tetramethylurea (119.6 μL, 0.95 mmol), and powdered AgOTf (218.4 mg, 0.85 mmol) in anhydrous CH_2Cl_2 (5 mL). The cooling was removed and, with continued stirring, the mixture was allowed to warm up to room temperature (~1 h), when TLC (8:1 toluene–acetone) indicated that all acceptor was consumed. Et_3N (0.5 mL) was added, the mixture was diluted with dichloromethane (5 mL), and filtered through a Celite® pad. The filtrate was washed successively with 0.5 M aq HCl, aq $NaHCO_3$, brine, and dried. After concentration, chromatography (15:1 toluene–acetone) gave the desired disaccharide **3** (357 mg, 91%), m.p. 203.5–204°C (*i*-PrOH), [α]$_D$ –20 (*c* 1, $CHCl_3$). ^{1}H NMR (600 MHz, $CDCl_3$): δ 7.49–7.37 (m, 5 H, Ph); 7.01 (d, 1 H, *J* 7.9 Hz, NH); 5.56 (s, 1 H, PhC*H*), 5.32 (dd, 1 H, $J_{3,4}$ 2.7 Hz, H-4^{II}), 5.19 (dd, 1 H, $J_{1,2}$ 7.9 Hz, $J_{2,3}$ 10.4 Hz, H-2^{II}), 5.10 (d, 1 H, $J_{1,2}$ 10.4 Hz, H-1^I), 4.92 (dd, 1 H, $J_{2,3}$ 10.4 Hz, $J_{3,4}$ 3.4 Hz, H-3^{II}), 4.76 (d, 1 H, $J_{1,2}$ 7.9 Hz, H-1^{II}), 4.48 (t, 1 H, *J* 9.5 Hz, H-3^I), 4.36 (dd, 1 H, *J* 4.9, 10.6 Hz, H-6^I_a), 4.11 (dd, 1 H, *J* 6.9, 11.3 Hz, H-6$^{II}_a$), 4.01 (dd, 1 H, *J* 6.5, 11.3 Hz, H-6$^{II}_b$), 3.82-3.73 (m, 3 H, H-6^I_b, H-4^I, H-5^{II}); 3.67 (m, 1 H, H-2^I), 3.57 (m, 1 H, H-5^I), 2.74 (m, 2 H, SC*H$_2$*), 2.12, 2.03, 1.95, 1.89 (4 s, 12 H, 4 COC*H$_3$*), 1.28 (t, 3 H, *J* 7.4 Hz, SCH$_2$C*H$_3$*). ^{13}C NMR (150 MHz, $CDCl_3$): δ 170.3–169.6 (4 O*C*OCH$_3$), 161.5 (N*C*OCCl$_3$), 136.9 (ipso Ph), 129.3, 128.3, 126.1 (Ph), 101.3 (Ph*C*H), 99.3 (C-1^{II}); 92.3 (*C*Cl$_3$), 83.1 (C-1^I), 78.7 (C-4^I), 77.1 (C-3^I), 70.9 (C-3^{II}), 70.7 (C-5^I), 70.6 (C-5^{II}), 68.8 (C-2^{II}), 68.5 (C-6^I), 66.8 (C-4^{II}), 61.3 (C-6^{II}), 57.5 (C-2^I), 24.9 (S*C*H$_2$), 20.7–20.5 (4 OCO*C*H$_3$), 15.1 (SCH$_2$*C*H$_3$). Anal. Calcd for $C_{31}H_{38}Cl_3NO_{14}S$: C, 47.31; H, 4.87; Cl, 13.51; N, 1.78; S, 4.07. Found: C, 47.23; H, 4.82; Cl, 13.66; N, 1.84; S, 3.88.

---

* Freshly crystallized[3] and thoroughly dried. Amorphous material or crystalline material that had been stored for a longer time may be partially hydrolyzed and contain a small amount of HBr. Introducing such material into the reaction mixture may eventually change the acidity of the reaction mixture, resulting in (partial) cleavage of the benzylidene group.

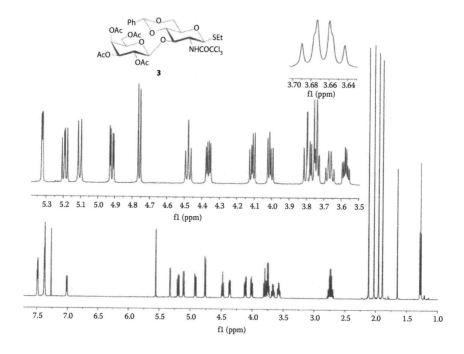

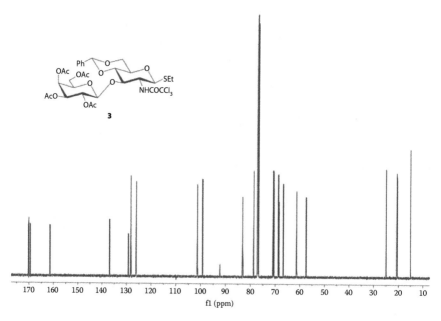

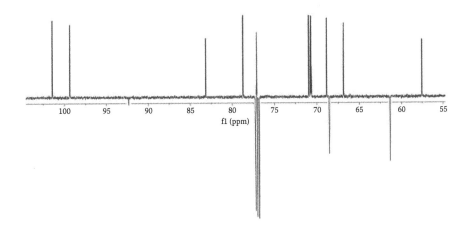

REFERENCES

1. Vibert, A.; Lopin-Bon, C.; Jacquinet, J. -C. *Tetrahedron Lett.,* 2010, *51*, 1867–1869.
2. Sherman, A. A.; Yudina, O. N.; Mironov, Y. V.; Sukhova, E. V.; Shashkov, A. S.; Menshov, V. M.; Nifantiev, N. E. *Carbohydr. Res.* 2001, *336*, 13–46.
3. Schroeder, L. R.; Counts, K. M.; Haigh, F. C. *Carbohydr. Res.* 1974, *37*, 368–372.
4. Hou, S. J.; Kováč, P. *Carbohydr. Res.* 2011, *346*, 1394–1397.
5. Banoub, J.; Bundle, D. R. *Can. J. Chem.* 1979, *57*, 2091–2097.
6. Garegg, P. J.; Norberg, T. *Acta Chem. Scand.* 1979, *B33*, 116–118.

26 Synthesis of α-D-Galactofuranosyl Phosphate

Carla Marino
Cecilia Attorresi
*Silvia Miranda**
Rosa M. de Lederkremer†

CONTENTS

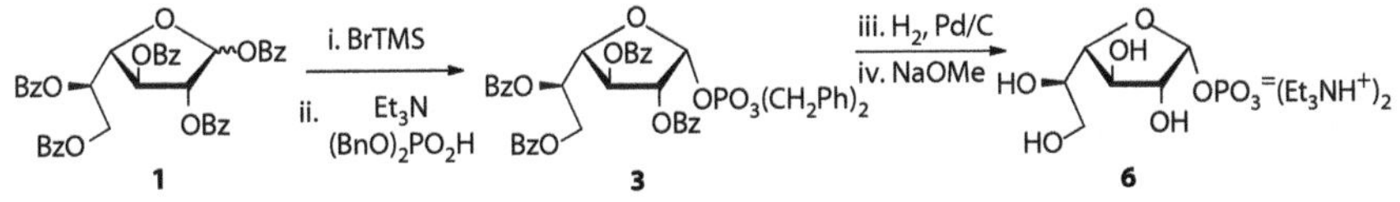

D-Galactofuranosyl units (Gal*f*), mainly in the β-configuration, are constituents of infectious microorganisms, such as the *Mycobacteria*,[1] trypanosomatids like *Trypanosoma cruzi*[2] and *Leishmania*,[3] and fungi like *Aspergillus fumigatus*.[4] Several chemical syntheses of the Gal*f* nucleotide have been reported.[5–7] Recently, an effective chemoenzymatic procedure for the preparation of UDP-Gal*f* was described in detail.[8] The chemical and enzymatic methods start with α-D-Gal*f* phosphate, prepared for the first time by Lederkremer et al.[9] Starting from α-D-[6-³H]-Gal*f* phosphate, the radioactive nucleotide, useful for metabolic studies, was chemically synthesized.[10]

---

* Checker under supervision of C. Lopez: jc.lopez@csic.es.
† Corresponding author: lederk@qo.fcen.uba.ar.

We have optimized the original procedure for the synthesis of α-D-Gal*f* phosphate (**6**), which is described in detail here, taking into account the exceptional lability of a furanose 1-phosphate. The starting compound, perbenzoylated galactofuranose **1**, was obtained crystalline in one step from galactose. The synthesis is fully described in Volume 1 of this Series.[11] Treatment of **1** with trimethylsilyl bromide gave tetra-*O*-benzoyl-β-D-galactofuranosyl bromide (**2**) in almost theoretical yield. The second step is the introduction of the phosphate by treatment of **2** with dibenzyl phosphate, to give an anomeric mixture of dibenzyl 2,3,5,6-tetra-*O*-benzoyl-D-galactofuranosyl phosphates (**3, 4**) from which the α-phosphate **3** was isolated by chromatography. After hydrogenolysis of the benzyl groups followed by debenzoylation, α-D-galactofuranosyl phosphate was obtained as the bis-triethylammonium salt **6**.

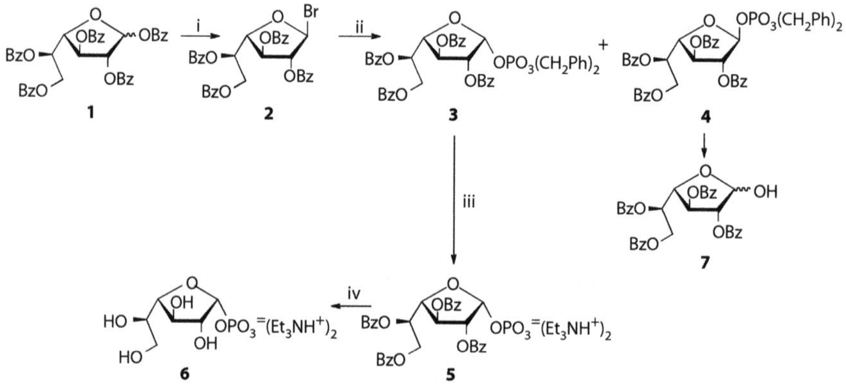

SCHEME 26.1 (i) BrTMS, CH$_2$Cl$_2$; (ii) triethylammonium dibenzylphosphate, toluene; (iii) H$_2$ (30 psi), Pd/C, EtOAc, NEt$_3$; (iv) 0.07 M NaOMe in 1:1 CH$_2$Cl$_2$/MeOH.

EXPERIMENTAL

GENERAL METHODS

Bromotrimethylsilane was purchased from Merck and dibenzyl phosphate from Sigma. Thin-layer chromatography (TLC) was performed on 0.2 mm Silica Gel 60 F254 (Merck) aluminium supported plates. Detection was effected by ultraviolet (UV) light and by charring with 10% (v/v) H$_2$SO$_4$ in EtOH. Column chromatography was performed on Silica Gel 60 (230–400 mesh, Merck). The ^{1}H, ^{13}C, and ^{31}P NMR (nuclear magnetic resonance) spectra were recorded with a Bruker AM 500 spectrometer. The chemical shift reference for ^{31}P was that of external phosphoric acid (85%) in D$_2$O set at 0.0 ppm. Melting points were determined with a Fisher-Johns apparatus and are uncorrected. Optical rotations were measured at 25°C with a Perkin-Elmer 343 polarimeter.

DIBENZYL 2,3,5,6-TETRA-*O*-BENZOYL-α-D-GALACTOFURANOSYL PHOSPHATE (3)

A dry 50 mL, single-neck round-bottom flask equipped with a magnetic stirring bar and a rubber septum is charged with 1,2,3,5,6-penta-*O*-benzoyl-α,β-D-galactofuranose

(**1**, 1.00 g, 1.43 mmol)* and anhydrous CH_2Cl_2 (7 mL).† The solution is stirred with external cooling in an ice-water bath, bromotrimethylsilane (8.0 mL, 60.6 mmol) is added in one portion and the reaction vessel is flushed with argon. After 10 min, the ice bath is removed and the mixture is stirred in the dark at room temperature for 24 h or until TLC examination shows total consumption of **1** [R_f ~0.7 and ~0.6 for β and α anomer, respectively, 9:1 toluene-EtOAc, R_f ~0.3 for 2,3,5,6-tetra-O-benzoyl-α,β-D-galactofuranose (**7**), formed as a result of the hydrolysis of bromide **2** on the silica gel plate]. If TLC analysis indicates the reaction is not complete, the solution is cooled again in the ice bath, an additional portion of bromotrimethylsilane is added (4.0 mL) and the stirring is continued for 6 h at room temperature.

Upon completion of the reaction, the magnetic stirring bar is removed and the solution is concentrated.‡ Several portions of CH_2Cl_2 and then toluene are successively added and co-evaporated under vacuum in order to remove the excess of reagent. The resulting 2,3,5,6-tetra-O-benzoyl-β-D-galactofuranosyl bromide (**2**,[9] 0.94 g, virtually theoretical yield), pure β-anomer (¹H NMR, H-1 broad singlet at 6.60 ppm), is used for the next step without further purification.

Crude **2** (0.94 g, 1.43 mmol) is dried under vacuum for 30 min and dissolved in anhydrous toluene (4 mL).† Triethylamine (0.22 g, 0.3 mL)§ is added followed by dibenzyl phosphate (0.60 g, to form 1.5 equiv. of triethylammonium dibenzyl phosphate)¶ and the flask is capped with a rubber septum. Triethylamine hydrobromide immediately begins to precipitate. The heterogeneous mixture is stirred at room temperature for 3 h, filtered, and the solid is washed with a small amount of toluene (1–2 mL). The filtrate is concentrated at 25°C. The residue, which shows two main spots by TLC (R_f 0.24 and 0.16, 9:1 toluene-EtOAc) is chromatographed,** with 13:1:0.1 toluene-EtOAc-Et_3N as eluent. Fractions of R_f 0.24 (0.68 g) contain a syrupy mixture of the unstable dibenzyl 2,3,5,6-tetra-O-benzoyl-β-D-galactofuranosyl phosphate (**4**), and mainly product of its hydrolysis 2,3,5,6-tetra-O-benzoyl-α,β-D-galactofuranose (**7**), as shown by ¹³C NMR spectroscopy.

Fractions with R_f 0.16 contain dibenzyl 2,3,5,6-tetra-O-benzoyl-α-D-galactofuranosyl phosphate (**3**, 0.41–0.48 g, 30–34%), which solidifies upon concentration: $[\alpha]_D$ + 54.6° (c 1, CHCl₃). Crystallization from a suitable solvent was not attempted because the compound is not stable. However, it can be stored for several months at −20°C without decomposition. ¹H NMR (CDCl₃) δ 8.12–7.04 (H-aromatic), 6.34 (dd, 1 H, $J_{1,2}$ 4.5 Hz, $J_{1,P}$ 5.7 Hz, H-1), 6.15 (apparent t, 1 H, J 7.4 Hz, H-3), 5.83 (m, 1 H, H-5), 5.73 (dddd, $J_{1,2}$ 4.5 Hz, $J_{2,3}$ 6.5 Hz, $J_{2,P}$ 1.8 Hz 1, H, H-2), 5.05–4.90

---

* 1,2,3,5,6-Penta-O-benzoyl-α,β-D-galactofuranose (**1**) is prepared from D-galactose according to the published procedure[11] and dried under vacuum (P_2O_5, toluene reflux).
† CH_2Cl_2 and toluene are freshly distilled from P_2O_5.
‡ The rotary evaporator is first dried by evaporation of some of CH_2Cl_2 and then toluene. A calcium chloride–filled drying tube is attached to the evaporator, allowing the evaporator to be depressurized under dry conditions. A medium-pressure pump is used (50 mm Hg).
§ Triethylamine is distilled from NaOH and stored over NaOH under N_2.
¶ Dibenzylphosphate is dried under vacuum (P_2O_5, room temperature).
** The column (9 × 1.5 cm) is prepared by suspending the silica in toluene with 0.1% of Et_3N. CH_2Cl_2 (2 mL) and silica gel (0.8 g) are added to the sample. The mixture is concentrated to afford a free-flowing powder that is placed at the top of the column. The column is eluted with 10 mL of toluene, followed by 13:1:0.1 toluene-EtOAc-Et_3N. After 120 mL of eluent is collected, compound **3** starts to elute.

(m, 3 H, H-benzylic), 4.82 (dd, 1 H, $J_{5,6}$ 7.7 Hz, $J_{6,6'}$ 11.8 Hz, H-6), 4.72 (m, 2 H, H-4, H-benzylic), 4.62 (dd, 1 H, $J_{5,6'}$ 6.2 Hz, $J_{6,6'}$ 11.8 Hz, H-6'); ^{13}C NMR (CDCl$_3$) δ 165.8–165.4 (4 CO), 135.4–127.6 (C-aromatic), 97.6 (C-1, $J_{C-1,P}$ 4.7 Hz), 79.9 (C-4), 76.5 (C-2, $J_{c-2,P}$ 7.7 Hz), 73.3 (C-3), 70.7 (C-5), 69.3 ($J_{C,P}$ 5.4 Hz, PhCH_2), 69.2 ($J_{C,P}$ 5.4 Hz, PhCH_2), 62.7 (C-6), ^{31}P NMR (CDCl$_3$) δ -3.7. Anal. Calcd for $C_{48}H_{41}O_{13}P$, 67.29; H, 4.82. Found: C, 67.11; H, 4.86.

2,3,5,6-TETRA-O-BENZOYL-α-D-GALACTOFURANOSYL PHOSPHATE BIS-TRIETHYLAMMONIUM SALT (5)

A solution of compound **3** (0.35 g, 0.41 mmol) in EtOAc (5 mL)* containing Et$_3$N (0.35 mL)* was treated with hydrogen in the presence of 10% Pd/C (30 mg),† at 30 psi and room temperature‡ using a Parr apparatus. After 2 h, when TLC shows that the reaction is complete, the catalyst is removed by filtration through a pad of Celite, and the pad is washed with several portions of EtOAc. The combined filtrate is concentrated, 1 mL of CH$_2$Cl$_2$ is added to the residue, and solvents are evaporated to remove EtOAc. The resulting syrup is dried under high vacuum to afford **5** as a white solid (0.33–0.35 g, 94–97%; R_f 0.6, 7:1:2 PrOH-NH$_3$-H$_2$O), mp 89–90°C (from EtOH–Et$_2$O); $[α]_D$ +68° (c 1, CHCl$_3$); ^{1}H NMR (CDCl$_3$ with D$_2$O exchange) δ 8.15–7.22 (H-aromatic), 6.16 (m, 2 H, $J_{1,2}$ 4.5 Hz, $J_{1,P}$ 7.2 Hz, H-1, H-3), 5.85 (m, 1 H, H-5), 5.65 (m, 1 H, H-2), 4.83 (dd, 1 H, $J_{5,6}$ 3.8 Hz, $J_{6,6'}$ 12.0 Hz, H-6), 4.71 (dd, 1 H, $J_{5,6'}$ 6.5 Hz, $J_{6,6'}$ 12.0 Hz, H-6'), 4.59 (dd, 1 H, $J_{3,4}$ 4.5 Hz, $J_{4,5}$ 6.5 Hz, H-4), 2.81 (q, 4 H, J 7.2 Hz, CH$_3$CH$_2$N), 1.05 (t, 6 H, J 7.2 Hz, CH$_3$CH$_2$N); ^{13}C NMR (CDCl$_3$) δ 165.9–165.5 (4 CO), 133.3–127.0 (C-aromatic), 96.0 (C-1, $J_{C-1,P}$ 3.8 Hz), 78.9 (C-4), 76.4 (C-2, $J_{C-2,P}$ 6.3 Hz), 74.0 (C-3), 71.2 (C-5), 63.0 (C-6), 45.4 ((CH$_3$CH$_2$)$_3$N), 8.3 ((CH$_3$CH$_2$)$_3$N). Anal. Calcd for $C_{34}H_{28}O_{13}P \cdot (C_2H_5)_3NH \cdot H_2O$, C, 60.37; H, 5.83. Found: C, 59.90; H, 5.90.

α-D-GALACTOFURANOSYL PHOSPHATE BIS-TRIETHYLAMMONIUM SALT (6)

A dry 25 mL, single-neck round-bottom flask with a magnetic stirring bar and a rubber septum is charged with **5** (0.25 g, 0.28 mmol) and dried under vacuum for 30 min at room temperature. Anhydrous CH$_2$Cl$_2$ (3 mL) is added and the solution is stirred with external cooling in an ice-water bath. A 0.07 M solution of NaOMe in anhydrous MeOH§ (2 mL)¶ is added in one portion. After 30 min, TLC analysis (7:1:2 PrOH-28% aq. NH$_3$-H$_2$O) shows complete conversion of the starting material into a more polar product (R_f 0.13), with mobility lower than a galactose standard (R_f 0.42).

---

* EtOAc was purchased from Baker and used as supplied.
† 10% Palladium on activated carbon (Degussa type) was purchased from Aldrich and employed as received.
‡ 5 mL tubes with aliquots of the solution of **3** in EtOAc with the catalyst are placed into a 250 mL Parr hydrogenation bottle. The vessel is connected to the Parr hydrogenator, and after alternate evacuation and flushing with hydrogen gas, it is shaken under 30 psi during 2 h.
§ Sodium methoxide is prepared by carefully reacting methanol with sodium.
¶ Reagent-grade methanol is dried by refluxing over magnesium turnings and a little iodine, distilling, and storing over 5Å MS.

The solution is diluted with 2 mL of MeOH and concentrated to a small volume, which is applied to a column with 1.0 mL of Bio-Rad AG 50 W-X12 resin, triethylammonium form[*] and eluted with 5 mL MeOH.[†] The solution is collected in a 25 mL round-bottom flask. Several portions of toluene (2 mL) and H_2O (2 mL) are added and co-concentrated at 25°C to remove methylbenzoate. Finally, the syrup obtained is freeze dried to afford **6** (0.11–0.12 g, 90–92%) as a white solid, ^1H NMR (D_2O) δ 5.51 (apparent t, 1 H, $J_{1,2} \approx J_{1,P}$ 4.7 Hz, H-1), 4.22 (apparent t, 1 H, J_1 4.7Hz, H-3), 4.14 (ddd, 1 H, $J_{1,2}$ 4.3 Hz, $J_{2,3}$ 8.1 Hz, $J_{2,P}$ 2.3 Hz, H-2), 3.82 (m, 1 H, H-4), 3.76 (m, 1 H, H-5), 3.70 (dd, 1 H, $J_{5,6}$ 4,4 Hz, $J_{6,6'}$ 11.7 Hz, Hz, H-6), 3.62 (dd, 1 H, $J_{5,6'}$ 6.2 Hz, $J_{6,6'}$ 11.7 Hz, H-6'), 3.19 (q, 2 H, J 7.2 Hz, CH_3CH_2N), 1.26 (t, 3 H, J 7.2 Hz, CH_3CH_2N); ^{13}C NMR (D_2O) δ 97.4 (C-1, $J_{C-1,P}$ 5.9 Hz), 82.0 (C-4), 76.9 (C-2, $J_{C-2,P}$ 8.0 Hz), 73.9 (C-3), 72.3 (C-5), 62.8 (C-6), 47.3 ((CH_3CH_2)$_3$N), 8.8 ((CH_3CH_2)$_3$N). Anal. Calcd for $C_6H_{11}O_9P\cdot2C_2H_5N$, C, 46.74; H, 9.37. Found: C, 46.69; H, 9.48.

ACKNOWLEDGMENTS

We are indebted to Agencia Nacional de Promoción Científica y Tecnológica and Universidad de Buenos Aires for financial support. C. Marino and R. M. de Lederkremer are research members of CONICET.

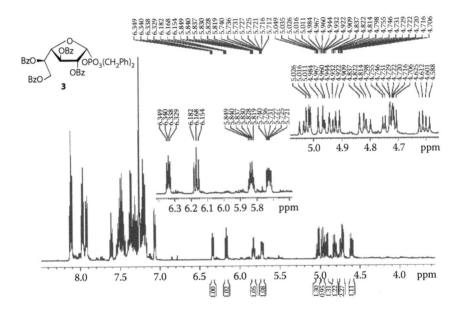

---

[*] Commercial resin is successively washed with water, triethylamine (7.5% in water), water until neutrality, and methanol.
[†] Alternatively, a Dowex 50W X8 resin may be used. In this case, compound **6** is obtained as the disodium salt along with sodium benzoate. Sodium benzoate can be removed by successive digestion with 9:1 nPrOH-25% aq NH_3 solution.

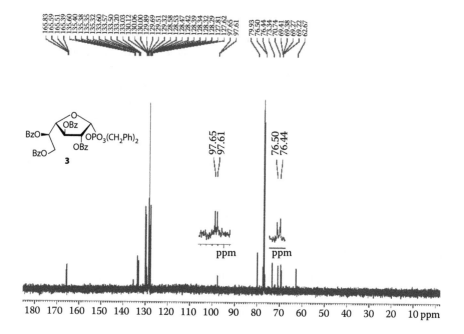

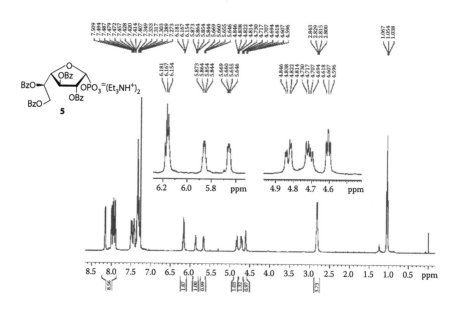

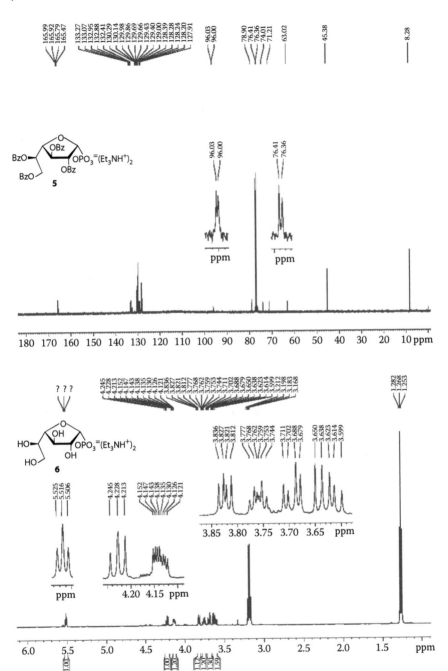

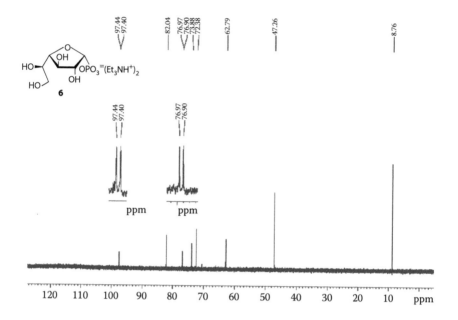

REFERENCES

1. (a) Crick, D. C.; Mahapatra, S.; Brennan, P. J. *Glycobiology* 2001, *11*, 107R–118R; (b) Alderwick, L. J.; Radmacher, E.; Seidel, M.; Gande, R.; Hitchen, P. G.; Morris, H. R.; Dell, A.; Sahm, H.; Eggeling, L.; Besra, G. S. *J. Biol. Chem.* 2005, *280*, 32362–32371; (c) Bhamidi, S.; Scherman, M. S.; Rithner C. D.; Prenni, J. E.; Chatterjee, D.; Khoo, K. H.; McNeil, M. R. *J. Biol. Chem.* 2008, *283*,12992–13000.

2. (a) Lederkremer, R. M.; Casal, O. L.; Alves, M. J. M.; Colli, W. *FEBS Lett.* 1980, *116*, 25–29; (b) Agusti, R.; Lederkremer, R. L. *Adv. Carbohydr. Chem. Biochem.* 2009, *62*, 311–366.

3. (a) Turco, S. J.; Descoteaux, A. *Annu. Rev. Microbiol.* 1992, *46*, 65–94; (b) McConville, M. J.; Collidge, T. A.; Ferguson, M. A.; Schneider, P. *J. Biol. Chem.* 1993, *268*, 15595–15604.

4. Latge, J. P. *Med. Mycol.* 2009, *47*, Supp. 1, 104–109.

5. Zhang, Q.; Liu, H.-w. *J. Am. Chem. Soc.* 2000, *122*, 9065–9070.

6. Tsvetkov, Y. E.; Nikolaev, A. V. *J. Chem. Soc. Perkin Trans.* 2000, *1*, 889–912.

7. Marlow, A. L.; Kiessling, L. L. *Organic Lett.* 2001, *3*, 2517–2519.

8. Poulin, M. B.; Lowary, T. L. *Methods Enzymol.* 2010, *478*, 389–411.

9. Lederkremer, R. M.; Nahmad, V.; Varela, O. *J. Org. Chem.* 1994, *59*, 690–692.

10. Mariño, K.; Marino, C.; Lima, C.; Baldoni, L.; Lederkremer, R. M. *Eur. J. Org. Chem.* 2005, 2958–2964.

11. Marino, C.; Gallo-Rodriguez, C.; Lederkremer R. M. in *Carbohydrate Chemistry: Proven Synthetic Methods*, Vol. 1, Kováč, P., Ed., 2012, *1*, 231–238.

27 Synthesis of 1,2:4,5-Di-O-(3,3-pentylidene) arabitol via Kinetic Acetal Formation

Konrad Hohlfeld
Solen Josse
Sylvain Picon*
Bruno Linclau†

CONTENTS

Both enantiomers of arabitol are commercially available at reasonable and similar cost, which makes them useful building blocks. Arabitol has *pseudo*-C_2-symmetry, with two pairs of diastereotopic alcohols and a non-stereogenic central carbon atom. It was found[1,2] that *pseudo*-C_2-symmetric *bis*-acetal **4** can be obtained under kinetic protection conditions using 3,3-dimethoxypentane **2** in THF (Scheme 27.1). The reflux temperature was employed to enhance solubility. Arabitol predominantly gave **4** when its refluxed suspension in THF was treated for 5 min with 30 mol% of CSA. *Bis*-acetals **4** and **5** were formed, together with a small amount of monoprotection products, which could be separated by careful chromatography. For large scale preparation, a more convenient scavenging procedure was developed in which the acetal mixture was subjected to reaction with succinic anhydride. Selective functionalization of the more reactive primary alcohol, which is only present in the undesired products, enabled facile separa-

---

* Checker under supervision of Dr. N. Tomkinson: nicholas.tomkinson@strath.ac.uk.
† Corresponding author: bruno.linclau@soton.ac.uk.

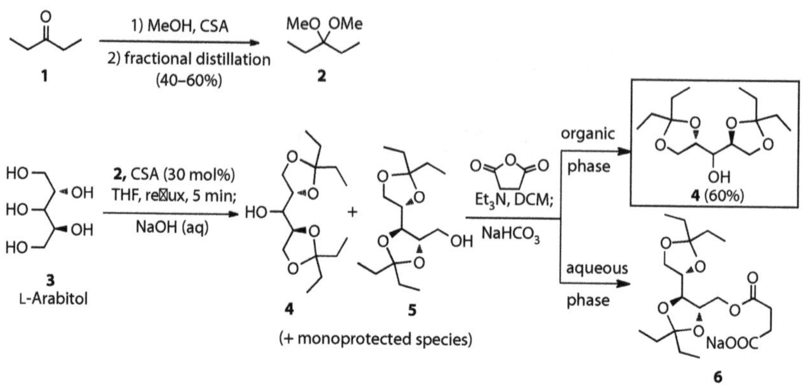

SCHEME 27.1   Kinetic protection of arabitol, followed by a scavenging workup.

tion by extraction. Finally, short-column chromatography afforded pure **4** in a reproducible yield of 60%.

A different regioselectivity is obtained when arabitol is protected under thermodynamic conditions (Scheme 27.2). With ketones, the desymmetrized 2,3:4,5-*bis*-acetonide **7** is obtained as the only isomer in good yield.[3] Acetal formation with benzaldehyde leads exclusively to the 1,3-benzylidene acetal **8**.[4]

EXPERIMENTAL

GENERAL METHODS

L-Arabitol was obtained from CMS chemicals and used without further purification. All other chemical reagents were obtained from Acros, Alfa Aesar, or Sigma-Aldrich and were used as supplied. Dry, high-performance liquid chromatography (HPLC)-grade MeOH was obtained from Fisher. THF was distilled from Na/benzophenone

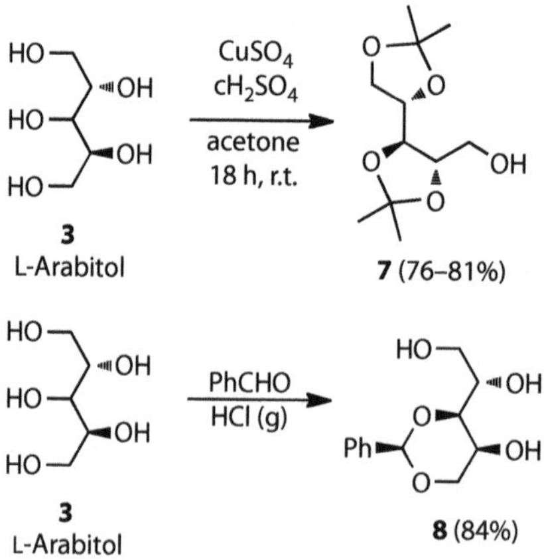

SCHEME 27.2   Thermodynamic protection of arabitol.

immediately prior to use. CH_2Cl_2 and Et_3N were distilled from CaH_2 immediately prior to use. All glassware was flame-dried under vacuum and cooled under N_2 prior to use. Acetal formation was carried out under inert gas and monitored by thin-layer chromatography (TLC) (Merck Kieselgel 60 F_{254}, aluminium sheet). Detection was carried out by spraying the plates with a solution of 5.1 mL *p*-anisaldehyde, 2.1 mL AcOH, and 6.9 mL H_2SO_4 in 186 mL EtOH followed by heating, which produced colored spots. Flash column chromatography was performed on silica gel (60 Å, particle size 35–70 μm).

Nuclear magnetic resonance (NMR) spectra were recorded with a Bruker *AV300* [300.13 MHz (^{1}H NMR) and 75.47 MHz (^{13}C NMR)] or a Bruker *DPX400* [400.13 MHz (^{1}H NMR) and 100.61 MHz (^{13}C NMR)] spectrometer. Chemical shifts (δ) are given in ppm relative to the residual solvent peak. Splitting of proton NMR signals are reported as follows: s, singlet, d, doublet, dd, doublet of doublets, ddd, doublet of doublets of doublets, t, triplet, dt, doublet of triplets, td, triplet of doublets, q, quartet, m, multiplet, and b, broad. Low-resolution electrospray mass spectra were recorded with a Waters ZMD single quadrupol system. High-resolution mass spectra (HRMS) were measured on a Bruker APEX III Fourier transform ion cyclotron resonance mass spectrometry (FTICR-MS) system. Solutions in organic solvents were concentrated at 40°C/30 mbar.

3,3-DIMETHOXYPENTANE (2)

A 2-L, round-bottomed flask equipped with a magnetic stirrer bar was charged with dry methanol (800 mL), pentan-3-one **1** (200 mL, 1.88 mol, 1 equiv.) and trimethyl orthoformate (248 mL, 2.26 mol, 1.2 equiv.). DL-10-Camphorsulfonic acid (4.50 g, 19 mmol, 0.01 equiv) was added in one portion. The flask was capped with a septum, which was pierced with a needle (inert gas is not required for this reaction). The reaction mixture was stirred at 30°C for 3 d. The septum was removed, and the flask was fitted with a Claisen condensor and a thermometer. Volatiles (some of the solvent and the methyl formate formed) were distilled off, until the reaction mixture became reddish-brown (ca 550 mL were removed). The remaining mixture was cooled to room temperature and neutralized by addition of solid anhydrous sodium methoxide (1.5 g), whereupon the color of the mixture changed from light red to bright yellow. Diethyl ether and water (800 mL each) were added and the phases were separated. The organic phase was washed with brine (400 mL) and dried over anhydrous potassium carbonate (20 g). After filtration and concentration, the crude product was distilled (Claisen condenser, no Vigreux head) in the presence of 1.5 g of K_2CO_3 at atmospheric pressure (128–130°C, Lit.[5] 120–124°C) to give two fractions of 3,3-dimethoxypentane **2** as a colorless liquid, 89.50 g (mixture 95:5 with starting ketone, second fraction) + 33.0 g (mixture 87:13 with starting ketone, first fraction). ^{1}H NMR ($CDCl_3$) δ 3.11 (s, 6 H, OCH_3), 1.55 (q, 4 H, *J* 7.6 Hz, CH$_3$CH_2), 0.78 (t, 6 H, *J* 7.6 Hz, CH_3CH$_2$C); ^{13}C NMR ($CDCl_3$) δ 104.1, 47.5, 24.2, 7.7. These ^{1}H and ^{13}C NMR data correspond to the reported data.[5]

(2S,4S)-1,2:4,5-Di-O-(3,3-PENTYLIDENE)ARABITOL (4)

A 500-mL, three-neck, round-bottomed flask equipped with a magnetic stirrer bar and a reflux condenser was charged with L-arabitol **3** (20.00 g, 131.5 mmol), 3,3-dimethoxypentane (mixture 95:5 with 3-pentanone, 76.46 g, 578.4 mmol) and dry THF (200 mL). The resulting suspension was heated to reflux and stirred at this temperature for 15 min. DL-10-Camphorsulfonic acid (9.16 g, 39.4 mmol, 0.3 equiv.) was added in one portion and the reaction mixture was stirred at reflux for exactly 5 min. NaOH (aq., 2 M, 40 mL) was added at reflux temperature, rapidly in one portion, to quench the reaction. After cooling to room temperature, Et$_2$O (70 mL) and H$_2$O (30 mL) were added and the layers were separated. The aqueous phase was extracted with Et$_2$O (3 × 50 mL), and the combined organic layers were dried (MgSO$_4$), filtered, and concentrated to give a pale yellow oil. The excess 3,3-dimethoxypentane was removed at 20°C/3 mbar, and (importantly) the last traces of the reagent were removed by keeping the product and the magnetic stirring bar in the rotating flask for ca. 4–5 h.

A solution of the crude product and dry Et$_3$N (20 mL) in dry CH$_2$Cl$_2$ (200 mL) was placed in a 500 mL, three-neck, round-bottomed flask equipped with a magnetic stirrer bar and a reflux condenser. The mixture was heated to reflux, and succinic anhydride (3.29 g, 33 mmol) was added in one portion. After stirring at reflux temperature for 1.5 h, the mixture was allowed to cool to rt and quenched by addition of NaHCO$_3$ (aq., sat., 200 mL). The layers were separated and the aqueous phase was extracted with CH$_2$Cl$_2$ (2 × 100 mL). The combined extracts were dried (MgSO$_4$), concentrated, and chromatography [(Ø 9 cm, L ~ 15 cm of silica (1 L); LP/EtOAc 90:10 (about 3 L) →85:15 (1 L)] afforded 1,2:4,5-di-O-(3,3-pentylidene)arabitol **4** as a pale yellow oil (22.7 g, 60%), [α]$_D$ −5.8 (c 0.9, CHCl$_3$, 20°C). ^{1}H NMR (CDCl$_3$) δ 4.20 (m, 1 H, OCH$_2$CHO), 4.14 (dd, 1 H, J 7.8, 5.8 Hz, OCHHCHO), 4.08 (dd, 1 H, J 8.0, 6.5 Hz, OCHHCHO), 3.98 (m, 1 H, OCH$_2$CHO), 3.93 (app. t, 1 H, J 7.3 Hz, OCHHCHO), 3.86 (app. t, 1 H, J 8.0 Hz, OCHHCHO), 3.46 (dt, 1 H, J 7.5, 5.3 Hz, CHOH), 2.39 (d, 1 H, J 5.5 Hz, CHOH), 1.74–1.56 (m, 8 H, OCH$_2$CH$_3$), 0.94–0.86 (m, 12 H, OCH$_2$CH$_3$); ^{13}C NMR (CDCl$_3$) δ 113.30 (C), 112.91 (C), 76.81 (OCH$_2$CHO), 76.45 (OCH$_2$CHO), 72.99 (CHOH), 67.87 (OCH$_2$CHO), 66.55 (OCH$_2$CHO), 29.54 (OCH$_2$CH$_3$), 29.52 (OCH$_2$CH$_3$), 29.05 (OCH$_2$CH$_3$), 28.96 (OCH$_2$CH$_3$), 8.19 (OCH$_2$CH$_3$), 8.17 (OCH$_2$CH$_3$), 8.04 (OCH$_2$CH$_3$), 8.03 (OCH$_2$CH$_3$); HRMS (EI) calcd for C$_{13}$H$_{23}$O$_5$ [M−C$_2$H$_5$]$^+$ 259.1546, found 259.1548. Anal. Calcd for C$_{15}$H$_{28}$O$_5$: C, 62.47; H, 9.79. Found: C, 62.33; H, 9.97.

ACKNOWLEDGMENTS

K.H. and S.J. thank Tibotec BVBA for funding.

3,3-Dimethoxypentane, 1st fraction (acetal : ketone = 87 : 13):

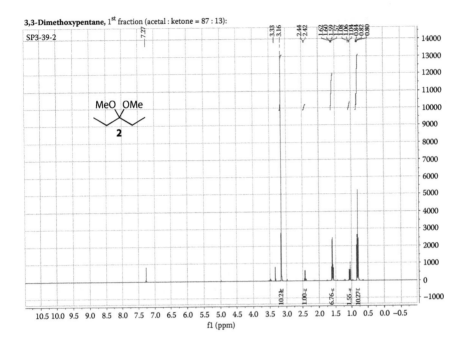

3,3-Dimethoxypentane, 2nd fraction (acetal : ketone = 95 : 5):

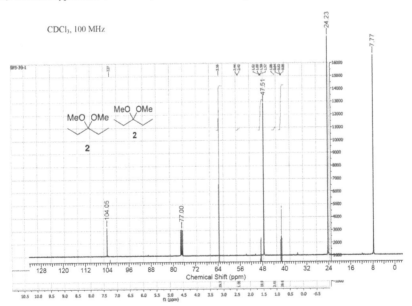

3,3-Dimethoxypentane, pure sample:

CDCl₃, 400 MHz

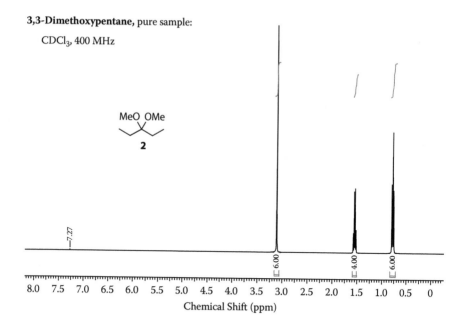

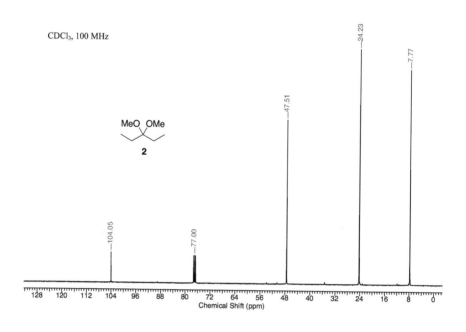

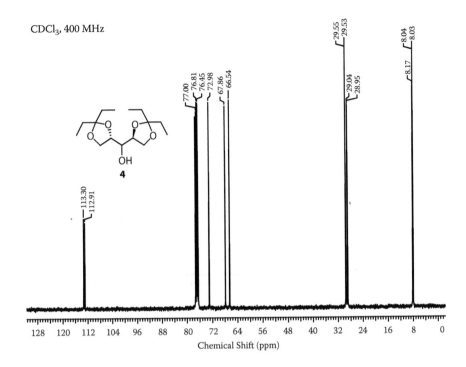

CDCl₃, 400 MHz

4

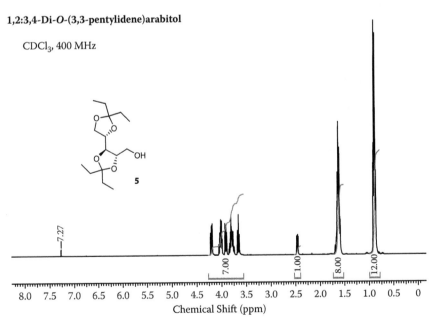

1,2:3,4-Di-O-(3,3-pentylidene)arabitol

CDCl₃, 400 MHz

5

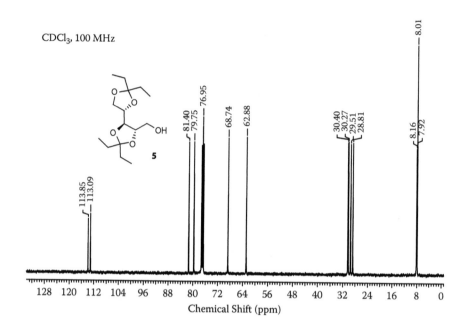

CDCl₃, 100 MHz

REFERENCES

1. Linclau, B.; Boydell, A. J.; Clarke, P. J.; Horan, R.; Jacquet, C. *J. Org. Chem.* 2003, *68*, 1821–1826.
2. Other reports describing symmetric arabitol protection, however without detailed experimental: (a) Maleczka, R. E. Jr.; Terrell, L. R.; Geng, F.; Ward, J. S. III *Org. Lett.* 2002, *4*, 2841–2844 (bis-3,3-pentylidene). (b) Terauchi, T.; Terauchi, T.; Sato, I.; Tsukada, T.; Kanoh, N.; Nakata, M. *Tetrahedron Lett.* 2000, *41*, 2649–2653 (*bis*-acetonide).
3. (a) Bukhari, M. A.; Foster, A. B.; Lehmann, J.; Webber, J. M.; Westwood, J. H. *J. Chem. Soc.* 1963, 2291–2295. (b) Nakagawa, T.; Tokuoka, H.; Shinoto, K.; Yoshimura, J.; Sato, T. *Bull. Chem. Soc. Jpn.* 1967, *40*, 2150–2154.
4. Haskins, W. T.; Hann, R. M.; Hudson, C. S. *J. Am. Chem. Soc.* 1943, *65*, 1663–1667.
5. Giersch, W.; Farris, I. *Helv. Chim. Acta* 2004, *87*, 1601–1606.

28 1,2-Anhydro-3,4,6-tri-O-benzyl-β-D-mannopyranose

*Shino Manabe**
Qingju Zhang[†]
*Yukishige Ito**

CONTENTS

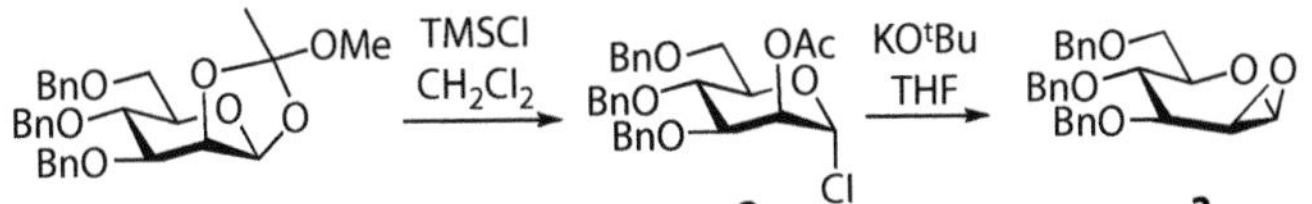

SCHEME 28.1

1,2-Anhydro pyranosides are useful intermediates for the preparation of O-glycosides, N-glycosides, and C-glycosyl compounds.[1] The epoxides can be activated under Lewis acidic conditions to give a cyclic cation. The cation captures nucleophiles to give glycosides. Danishefsky demonstrated the utility of 1,2-anhydro pyranosides in solid-phase oligosaccharide synthesis.[2] Since 1,2-epoxides are extremely sensitive to the acid, the typical epoxidation conditions of olefins using m-chloroperbenzoic acid[‡] cannot be used. The neutral oxidation reagent 3,3-methyldioxirane[3] was successfully used for preparation of 1,2-anhydro pyranosides from glycals.[4] The disadvantage of the latter method for epoxidation is the instability of

---

[*] Corresponding authors: smanabe@riken.jp; yukito@riken.jp
[†] Checker under supervision of B. Yu: byu@mail.sioc.ac.cn
[‡] m-Chloroperbenzoic acid is explosive

the reagent. Furthermore, the concentration of the reagent is limited. The disadvantage of the latter method for epoxidation is the instability of the reagent, which can be used only in limited concentration. For large-scale preparation, the method involving epoxide formation by nucleophilic attack of alkoxide generated in situ is reliable.[5,6]

EXPERIMENTAL

GENERAL METHODS

Dichloromethane (dehydrated, stabilized with amylene, Kanto Chemical Co.) and THF (dehydrated, stabilized with BHT, Kanto Chemical Co.) were used as received. Reagent-grade ether and technical-grade ethyl acetate (EtOAc) and hexane were used. Thin-layer chromatography (TLC) analysis was conducted using silica gel–coated glass plates (Merck, silica gel 60, F_{254}). Compounds were visualized by ultraviolet (UV) light (254 nm), and dipping TLC plates in a solution of $H_3(PMo_{12}O_{40}) \cdot nH_2O$ (24.2g/L), phosphoric acid (85%, 15 mL), H_2SO_4 (conc. 50 mL/L) in water, followed by charring at ~210°C. Flash chromatography was performed using silica gel 60N (spherical, neutral, 100–210 μm). TMSCl and ^{t}BuOK were purchased from Nacalai Tesque, Inc. (Japan). Nuclear magnetic resonance (NMR) spectra were recorded for solutions in $CHCl_3$ with a JEOL AL400 spectrometer and the chemical shifts (δ) are reported relative to $CHCl_3$ (^{1}H for $\delta = 7.24$, ^{13}C for $\delta = 77.0$ ppm). Peak assignments were made by HH-COSY (correlation spectroscopy) and CH-COSY measurements. Optical rotation was measured using JASCO DIP-310 polarimeter. Glycosyl chloride **2** was prepared in high yield from orthoester $\mathbf{1}^{7,*}$ by action of TMSCl.[8]

2-*O*-ACETYL-3,4,6-TRI-*O*-BENZYL-α-D-MANNOPYRANOSYL CHLORIDE (2)

To a solution of 3,4,6-tri-*O*-benzyl-1,2-*O*-(1-methoxyethylidene)-β-D-mannopyranose (**1**)[7] (10.0 g, 19.76 mmol) in CH_2Cl_2 (100 mL), TMSCl (3.4 mL, 26.79 mmol) was added at room temperature, and the mixture was refluxed under N_2 atmosphere for 2 h. After concentration, the chloride was obtained as colorless oil in nearly theoretical yield, which was sufficiently pure for the next step. The compound can be purified by silica gel column chromatography (7:3 hexane:EtOAc) to give pure **2**, 8.99g (89%), $[\alpha]^{26}_D$ +67.0 (c 1.0, $CHCl_3$). ^{1}H NMR (400 MHz, $CDCl_3$): δ 7.18–7.10 (m, 13H, aromatic), 7.02–7.00 (m, 2H), 5.91 (d, $J_{1,2}$ 2.0 Hz, 1H, H-1), 5.31 (dd, $J_{1,2}$ 2.0 Hz, $J_{2,3}$ 3.2 Hz,1H, H-2), 4.71 (d, J 10.8 Hz, C*H*HPh), 4.55 (d, J 11.2 Hz, 1H, C*H*HPh), 4.51 (d, J 12.0 Hz, 1H, C*H*HPh), 4.42 (d, J 11.2 Hz, 1H, C*H*HPh), 4.35 (d, J 10.8 Hz, C*H*HPh), 4.35 (d, J 12.0 Hz, 1H, C*H*HPh), 4.10

---

* The orthoester **1** is available from Carbosynth Limited.

(dd, $J_{3,4}$ 9.2 Hz, 1H, H-3), 3.92 (m, 1H, H-5), 3.83 (t, 1H, H-4), 3.68 (dd, J_{gem} 11.2 Hz, $J_{5,6}$ 4.0 Hz, 1H, H-6), 3.54 (dd, $J_{5,6}$ 2.0 Hz, 1H, H-6), 2.01 (CH$_3$); ^{13}C NMR (100 MHz, CDCl$_3$): δ 169.8 (CO), 137.8 (aromatic), 137.7 (aromatic), 137.3 (aromatic), 128.3 (CHarom), 128.2 (CHarom), 128.0 (CHarom), 127.8 (CHarom), 127.7 (CHarom), 127.7 (CHarom), 127.6 (CHarom), 127.6 (CHarom), 90.3 (C-1), 76.6 (C-3), 75.3 (CH$_2$Ph), 74.2 (C-5), 73.5 (C-4), 73.5 (CH$_2$Ph), 72.1 (CH$_2$Ph), 71.0 (C-2), 68.0 (C-6), 21.1 (CH$_3$, Ac).

1,2-ANHYDRO-3,4,6-TRI-O-BENZYL-β-D-MANNOPYRANOSE (3)*

To a solution of **2** (8.99g, 17.62 mmol) in THF (500 mL), commercially available potassium *tert*-butoxide (2.37 g, 21.15 mmol, 1.2 equiv.)† was added as a solid in some portions at room temperature. The mixture was refluxed under N$_2$ atmosphere for 1 h. After cooling to room temperature, the mixture was concentrated to ca. 50 mL, EtOAc (500 mL) and brine (100 mL) were added, and the aqueous layer was extracted with EtOAc. The combined extracts were washed with brine and dried over Na$_2$SO$_4$. After filtration and concentration, the product was crystallized from ether-hexane. After filtration, the solid was washed with hexane, to give pure epoxide **3** (6.47 g, 85%). $[\alpha]^{24}_D$ −6.5 (c 0.56, CHCl$_3$); ^{1}H NMR (400 MHz, CDCl$_3$): δ 7.39–7.15 (m, 15H, aromatic), 4.95 (d, $J_{1,2}$ 2.8 Hz, 1H, H-1), 4.81 (d, J 11.6 Hz, 1H, CHHPh), 4.81–4.75 (m, 2H, CHHPh), 4.60 (d, J 12.0 Hz, 1H, CHHPh), 4.54 (d, J 10.4 Hz, 1H, CHHPh), 4.52 (d, J 12.0 Hz, 1H, CHHPh), 3.93–3.89 (m, 2H, H-2, H-3), 3.72 (m, 1H, H-5), 3.67–3.59 (m, 2H, H-6), 3.33 (m, 1H, H-2); ^{13}C NMR (100 MHz, CDCl$_3$): δ 137.9 (Cq Bn), 128.4 (CHarom), 128.3 (CHarom), 127.9 (CHarom), 127.7 (CHarom), 127.6 (CHarom), 79.2 (C-3 or C-4), 78.7 (C-1), 78.1 (C-5), 76.0 (C-3 or C-4), 75.1 (CH$_2$Ph), 73.6 (CH$_2$Ph), 72.1 (CH$_2$Ph), 68.7 (C-6), 54.4 (C-2); Mp 88°C; Anal. Calcd for C$_{27}$H$_{28}$O$_5$, C, 74.98; H, 6.53; found C, 74.89, H, 6.50.

ACKNOWLEDGMENTS

S. M. thanks Grant-in-Aid for Scientific Research (C) (Grant Nos. 21590036) from the Japan Society for the Promotion of Science. We thank Ms. Akemi Takahashi for technical assistance.

---

* Compound **3** decomposes on silica gel. Monitoring reactions by TLC or purification by silica gel column chromatography is not possible.
† The amount of *tert*-BuOK is crucial for reproducibility. Although 2.3 equivalents of *tert*-BuOK was used in reference 6, an increased amount of *tert*-BuOK reduces the yield of **3**.

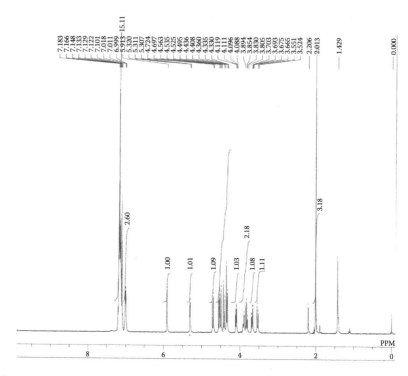

2-O-Acetyl-3,4,6-tri-O-benzyl-α-ᴅ-mannopyranoside chloride 2.

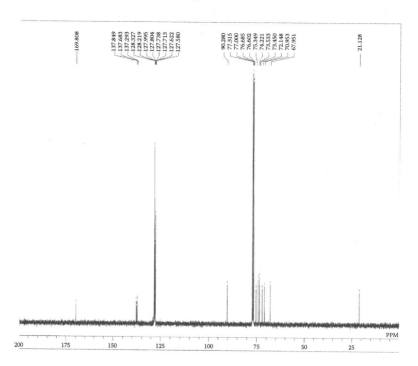

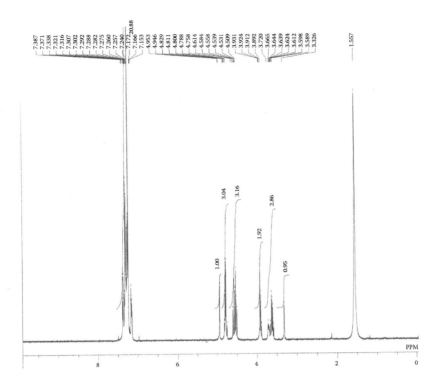

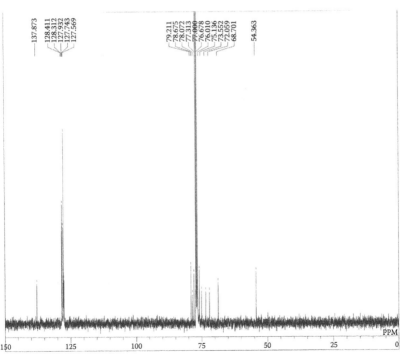

REFERENCES

1. (a) Danishefsky, S. J.; Bilodeau, M. T. *Angew. Chem. Int. Ed.* 1996, *35*, 1380–1419. (b) Chiara, J. L.; Sesmilo, E. *Angew. Chem. Int. Ed.* 2002, *41*, 3242–3246. (c) Manabe, S.; Ito, Y. *J. Am. Chem. Soc.* 1999, *121*, 9754–9755. (d) Manabe, S.; Ito, Y. *Synlett,* 2008, 880–882.
2. Roberge, J. Y.; Beebe, X.; Danishefsky, S. J. *Science,* 1995, *269*, 202.
3. Murray, R. W.; Singh, M. *Org. Syn. Coll.* 1998, *9*, 288.
4. (a) Halcomb, R. L.; Danishefsky, S. J. *J. Am. Chem. Soc.* 1989, *111*, 6661–6666. (b) Bellucci, G.; Catelani, G.; Chiappe, C.; D'Andrea, F. *Tetrahedon Lett.* 1994, *35*, 8433–8436.
5. Sondheimer, S. J.; Yamaguchi, H.; Schuerch, C. *Carbohydr. Res.,* 1979, *74*, 327–332.
6. Du, Y.; Kong, F. *J. Carbohydr. Chem.,* 1995, *14*, 341–352.
7. Franks, N. E.; Montgomery, R. *Carbohydr. Res.* 1968, *6*, 286–298.
8. Ogawa, T.; Katano, K.; Matsui, M. *Carbohydr. Res.* 1978, *64*, C3–C9.

29 Synthesis of Glucosyl C-1 Dihalides from Tetra-*O*-acetyl-β-D-glucopyranosyl Chloride

*Jean-Pierre Praly**
Tze Chieh Shiao[†]

CONTENTS

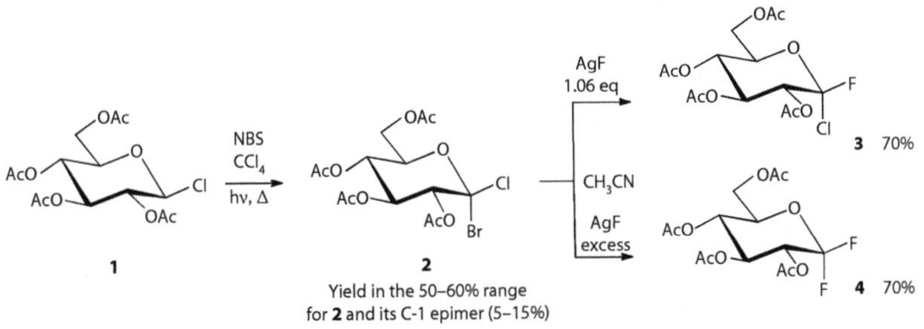

SCHEME 29.1

Sugar halides are important glycosyl donors. They are usually activated by means of heavy metal salts in glycosylations referred to as the Koenigs-Knorr reaction.[1] The first sugar anomeric dihalide reported was 2,3,4,6-tetra-*O*-acetyl-1-bromo-D-glucopyranosyl bromide isolated in a 17% yield by Blattner and Ferrier from products of photobromination of β-D-glucopyranose pentacetate with bromine in the absence of base.[2] This method is synthetically useful, as a C-5 brominated

---

[*] Corresponding author: jean-pierre.praly@univ-lyon1.fr.
[†] Checker under supervision of R. Roy: roy.rene@uqam.ca.

compound can be isolated in 82–90% yield when the reaction is carried out under non acidic conditions. Investigations with tetrahydropyranyl ethers and sugar derivatives have demonstrated that axial C–H bonds next to an oxygen atom [as in ethers, acetals, including the C-1, the C-4 (furanose) and the C-5 (pyranose) positions in cyclic sugars] are more reactive toward homolysis, as compared to equatorial C–H bonds.[3] This accounts for the higher reactivity of β-configured D-glycopyranosyl derivatives, as generally observed for reactions which involve homolysis of anomeric C–H bond as the key step. Similarly, axial C–H bonds next to the oxygen atom in cyclic sugar derivatives at the C-4 or C-5 positions are prone to homolysis. Bromination selectivity depends on steric and electronic factors and is markedly influenced by the structural elements (atoms, substituents) at or near C-1 and C-4/C-5: esterified D-glycopyranosyluronic derivatives readily undergo C-5 bromination due to the additional activating effect of the nearby ester group.[4] Under radical-based bromination conditions, 5-thio-sugar derivatives show a high reactivity with formation of various products that are brominated at C-1 and/or C-5.[5] Protected glycopyranosyl cyanides[6] are also valuable substrates that undergo bromination at C-1. Radical-mediated brominations at ring positions of carbohydrates have been reviewed by Somsák and Ferrier.[7]

In this context, we showed that photobromination of acetylated β-D-glucopyranosyl chloride **1**, available commercially or in high yield from β-D-glucose pentacetate upon treatment with AlCl$_3$[8] or PCl$_5$,[9] afforded the corresponding C-1 *gem*-dihalides in high yield.[10–12] Thus, β-chloride **1** undergoes photobromination with *N*-bromosuccinimide (NBS) giving predominantly the C-1 *gem*-dihalide **2** (Scheme 29.1). Dihalide **2**[*] is a crystalline solid, which can be stored at –20°C indefinitely. It has to be purified by chromatography and/or crystallization from the regioisomeric β-chloride brominated at C-5 (~15% yield), and two byproducts (~5% each), identified as the α-anomer of **1**, and 2,3,4,6-tetra-*O*-acetyl D-glucopyranose formed by hydrolysis, in particular from a trace amount of unchanged **1**. Photobromination proceeds similarly with D-*manno*-[10,11] and D-*galacto*[12] analogues of **1**.

Halogen exchange in dihalide **2** is chemoselective,[†] as exemplified by the conversion of **2** into the corresponding chlorofluoride and difluoride **3** and **4** upon treatment with a stoichiometric amount or excess AgF, respectively. *O*-Unprotected **4**[13] and its D-*galacto* analogue[14] have been investigated as glycosidase inhibitors.

---

[*] Dihalide **2** is difficult to visualize on thin-layer chromatography (TLC) plates upon charring, but fluorescein staining can reveal its presence as a pink-colored spot (see General Methods). Purity of samples of **2** that appear homogeneous by TLC should be verified by nuclear magnetic resonance (NMR) and mass spectrometry (MS) as they may contain the analogous C-1 *gem*-dichloride (<5%, not detected by NMR) and, presumably, the C-1 epimer of **2** (up to ~20%, based on NMR).[11] Recrystallized samples of **2** usually contain between 5 and 15% of the epimer. Being more stable than **2**, the dichloride can be recovered from reaction mixtures produced from **2**. The C-1 *gem*-dichloride is accessible by photochlorination of **1**, preferably with NCS.[11]

[†] The presence of a minor amount of the C-1 epimer of **2** seems to have no influence on the outcome of this reaction, which presumably occurs with neighboring group participation.[16,17]

EXPERIMENTAL

GENERAL METHODS

Technical-grade, brownish NBS can be used. Technical-grade CCl_4 was washed several times with water, dried over $CaCl_2$, and then distilled. Acetonitrile was distilled over calcium hydride. Bromine-containing compounds were visualized on TLC plates as pink-colored spots upon spraying successively with a fluorescein solution in absolute methanol (0.1% w/v), and a mixture of H_2O_2 (30% in water) and AcOH (1:1 v/v), followed by charring. Alternatively, spraying with 2% aqueous $AgNO_3$ and charring helps visualizing spots. NMR spectra were recorded for solutions in $CDCl_3$ with Bruker (WP-80, AM-300), Cameca (Cameca 350), or Varian XL-100 spectrometers, with tetramethylsilane as the reference (δ in ppm; J in Hz). The reference for ^{19}F NMR spectra was $CFCl_3$. Proton assignments were based on selective irradiations. Carbon assignments were tentatively based on analysis of chemical shifts, DEPT spectra, and, if applicable, on measured heteronuclear C–F coupling constants.[15] Mass spectra were measured with a NERMAG R10.10S spectrometer (ionization energy: 70 eV).

2,3,4,6-TETRA-*O*-ACETYL-1-BROMO-β-D-GLUCOPYRANOSYL CHLORIDE (2)

2,3,4,6-Tetra-*O*-acetyl-β-D-glucopyranosyl chloride **1** (10.0 g, 27.3 mmol) and NBS (12.5 g, 70 mmol) were added to a mixture of CCl_4 (350 mL)* contained in a 1 L Erlenmeyer flask equipped with a condenser and mounted above a 250 W tungsten lamp. The mixture was refluxed for 50 min, utilizing the heat generated by the lamp, whereupon the reaction was complete, as shown by TLC (1:1 EtOAc–*n*-hexane). After cooling with ice water, the dark-red mixture was filtered through Celite®, and the filtrate was concentrated under reduced pressure. A solution of the residue in Et_2O was washed with a 5% aqueous $Na_2S_2O_3$ solution, and water (3 × 150 mL), dried (Na_2SO_4), and concentrated. The syrup obtained was dissolved in CH_2Cl_2 (~5 mL), and crystallization was effected by portionwise addition of Et_2O and, if needed, petroleum ether, until the mixture became turbid, and storage at ~4°C. Seed crystals may be obtained as indicated above from homogeneous fractions collected after column chromatography. The first crop obtained (6.0 g) was recrystallized similarly twice, to give **2** (4.08 g, mp 107°C). Concentration of the combined mother liquors afforded a residue, which was separated by chromatography on silica gel, with 3.5:6.5 EtOAc–*n*-hexane or with a 1:4→3:7 gradient under flash conditions. The products were eluted in the following order: the C-5 brominated β-chloride, **2**, followed by the α-anomer of **1** and then the 2,3,4,6-tetra-*O*-acetyl D-glucopyranose. Homogeneous fractions containing **2** were pooled and concentrated under reduced pressure to afford 2.37 g of the title compound (Total

---

* Variations of this protocol with use of a 9:1 solvent mixture of CCl_4 and $CBrCl_3$ and addition of a small amount of benzoyl peroxide were tested, but found to have no significant influence.

yield: 14.47 mmol, 53%).* Colorless prisms, mp 102–109°C (Et$_2$O-petroleum ether); R_f = 0.57 (EtOAc/n-hexane 1:1). [α]$_D$ +136 (c 1.4, acetone). ^{1}H NMR (300.13 MHz) δ 5.37 (t, 1H, $J_{3,4}$ 9.5 Hz, H-3), 5.27 (t, 1H, $J_{4,5}$ 9.8 Hz, H-4), 5.19 (d, 1H, $J_{2,3}$ 9.5 Hz, H-2), 4.38 (dd, 1H, $J_{5,6}$ 4.1 Hz, $J_{6,6'}$ 12.7 Hz, H-6), 4.29 (dq, 1H, H-5), 4.20 (dd, 1H, $J_{5,6'}$ 1.9 Hz, H-6'), 2.18, 2.13, 2.05, 2.01 (4s, 12H). ^{13}C NMR (75.47 MHz) δ 170.1, 169.3, 168.8, 168.5 (C = O), 104.2 (C-1), 76.2, 76.1 (C-2, C-5), 71.6 (C-3), 66.3 (C-4), 60.5 (C-6), 20.6, 20.5, 20.4, 20.4 (CH$_3$). The following resonances were attributed to the epimer of **2**, present in 5–15% amount, depending on the sample grade: ^{1}H NMR (300.13 MHz) δ 5.55 (d, $J_{2,3}$ 9.5 Hz, H-2); ^{13}C NMR (75.47 MHz) δ 99.5, 77.0, 74.6, 70.5, 66.6, 60.7.

2,3,4,6-Tetra-O-acetyl-1-chloro-β-d-glucopyranosyl Fluoride (3)

2,3,4,6-Tetra-O-acetyl-1-bromo-β-d-glucopyranosyl chloride **2** (mp 107°C)† (1.78 g, 4 mmol) was dissolved in freshly distilled CH$_3$CN (16 mL), and the medium was cooled to 0°C. Anhydrous powdered AgF (85% commercial grade, 0.635 g, 1.06 eq. based on pure AgF) was added under stirring. After warming up to room temperature (30°C), the mixture maintained under normal atmosphere in a stoppered flask was stirred for 24 h. The initially formed white solid darkened gradually. Conversion of **2** can be visualized by TLC (1:1 Et$_2$O–petroleum ether) analysis through the disappearance of the pink-colored spot. After addition of 1,4-diazabicyclo[2,2,2]-octane (0.112 g, 1 mmol) to the mixture, stirring was continued for 3 h. After adding a few mL of a saturated aqueous NaCl solution, and stirring for 30 min, the mixture was filtered to remove the solids. The filtrate was concentrated under reduced pressure, and the residue was dissolved in 60 mL Et$_2$O. After aqueous washings (25 mL × 3), the ethereal phase was dried (Na$_2$SO$_4$), filtered, and concentrated under reduced pressure to afford a residue, which was purified by column chromatography (1:1 Et$_2$O–petroleum ether) on silica gel, to give **3** (1.09 g, 70% yield), which solidified. Crystallization (concentrated solution in hot absolute EtOH and cooling) gave material mp 57°C. R_f = 0.4 (1:1 Et$_2$O–petroleum ether). [α]$_D$ +101.5 (c 0.5, acetone). ^{1}H NMR (300.13 MHz) δ 5.45–5.23 (m, 3H, H-2, H-3, H-4), 4.42–4.18 (m, 3H, H-5, H-6, H-6'), 2.14, 2.10, 2.04, 2.01 (4s, 12H, acetyl). ^{13}C NMR (75.47 MHz) δ 170.3, 169.6, 169.2, 168.8 (C = O), 123.7 (d, $^1J_{C,F}$ −268.3 Hz, C1), 74.1 (d, $^3J_{C,F}$ 3.0 Hz, C5), 71.9 (d, $^2J_{C,F}$ 25.9 Hz, C2), 71.0 (d, $^3J_{C,F}$ 8.9 Hz, C3), 66.8 (s, C4), 60.6 (s, C6), 20.5, 20.4, 20.3, 20.3 (CH$_3$). ^{19}F NMR (75.2 MHz) δ −68.25 (d, 1F, $^3J_{F,H2}$ 6.2 Hz). Anal. Calcd for C$_{14}$H$_{18}$O$_9$Cl$_1$F$_1$ (384.74): Cl, 9.21; F, 4.94. Found: Cl, 9.69; F, 4.63.

---

* Reaction at a smaller scale (0.2–2 g of **1**) resulted in a somewhat more selective reaction (~60% yield for **2** within 30 min).

† Use of high-grade samples of **2** with a reduced content of C-1 gem-dichloride is recommended, as the more stable dichloride which is unaltered under these conditions proved very difficult to separate from the chlorofluoride **3** either by chromatography or crystallization.

2,3,4,6-TETRA-O-ACETYL-1-FLUORO-D-GLUCOPYRANOSYL FLUORIDE (4)

Powdered AgF (85% commercial grade, 0.6 g, 4.0 mmol based on pure AgF) was added to an ice-cooled solution of **2** (0.446 g, 1 mmol) in freshly distilled acetonitrile (4 mL) (protection from moisture is not required). Temperature was allowed to rise to ambient, ~30°C, and stirring was maintained for 4 days while the stoppered flask was protected from light with aluminium foil. After adding a few mL of a saturated aqueous NaCl solution, the mixture was stirred for 30 min and filtered. The solution was concentrated under reduced pressure and the residue was dissolved in 50 mL Et_2O. After aqueous washings (20 mL × 3), the ethereal phase was dried (Na_2SO_4), filtered, concentrated under reduced pressure, and chromatography on silica gel (1:1 Et_2O–petroleum ether) yielded difluoride **4** (0.26 g, 70% yield), which crystallized from Et_2O–petroleum ether as long, colorless needles, mp 97–98°C; $R_f = 0.54$ (EtOAc/n-hexane 1:1). $[\alpha]_D$ +42 (c 0.5, acetone). ^{1}H NMR (350 MHz) δ 5.41–5.31 (m, 2H, H-2, H-3), 5.26 (t, 1H, $J_{3,4}$ 9.2 Hz, $J_{4,5}$ 9.2 Hz, H-4), 4.32 (br dd, 1H, $J_{5,6}$ 4.3 Hz, $J_{6,6'}$ 13.3 Hz, H-6), 4.19 (m, 2H, H-5, H-6′) 2.14, 2.12, 2.05, 2.03 (4s, 12H, acetyl). ^{13}C NMR (25.2 MHz) δ 120.3 (dd, $^1J_{C,F}$ −256.0 Hz, $^1J_{C,F}$ −271.7 Hz, C1), 72.6 (dd, $^3J_{C,F}$ 2.8 Hz, $^3J_{C,F}$ 4.0 Hz, C5), 70.7 (d, $^3J_{C,F}$ 9.4 Hz, C3), 68.7 (t, $^2J_{C,F}$ 30.6 Hz, $^2J_{C,F}$ 30.6 Hz, C2), 66.8 (s, C4), 60.6 (s, C6), 170.1, 169.4, 168.9, 168.7 (C = O), 20.5, 20.3, 20.3, 20.2 (CH_3). ^{19}F NMR (75.2 MHz) δ −86.3 (dd, 1F, $^3J_{Fax,H2}$ 17.5 Hz, $^2J_{F,F}$ 148.8 Hz, Fax), −82.7 (dd, 1F, $^3J_{Feq,H2}$ 3.4 Hz, $^2J_{F,F}$ 148.8 Hz, Feq). Anal. Calcd for $C_{14}H_{18}O_9F_2$ (368.29): F, 10.32. Found: F, 10.28.

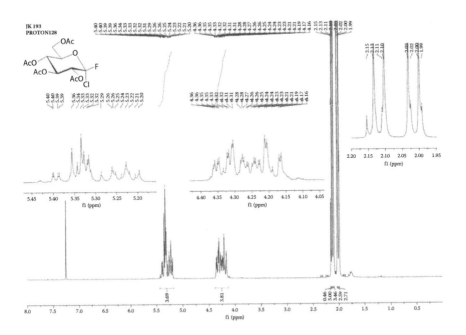

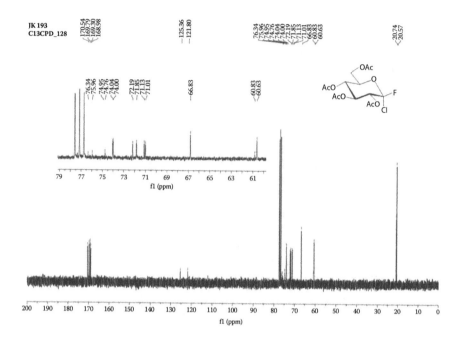

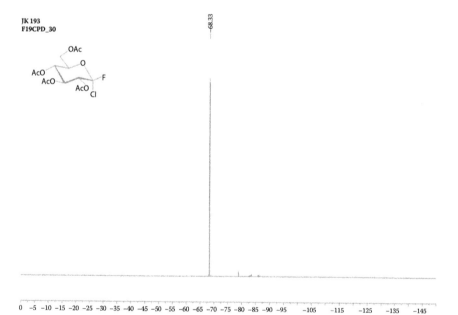

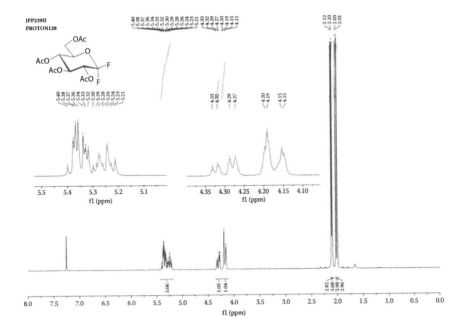

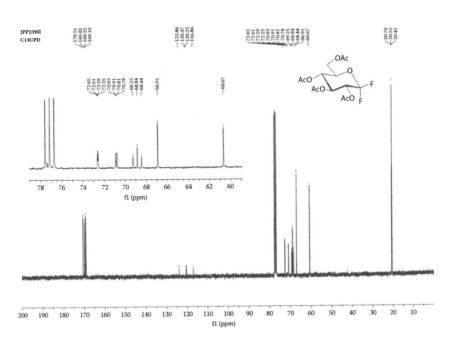

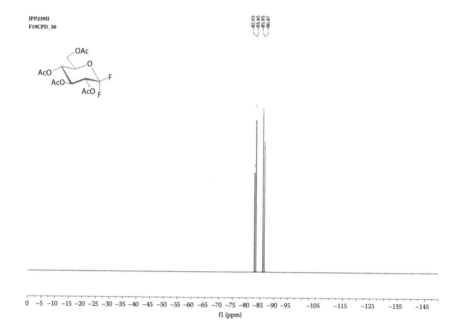

JPP239II
F19CPD_30

REFERENCES

1. Fügedi, P. in *The Organic Chemistry of Sugars,* Levy, D. E.; Fügedi, P. Eds., CRC Press, 2006, chap. 4, Glycosidation Methods, pp. 89–179, and chap. 5, Oligosaccharide Synthesis, pp. 181–268.
2. Blattner, R.; Ferrier, R. J. *J. Chem. Soc. Perkin Trans. 1,* 1980, *7*, 1523–1527.
3. (a) Bernasconi, C.; Descotes, G. *CR Hebd. Seances Acad. Sci. Ser. C.,* 1975, *280*, 469–472; (b) Hayday, K.; McKelvey, R. D. *J. Org.* Chem., 1976, *41*, 2222–2223.
4. Ferrier, R. J.; Furneaux, R. H. *J. Chem. Soc. Perkin Trans. 1,* 1977, 1996–2000.
5. Baudry, M.; Bouchu, M.-N.; Descotes, G.; Praly, J.-P.; Bellamy, F.; *Carbohydr. Res.,* 1996, *282*, 237–246.
6. (a) Somsák, L.; Batta, G.; Farkas, I. *Carbohydr. Res.,* 1982, *106*, c4–c5; (b) Lichtenthaler, F. W.; Jarglis, P. *Angew. Chem. Suppl.,* 1982, *21(S8)*, 1449–1459.
7. Somsák, L.; Ferrier, R. J. *Adv. Carbohydr. Chem. Biochem.,* 1991, *49*, 37–92.
8. Lemieux, R. U. In *Methods in Carbohydrate Chemistry;* Whistler, R. L., Wolfrom, M. L., Eds.; Academic Press: New York, 1963; Vol. 2, pp. 224–225.
9. Ibatullin, F. M.; Selivanov, S. I. *Tetrahedron Lett.,* 2002, *43*, 9577–9580.
10. Praly, J.-P.; Descotes, G. *Tetrahedron Lett.,* 1987, *28*, 1405–1408.
11. Praly, J.-P.; Brard, L.; Descotes, G.; Toupet, L. *Tetrahedron,* 1989, *45*, 4141–4152.
12. Praly, J.-P.; Brendle, J.-C.; Klett, J.; Pequery, F. *Compt. Rend. Serie IIc: Chimie,* 2001, *4*, 611–617.
13. Konstantinidis, A.; Sinnott, M. L. *Biochem. J.,* 1991, *279*, 587–593.
14. K, S.; Konstantinidis, A.; Sinnott, M. L.; Hall, B. G. *Biochem. J.,* 1993, *291*, 15–17.
15. Praly, J.-P.; Descotes, G. *C. R. Acad. Sci. Paris,* Série II, 1988, *307*, 1637–1639.
16. Praly, J.-P.; El Kharraf, Z.; Corringer, P.-J.; Brard, L.; Descotes, G. *Tetrahedron,* 1990, *46*, 65–75.
17. Praly, J.-P.; Chen, G.-R.; Gola, J.; Hetzer, G. *Eur. J. Org. Chem.,* 2000, 2831–2838.

30 Synthesis of 3,4,6-Tri-O-acetyl-D-galactal

Tze Chieh Shiao
Jacques Rodrigue
*Olivier Renaudet**
René Roy[†]

CONTENTS

1. Zn Dust, CuSO$_4$
AcOH/0.1 M NaH$_2$PO$_4$
pH 4.21 4:1; rt, 30 min

2. EtOAc, rt, 30 min
86%

1   **2**

SCHEME 30.1

Glycals have been extensively used for the synthesis of a wide range of carbohydrate derivatives such as O-,[1,2] C-,[3] S-,[4] and N-glycosides,[5] cyclopropanated carbohydrates,[6] and natural products.[7] In particular, the Ferrier[2] and the azidonitration reactions,[8] together with Danishefsky's glycal methodology[1a] (through epoxides), represent the most widely exploited applications.

The original Fischer–Zach method of preparation by reductive elimination[9] (i.e., the treatment of acetobromosugars with zinc dust in [buffered] aqueous acetic acid at a temperature ranging from –20°C to room temperature[10]) has undergone countless modifications. Over the years, some of these alterations included the reduction of protected glycosyl halides by Na, lithium naphthalenide, Li-NH$_3$, Zn-Ag, (Cp$_2$TiCl)$_2$, Cr(II), Al-Hg, K-graphite, SmI$_2$,[11,12] or using thiophenyl glycoside, glycosyl sulfones, and electrochemical approaches.[13,14] The yields usually obtained ranged from 58–95%.

---

* Checker: olivier.renaudet@ujf-grenoble.fr.
† Corresponding author: roy.rene@uqam.ca.

For a comprehensive review, together with reaction mechanisms, see the excellent review by Somsák.[11] We found that the slightly modified Fischer–Zach method, using freshly prepared and ideally crystalline acetobromogalactose (**1**) with $CuSO_4$-activated zinc dust in buffered sodium dihydrogen phosphate[14b] acetic acid conditions, provided high yields reproducibly (>85%). This protocol is simple, uses inexpensive reagents and a readily available starting material (**1**), and does not require expensive or environnementally noxious metals, such as chloroplatinic acid.[10] Purity of galactal **2** (Scheme 30.1), synthesized in this way and obtained after flash chromatography, is high, and **2** is suitable for further transformations. Although **2** was reported as a syrup after distillation (bp. 134°C at 0.01mm[15a] or 138–140°C at 0.2–0.3 mm[15b]), which crystallizes upon long standing (14 days)[10a] or in a mixture of ether-petroleum of ether (mp 30°C),[15a] we found it difficult to obtain suitable crystals. Identical nuclear magnetic resonance (NMR) spectra were obtained before and after distillation, and the $[\alpha]_D$ became more negative and constant after distillation. Hence, the dryness of the syrupy sample may explain the variations in the reported values $[\alpha]_D$.

EXPERIMENTAL

GENERAL METHODS

Solutions in organic solvents were dried over anhydrous Na_2SO_4. Reactions were monitored by thin-layer chromatography (TLC) using silica gel 60 F_{254} coated plates (E. Merck). Optical rotations were measured with a JASCO P-1010 polarimeter. NMR spectra were recorded on Varian Inova AS600 spectrometer. Proton and carbon chemical shifts (δ) are reported in ppm relative to the signals of the residual $CHCl_3$, which were set at 7.27 and 77.0 ppm for ^1H and ^{13}C, respectively. Coupling constants (J) are reported in Hertz (Hz), and the following abbreviations are used for peak multiplicities: singlet (s), doublet (d), doublet of doublets (dd), triplet (t), multiplet (m), and broad (b). Analysis and assignments were made using COSY (COrrelated SpectroscopY), DEPT (Distortionless Enhancement by Polarization Transfer), and HETCOR (Heteronuclear Chemical Shift Correlation) experiments; coupling constants were obtained from a J-resolved spectrum. High-resolution mass spectrum (HRMS) was measured on a LC-MS-TOF (Liquid Chromatography Mass Spectrometry Time Of Flight) instrument from Agilent Technologies in positive electrospray mode.

3,4,6-TRI-*O*-ACETYL-D-GALACTAL (2)

The reaction was carried out in a round-bottom flask (50 mL). To a suspension of 2,3,4,6-tetra-*O*-acetyl-α-D-galactopyranosyl bromide **1**[16] (411 mg, 1.0 mmol, 1.0 equiv.) and zinc powder (1.05 g, 16.0 mmol, 16.0 equiv.) in acetic acid (8 mL) was added a solution of $CuSO_4$ (72 mg, 0.29 mmol, 0.29 equiv.) in sodium dihydrogen phosphate solution (NaH_2PO_4, 0.1M, pH 4.21, 2 mL). The mixture was stirred vigorously for 30 min. at room temperature, and EtOAc (10 mL) was added. After 30 min., the suspension was filtered through Celite®, and the solids were washed with EtOAc. The filtrate was washed successively with ice water (2×), saturated solution of

NaHCO$_3$ (3×), dried, concentrated, and chromatographed (4:1→7:3 hexane-AcOEt) to give **2** as a colorless oil. Yield: 234 mg, 0.86 mmol (86%). R_f = 0.21, 4:1 hexane-EtOAc; $[\alpha]_D$ −8.8 (c 1.1, CHCl$_3$) before distillation and $[\alpha]_D$ −14.6 (c 0.5, CHCl$_3$) after distillation (heating at 225°C at 16 mmHg); the checker found: $[\alpha]_D$ −19 (c 0.8, CHCl$_3$). [Lit.[14] $[\alpha]_D$ −9 (c 1.0, EtOAc); Lit.[10b,15a,b] $[\alpha]_D$ −12.4 (c 2.5 in CHCl$_3$); Lit.[10a] $[\alpha]_D$ −16.5 (c 3.0, CHCl$_3$); Lit.[17] $[\alpha]_D$ −16.9 (c 1.1, CHCl$_3$);]; NMR data (^{1}H and ^{13}C, in agreement with reported literature data)[14,18]: ^{1}H NMR (CDCl$_3$) δ 6.45 (dd, 1H, $J_{1,3}$ 1.4 Hz, $J_{1,2}$ 6.3 Hz, H-1), 5.55 (m, 1H, $J_{2,3}$ 1.7 Hz, $J_{3,4}$ 4.6 Hz, H-3), 5.42 (m, 1H, $J_{4,5}$ 1.7 Hz, H-4), 4.72 (m, 1H, H-2), 4.31 (m, $J_{5,6a}$ 5.5 Hz, $J_{5,6b}$ 2.0 Hz, 1H, H-5), 4.26 (m, 1 H, $J_{6a,6b}$ 11.2 Hz, H-6a), 4.21 (m, 1H, H-6b), 2.14, 2.08, 2.02, 1.96 ppm (4 × s, 12H, CH$_3$); ^{13}C NMR (CDCl$_3$) δ 170.5, 170.2, 170.1 (CO), 145.3 (C-1), 98.8 (C-2), 72.7 (C-5), 63.8 (C-3), 63.7 (C-4), 61.8 (C-6), 20.7, 20.7, 20.6 ppm (CH$_3$). Electrospray ionization high-resolution mass spectrometry (ESI$^+$-HRMS): [M + NH$_4$]$^+$ calcd for C$_{12}$H$_{20}$NO$_7$, 290.1234; found, 290.1228.

ACKNOWLEDGMENTS

This work was supported from Natural Sciences and Engineering Research Council of Canada (NSERC) and a Canadian Research Chair in Therapeutic Chemistry to R.R. The authors would like to thank Alexandre A. Arnold for assistance with NMR.

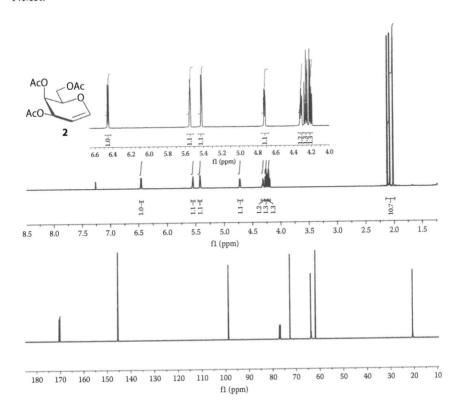

REFERENCES

1. (a) Danishefsky, S. J.; Bilodeau, M. T. *Angew. Chem. Int. Ed. Engl.* 1996, *35*, 1380–1419; (b) Honda, E.; Gin, D. Y. *J. Am. Chem. Soc.* 2002, 124, 7343–7352; (c) Tiwari, P.; Misra, A. K. *J. Org. Chem.* 2006, *71*, 2911–2913.

2. (a) Ferrier, Robert J. *J. Chem. Soc., Perkin Trans. 1,* 1979, 1455–1458; (b) Ferrier, Robert J. *Adv. Carbohydr. Chem. Biochem.*, 1969, *24*, 199–266.

3. (a) Rainier, J. D.; Cox, J. M. *Org. Lett.* 2000, *2*, 2707–2709; (b) Parrish, J. D.; Little, R. D. *Org. Lett.* 2002, *4*, 1439–1442; (c) Lehmann, U.; Awasthi, S.; Minehan, T. *Org. Lett.* 2003, *5*, 2405–2408; (d) Saeeng, R.; Isobe, M. *Org. Lett.* 2005, *7*, 1585–1588; (e) Yin, J.; Spindler, J.; Linker, T. *Chem. Commun.* 2007, *26*, 2712–2713.

4. (a) Seeberger, P. H.; Eckhard, M.; Gutteridge, C. E.; Danishefsky, S. J. *J. Am. Chem. Soc.* 1997, *119*, 10064–10072; (b) Boulineau, F. P. Wei, A. *Org. Lett.* 2004, *6*, 119–121.

5. (a) Li, B. Q.; Franck, R. W.; Capozzi, G.; Menichetti, S.; Nativi, C. *Org. Lett.* 1999, *1*, 111–114; (b) Colinas, P. A.; Bravo, R. D. *Org. Lett.* 2003, *5*, 4509–4511; (c) Dahl, R. S.; Finney, N. S. *J. Am. Chem. Soc.* 2004, *126*, 8356–8357.

6. (a) Cousins, G. S.; Hoberg, J. O. *Chem. Soc. Rev.*, 2000, *29*, 165–174; (b) Haveli, S. D.; Sridhar, P. R.; Suguna, P.; Chandrasekaran, S. *Org. Lett.* 2007, *9*, 1331–1334.

7. (a) Zakarian, A.; Batch, A.; Holton, R. A. *J. Am. Chem. Soc.* 2003, *125*, 7822–7824; (b) Takahashi, K.; Matsumura, T.; Corbin, G. R. M.; Ishihara, J.; Hatakeyama, S. *J. Org. Chem.* 2006, *71*, 4227–4231; (c) Denmark, S. E.; Regens, C. S.; Kobayashi, T. *J. Am. Chem. Soc.* 2007, *129*, 2774–2776.

8. Lemieux, R. U.; Ratcliffe, R. M. *Can. J. Chem.* 1979, *57*, 1244–1251.

9. Fischer, E.; Zach, K. *Sitzberg Kgl. preuss. Akad. Wiss*, 1913, *16*, 311–317.

10. (a) Shafizadeh, F. *Methods Carbohydr. Chem.*, 1963, *2*, 409–410; (b) Overrend, W. G.; Shafizadeh, F.; Stacey, M. *J. Chem. Soc.* 1950, 671–677.

11. Somsák, L. *Chem. Rev.*, 2011, *101*, 81–135.

12. (a) Li, B. Q.; Franck, R. W.; Capozzi, G.; Menichetti, S.; Nativi, C. *Org. Lett.* 1999, *1*, 111–114; (b) Eitelman, S. J.; Jordaan, A. *J. Chem. Soc., Chem. Commun.* 1977, 552–553; (c) Cavallaro, C. L.; Schwartz, J. *J. Org. Chem.* 1995, *60*, 7055–7057; (d) Spencer, R. P.; Cavallaro, C. L.; Schwartz, J. *J. Org. Chem.* 1999, *64*, 3987–3995; (e) De Pouilly, P.; Chénedé, A.; Mallet, J. -M.; Sinaÿ, P. *Tetrahedron Lett.* 1992, *33*, 8065–8068; (f) Kovács, G.; Micskei, K.; Somsák, L. *Carbohydr. Res.* 2001, *336*, 225–228.

13. (a) Swallen, L. C.; Boord, C. E. *J. Am. Chem. Soc.*, 1930, *52*, 651–660; (b) Parrish, J. D.; Little, R. D. *Tetrahedron Lett.* 2001, *42*, 7371–7374; (c) Micskei, K.; Juhász, Z.; Ratković, Z. R.; Somsák, L. *Tetrahedron Lett.* 2006, *47*, 6117–6120.

14. (a) Zhao, J.; Wei, S.; Ma, X.; Shao, H. *Green Chem.* 2009, *11*, 1124–1127; (b) Zhao, J.; Wei, S.; Ma, X.; Shao, H. *Carbohydr. Res.*, 2010, *345*, 168–171.

15. (a) Levene, P. A.; Tipson, S. *J. Biol. Chem.*, 1931, *93*, 631–644; (b) Komada, T. *Bull. Jap. Chem. Soc.* 1932, *7*, 211–216.

16. Shiao, T. C.; Papadopoulos, A.; Renaudet, O.; Roy, R. Preparation of *O*-β-D-Galactopyranosylhydroxylamine. Kovac, P. Ed.; Taylor & Francis, Boca Raton, FL, *Carbohydrate Chemistry: Proven Synthetic Methods*, 2011, *1*, 289–293.

17. Shull, B. K.; Wu, Z.; Koreeda, M. *J. Carbohydr. Chem.*, 1996, *15*, 955–964.

18. (a) Pilgrim, W.; Murphy, P. V. *Org. Lett.*, 2009, *11*, 939–942. (b) Bukowski, R.; Morris, L. M.; Woods, R. J.; Weimar, T. *Eur. J. Org. Chem.*, 2001, 2697–2705.

31 Efficient Synthesis of Hepta-**O**-acetyl-β-lactosyl Azide via Phase Transfer Catalysis

Tze Chieh Shiao
Denis Giguère
Nicolas Galanos[*]
René Roy[†]

CONTENTS

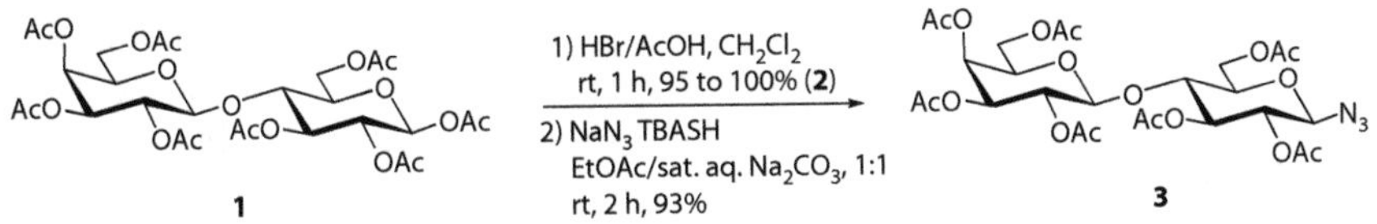

SCHEME 31.1

Glycosyl azides are an important family of carbohydrate derivatives.[1] They have been used as precursors of glycopeptides[2] and heterocyclic *N*-glycosides.[3] The corresponding glycosylamines, which can be obtained by reduction of the azide group, can be used as precursors to glycopolymers or hyper-branched dendrimers, after *N*-acryloylation.[4] Furthermore, azide functionalities can be easily converted into

---

[*] Checker under supervision of S. Vidal: sebastien.vidal@univ-lyon1.fr.
[†] Corresponding author: roy.rene@uqam.ca.

isothiocyanates for the synthesis of novel glycosyl intermediates.[5] Recent applications of the Cu(I)-catalyzed Huisgen azide-alkyne [1,3]-dipolar cycloaddition[6] have renewed interest in glycosyl azides.[7,8]

We present herein an efficient synthesis of hepta-*O*-acetyl-β-lactosyl azide **3** via phase transfer catalysis under conditions developed by our group (Scheme 31.1).[9] Firstly, β-lactose octaacetate was transformed into acetobromo-α-lactose **2**, using a commercially available solution of 33% HBr in AcOH. Without further purification, bromide **2** was converted under mildly basic, phase transfer catalyzed conditions into compound **3**[10] in 88 to 93% yields in two steps. Other methods for the synthesis of glycosyl azides require strong Lewis acid–catalyzed treatment with TMSN₃ for several hours[11] or the use of sonication.[12] These methods can be lengthy and expensive, and often provide anomeric mixtures of azides.

EXPERIMENTAL

GENERAL METHODS

The reaction in organic media was carried out using solvents for high-performance liquid chromatography (HPLC). After workup, organic phases were dried over anhydrous Na_2SO_4. Progress of reactions was monitored by thin-layer chromatography (TLC) using silica gel 60 F_{254} coated plates (E. Merck). Optical rotations were measured with a JASCO P-1010 polarimeter. Melting points were measured on a Fisher Jones apparatus. Nuclear magnetic resonance (NMR) spectra were recorded on Varian Inova AS600 spectrometers. Proton and carbon chemical shifts (δ) are reported in ppm relative to the signals of the residual $CHCl_3$, which were set at 7.27 and 77.0 ppm for ^1H and ^{13}C, respectively. Coupling constants (*J*) are reported in Hertz (Hz), and the following abbreviations are used for peak multiplicities: singlet (s), doublet (d), doublet of doublets (dd), triplet (t), multiplet (m), and broad (b). Analysis and assignments were made using COSY (COrrelated SpectroscopY), DEPT (Distortionless Enhancement by Polarization Transfer), and HETCOR (Heteronuclear Chemical Shift Correlation) experiments. Fourier transform infrared (FTIR) spectra were obtained with a NICOLET 6700 FT-IR (Thermo Scientific) spectrometer. The absorptions are given in wave numbers (cm⁻¹). The intensity of the bands is described as *s* (strong), *m* (medium), or *w* (weak). High-resolution mass spectra (HRMS) were recorded at UQAM (Université du Québec à Montréal, Canada).

2,3,4,6-TETRA-*O*-ACETYL-β-D-GALACTOPYRANOSYL-(1 → 4)-2,3,6-TRI-*O*-ACETYL-α-D-GLUCOPYRANOSYL BROMIDE (2)

A 33% solution of HBr in AcOH (2.28 mL, 13.0 mmol, 13.0 equiv.) was added in one portion to a solution of lactose octaacetate[13] (679 mg, 1.0 mmol, 1.0 equiv.) in dry CH_2Cl_2 (3.0 mL, 0.33 M). The resulting mixture was stirred at room temperature for 1 hour. Upon completion of the reaction (TLC 3:2 EtOAc-Hexanes), the mixture was poured slowly into a cold, saturated aqueous solution of $NaHCO_3$ (50 mL), the flask was washed with CH_2Cl_2 (3 × 10 mL) and the contents were added to the mixture.

The two phases were stirred until effervescence ceased.[*] The aqueous layer was decanted, and the organic phase was washed with a saturated aqueous solution of $NaHCO_3$ (3 × 50 mL) and brine (50 mL). Drying over Na_2SO_4, filtration, and concentration under reduced pressure gave the bromide intermediate **2** as a light yellow oil. Compound **2** could be used for the next step without further purification or be recrystallized from DCM-Et_2O-petroleum ether.[†] $R_f = 0.27$ hexanes/EtOAc (1:1); mp 138.5–139.0°C; $[\alpha]_D$ +98.6 (c 1, $CHCl_3$); [Ref[14] mp 141–142°C; $[\alpha]_D$ +105.1]; ^{1}H NMR ($CDCl_3$) δ 6.52 (d, 1H, $J_{1,2}$ 4.0 Hz, H-1[I]), 5.55 (dd, 1H, $J_{2,3} = J_{3,4}$ 9.7 Hz, H-3[I]), 5.35 (dd, 1H, $J_{3,4} = J_{4,5}$ 3.4 Hz, H-4[II]), 5.12 (dd, 1H, $J_{1,2}$ 8.0 Hz, $J_{2,3}$ 10.3 Hz, H-2[II]), 4.96 (dd, 1H, H-3[II]), 4.76 (dd, 1H, H-2[I]), 4.51 (d, 2H, H-1[II]), 4.51 (m, 1H, H-6a[I]), 4.18 (m, 3H, H-6a[II], H-6b[I] and H-5[II]), 4.08 (dd, 1H, $J_{6a,6b}$ 11.2 Hz, $J_{5,6b}$ 7.3 Hz, H-6b[II]), 3.87 (m, 2H, H-5[I] and H-4[I]), 2.16, 2.13, 2.09, 2.06, 2.06, 2.05 and 1.96 ppm (7s, 21H, $COCH_3$); ^{13}C NMR ($CDCl_3$) δ 170.3, 170.1, 170.1, 170.0, 170.0, 169.2, 168.9 (CO), 100.8 (C-1[II]), 86.4 (C-1[I]), 75.0 (C-4[I]), 72.9 (C-5[I]), 71.0 (C-3[I]), 70.8 (C-5[II]), 70.8 (C-3[II]), 69.6 (C-2[I]), 69.0 (C-2[II]), 66.6 (C-4[II]), 61.0 (C-6[II]), 60.8 (C-6[I]), 20.8, 20.8, 20.7, 20.6, 20.6, 20.6 and 20.5 ppm ($COCH_3$). ESI[+]-HRMS: $[M + NH_4]^+$ calcd for $C_{26}H_{39}NO_{17}Br$, 716.1396; found, 716.1383.

2,3,4,6-Tetra-O-acetyl-β-D-galactopyranosyl-(1 → 4)-2,3,6-tri-O-acetyl-β-D-glucopyranosyl azide (3)

To a solution of bromide **2** (515 mg, 0.736 mmol, 1.0 equiv.) in ethyl acetate (1.0 mL/100 mg of sugar, 5.2 mL) was added NaN_3 (239 mg, 3.68 mmol, 5.0 equiv.), tetrabutylammonium hydrogen sulfate (255 mg, 0.736 mmol, 1.0 equiv.) and a saturated aqueous solution of Na_2CO_3[‡] (1.0 mL/100 mg of sugar, 5.2 mL). The mixture was vigorously stirred for 2 hours when TLC (2:3 hexane–EtOAc) showed that the starting material was consumed. The organic layer was separated, washed with a saturated aqueous $NaHCO_3$ solution (3 × 15 mL), H_2O (15 mL), brine (15 mL), dried over Na_2SO_4, filtered,[§] and concentrated under reduced pressure. The crude product was then crystallized[¶,**] (EtOH–petroleum ether at –20°C for 16 hours) to give hepta-O-acetyl-β-lactosyl azide **3** (453 mg, 93%). $R_f = 0.34$ hexanes/EtOAc (1:1); mp 73.5–75.0°C; $[\alpha]_D$ –19.4 (c 1, $CHCl_3$); [Ref.[10] mp 72–74°C (Et_2O-light petroleum)]; $[\alpha]_D$ –22 (c 0.6, $CHCl_3$)]. ^{1}H NMR ($CDCl_3$) δ 5.33 (dd, 1H, $J_{3,4} = J_{4,5}$

---

[*] *Caution*: A large amount of CO_2 is produced during this operation, and slow addition of the reaction mixture is particularly important when working on a large scale.

[†] When crystallized, compound **2** is relatively stable when kept in a freezer (1+ year). Approximately 30 mL of the ternary mixture of solvents was used in equal volumes (1:1:1, v/v/v).

[‡] Changing the aqueous phase from saturated Na_2CO_3 to 1M Na_2CO_3 or 2M Na_2CO_3 resulted in slight decrease of the reaction rate.

[§] Filtration over silica gel pad and washing with EtOAc facilitated direct crystallization.

[¶] Solubilize in minimum hot EtOH and then slowly add petroleum ether until slight turbidity. Let cool to room temperature and keep overnight at –20°C.

[**] If crude **2** is used as starting material, purification by silica gel column chromatography (3:2 hexanes–EtOAc) is required.

3.4 Hz, H-4^{II}), 5.19 (dd, 1H, $J_{2,3}$ = $J_{3,4}$ 9.3 Hz, H-3^{I}), 5.09 (dd, 1H, $J_{1,2}$ 7.9 Hz, $J_{2,3}$ 10.4 Hz, H-2^{II}), 4.94 (dd, 1H, H-3^{II}), 4.84 (dd, 1H, $J_{1,2}$ = 8.9 Hz, $J_{2,3}$ 9.4 Hz, H-2^{I}), 4.62 (d, 1H, H-1^{I}), 4.49 (dd, 1H, $J_{6a,6b}$ 12.1 Hz, $J_{5,6a}$ 2.1 Hz, H-6a^{I}), 4.47 (d, 1H,H-1^{II}), 4.13–4.047 (m, 3H, H-6a^{II}, H-6b^{I} and H-6b^{II}), 3.86 (m, 1H, H-5^{II}), 3.77 (dd, 1H, $J_{4,5}$ 9.2 Hz, H-4^{I}), 3.69 (ddd, 1H, $J_{5,6a}$ 2.1 Hz, $J_{5,6b}$ 5.1 Hz, H-5^{I}), 2.13, 2.12, 2.06, 2.05, 2.03, 2.03 and 1.95 ppm (7s, 21H, COCH_3); ^{13}C NMR (CDCl$_3$) δ 170.3, 170.2, 170.1, 170.0, 169.6, 169.4, 169.0 (CO), 101.0 (C-1^{II}), 87.6 (C-1^{I}), 75.7 (C-4^{I}), 74.7 (C-5^{I}), 72.4 (C-3^{I}), 70.9 (C-2^{I}), 70.9 (C-3^{II}), 70.7 (C-5^{II}), 69.0 (C-2^{II}), 66.5 (C-4^{II}), 61.7 (C-6^{I}), 60.7 (C-6I^{I}), 20.7, 20.7, 20.6, 20.6, 20.5, 20.5 and 20.4 ppm (COCH_3). IR ν_{max} (CHCl$_3$) cm^{-1}: 2120s (N$_3$). ESI$^+$-HRMS: [M + NH$_4$]$^+$ calcd for C$_{26}$H$_{39}$N$_4$O$_{17}$, 679.2305 ; found, 679.2316.

ACKNOWLEDGMENTS

This work was supported by a grant from the Natural Science and Engineering Research Council of Canada (NSERC) to R.R. D.G. is thankful to the FQRNT for a post-graduate fellowship. N.G. acknowledges financial support from the Région Rhône-Alpes, Université Lyon 1 and CNRS.

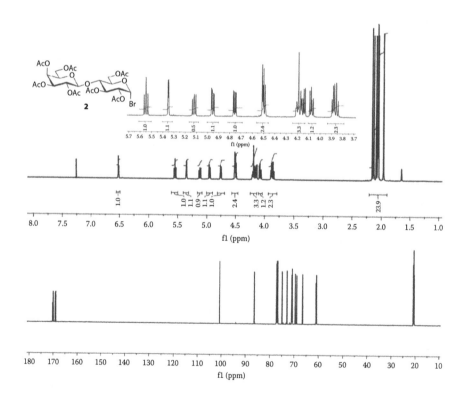

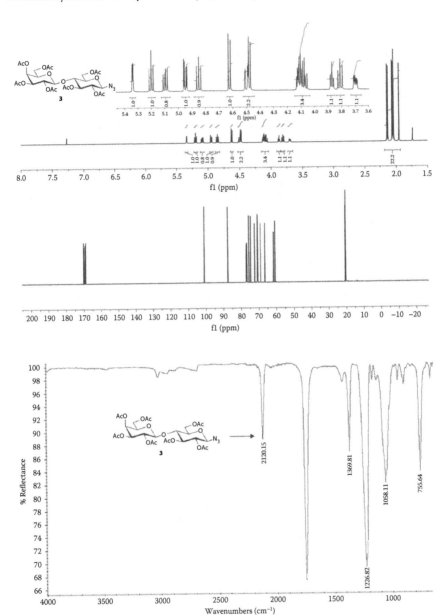

REFERENCES

1. Micheel, F.; Klemer, A. *Adv. Carbohydr. Chem. Biochem.*1961, *16*, 85–103.
2. (a) Ying, L.; Gervay-Hague, J. *Carbohydr. Res.* 2004, *339*, 367–375; (b) Ying, L.; Gervay-Hague, J. *Carbohydr. Res.* 2003, *338*, 835–841; (c) Garg, H. G.; Jeanloz, R. W. *Adv. Carbohydr. Chem. Biochem.* 1985, *43*, 135–201.

3. Baum, G.; Micheel, F. *ChemBer.* 1957, *90*, 1595–1596.

4. (a) Kallin, E.; Lonn, H.; Norberg, T.; Elofsson, M. *J. Carbohydr. Chem.* 1989, *8*, 597–611; (b) Roy, R.; Laferrière, C. A. *J. Chem. Soc., Chem. Commun.*1990, 1709–1711; (c) Roy, R.; Park, W. K. C.; Wu, Q.; Wang, S.-N. *Tetrahedron Lett.* 1995, *36*, 4377–4380; (d) Chabre, Y. M.; Contino-Pépin, C.; Placide, V.; Shiao, T. C.; Roy, R. *J. Org. Chem.* 2008, *73*, 5602–5605.

5. (a) Jiménez Blanco, J. L.; Sylla, B.; Ortiz, Mellet, C.; Garcia Fernandez, J. M. *J. Org. Chem.* 2007, *72*, 4547–4550; (b) Menuel, S.; Porwanski, S.; Marsura, A. *New J. Chem.* 2006, *30*, 603–608.

6. (a) Himo, F.; Lovell, T.; Hilgraf, R.; Rostovtsev, V. V.; Noodleman, L.; Sharpless, K. B.; Fokin, V. V. *J. Am. Chem. Soc.* 2005, *127*, 210–216; (b) Kolb, H. C.; Finn, M. G. Sharpless, K. B. *Angew. Chem. Int. Ed.* 2001, *40*, 2004–2021.

7. For reviews on click chemistry including carbohydrates: (a) Meldal, M.; Tornoe, C. W. *Chem. Rev.* 2008, *108*, 2952–3015; (b) Dedola, S.; Nepogodiev, S. A.; Field, R. A. *Org. Biomol. Chem.* 2007, *5*, 1006–1017; (c) Santoyo-Gonzalez, F.; Hermandez-Mateo, F. *Top. Heterocycl. Chem.* 2007, *7*, 133–177.

8. For the use of hepta-*O*-acetyl-β-lactosylazide[2] as precursor for click chemistry: (a) Giguère, D.; Bonin, M.-A.; Cloutier, P.; Patnam, R.; St-Pierre, C.; Sato, S.; Roy, R. *Bioorg. Med. Chem. Lett.* 2008, *16*, 7811–7823; (b) Geng, J.; Lindqvist, J.; Mantovani, G.; Chen, G.; Sayers, C.; Clarkson, G. J.; Haddleton, D. M. *QSAR Comb. Sci.* 2007, *26*, 1220–1228; (c) Zhan, J.; Garrossian, M.; Gardner, D.; Garrossian, A.; Chang, Y.-T.; Kim, Y. K.; Chang, C.-W. T. *Bioorg. Med. Chem. Lett.* 2008, *18*, 1359–1363; (d) Giguère, D.; Patnam, R.; Bellefleur, M.-A.; St-Pierre, C.; Sato, S.; Roy, R. *Chem. Commun.* 2006, *22*, 2379–2381; (e) Joosten, J. A. F.; Tholen, N. T. H.; El Maate, F. A.; Brouwer, A. J.; van Esse, G. W.; Rijkers, D. T. S.; Liskamp, R. M. J.; Pieters, R. J. *Eur. J. Org. Chem.* 2005, *15*, 3182–3185; (f) Ermolat'ev, D.; Dehaen, W.; Van der Eycken, E. *QSAR Comb. Sci.* 2004, *23*, 915–918.

9. (a) Roy, R.; Tropper, F. D.; Cao, S.; Kim, J. M. *Phase Transfer Catalyst Mechanism and Syntheses.* ACS Symposium Series. 1997, *569*, 163–180; (b) Carrière, D.; Meunier, S. J.; Tropper, F. D.; Cao, S.; Roy, R. *J. Mol. Catal. A: Chem.* 2000, *154*, 9–22; (c) Meunier, S. J.; Andersson, F. O.; Letellier, M.; Roy, R. *Tetrahedron: Asymm.* 1994, *5*, 2303–2312; (d) Tropper, F. D.; Andersson, F. O.; Braun, S.; Roy, R. *Synthesis*, 1992, *7*, 618–620.

10. Dunstan, D.; Hough, L. *Carbohydr. Res.* 1972, *23*, 17–21.

11. For example, see: (a) Malkinson, J. P.; Falconer, R. A.; Toth, I. *J. Org. Chem.* 2000, *65*, 5249–5252; (b) Soli, E. D.; DeShong, P. *J. Org. Chem.* 1999, *64*, 9724–9726.

12. Deng, S.; Gangadharmath, U.; Chang, C.-W. T. *J. Org. Chem.* 2006, *71*, 5179–5185.

13. Šardzík, R.; Noble, G. T.; Weissenborn, M. J.; Martin, A.; Webb, S. J.; Flitsch, S. L. *Beilstein J. Org. Chem.*, 2010, *6*, 699–703.

14. Fischer, E.; Fischer, H. *Ber.*, 1910, *43*, 2521–2536.

32 Synthesis of 2,3,4,6-Tetra-*O*-acetyl-α-D-mannopyranosyl Azide

Tze Chieh Shiao
Yoann M. Chabre
*Myriam Roy**
René Roy†

CONTENTS

SCHEME 32.1

Among natural glycosylated structures, glycoproteins and glycopeptides are of significant biological interest due to their widespread roles in cell recognition and cell adhesion, and improving absorption of poorly bioavailable drugs and peptides by enhancing membrane transport.[1] The azido function usually serves as a latent amino group and therefore azido glycosides represent key intermediates in the synthesis of corresponding glycosyl amino acids.[2] Another useful application of azido glycosides has recently been highlighted by the use of the Cu[I]-catalyzed Huisgen azide-alkyne 1,3-dipolar cycloaddition reaction to efficiently afford various *N*-glycosyl-triazole

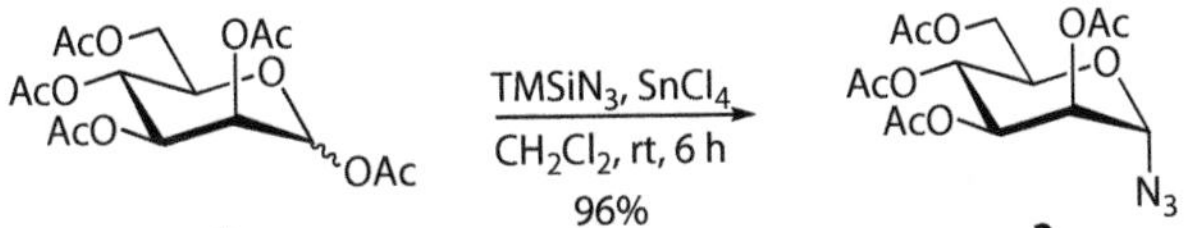

---

* Checker under supervision of M. Gingras: marc.gingras@univ-amu.fr.
† Corresponding author: roy.rene@uqam.ca.

mimetics.[3] In this context, various synthetic methods have been developed to introduce the azide function at the anomeric position, using corresponding glycosyl halides,[4] phosphates,[5] or 1,2 cyclic sulfites.[6] Unlike strategies that involve low stability precursors, modest yields, or formation of anomeric mixture, 2,3,4,6-tetra-O-acetyl-α-D-mannopyranosyl azide was obtained[7] in (80% to nearly theoretical) yields from the corresponding pentaacetate and azidotrimethylsilane (TMSN$_3$) by a Lewis acid–catalyzed reaction. Here we describe the high yielding synthesis of 2,3,4,6-tetra-O-acetyl-α-D-mannopyranosyl azide from 1-O-acetyl derivative **1** and TMSN$_3$ through a SnCl$_4$-catalyzed reaction in CH$_2$Cl$_2$.

EXPERIMENTAL

GENERAL METHODS

The reaction was carried out under nitrogen atmosphere using dichloromethane which was freshly distilled from P$_2$O$_5$. After workup, the organic phase was dried over anhydrous Na$_2$SO$_4$. The reaction was monitored by thin-layer chromatography (TLC) using silica gel 60 F$_{254}$ coated plates (E. Merck). Optical rotation was measured with a JASCO P-1010 polarimeter. Nuclear magnetic resonance (NMR) spectra were recorded on Varian Inova AS600 spectrometer. Proton and ^{13}C chemical shifts (δ) are reported in ppm relative to the signal of the residual CHCl$_3$, which was set at 7.27 ppm. Coupling constants (J) are reported in Hertz (Hz), and the following abbreviations are used for peak multiplicities: singlet (s), doublet (d), doublet of doublets (dd), triplet (t), multiplet (m), and broad (b). Analysis and assignments were made using COSY (COrrelated SpectroscopY), DEPT (Distortionless Enhancement by Polarization Transfer), and HETCOR (Heteronuclear Chemical Shift Correlation) experiments. The α-stereochemistry of the glycosidic linkage in the product azide followed from $^1J_{C-1,H-1}$ coupling constant, determined by a coupled HSQC (Heteronuclear Simple Quantum Correlation) experiment (*ca.* 170 Hz).[8] Fourier transform infrared (FTIR) spectra were obtained with a Bomem (Hartmann-Braun) MB series Michelson FTIR spectrometer and spectra were measured for neat substance on NaCl. The absorptions are given in wave numbers (cm^{-1}). The intensity of the bands is described as s (strong), m (medium), or w (weak). High-resolution mass spectra (HRMS) were measured with a LC-MS-TOF (Liquid Chromatography Mass Spectrometry Time Of Flight) instrument (Agilent Technologies) in positive electrospray mode by Plateforme analytique pour molécules organiques de l'UQAM (Université du Québec à Montréal, Canada).

2,3,4,6-TETRA-O-ACETYL-α-D-MANNOPYRANOSYL AZIDE (2)

TMSN$_3$ (0.43 mL, 3.07 mmol, 4.00 equiv.) and SnCl$_4$ (1M in CH$_2$Cl$_2$, 0.20 mL, 0.20 mmol, 0.26 equiv.) were added under nitrogen atmosphere to a solution of penta-O-acetyl-α,β-D-mannopyranose **1**[4e] (300 mg, 0.77 mmol, 1.00 equiv.) in dry CH$_2$Cl$_2$ (3 mL), and the mixture was stirred at room temperature. The progress of the reaction was monitored by TLC (3:3:4 hexane–toluene–AcOEt) until the starting material was consumed (~6 h). CH$_2$Cl$_2$ (15 mL) was added and the solution was washed with a saturated aqueous NaHCO$_3$ (10 mL), water (10 mL), and brine (10 mL). The organic phase was dried and concentrated and chromatographed (3:1 hexane–AcOEt) to give **2** as a colorless oil (276 mg, 96%, 0.74 mmol). R_f = 0.4, 3:3:4 hexane–toluene–AcOEt; [α]$_D$ +102 (c 0.1, CHCl$_3$) [Ref.[4a] [α]$_D$ +105 (c 0.1, CHCl$_3$); Ref. [4f] [α]$_D$ +103 (c 0.1, CHCl$_3$)]; ^{1}H NMR (CDCl$_3$) δ 5.37 (d, 1H, $J_{1,2}$ 1.9 Hz, H-1), 5.25 (dd, 1H, $J_{3,4}$ = $J_{4,5}$ 9.9 Hz, H-4), 5.21 (dd, 1H, $J_{2,3}$ 2.3 Hz, H-3), 5.12 (dd, 1H, H-2), 4.25 (dd, 1H, $J_{6a,6b}$ 12.4 Hz, $J_{5,6}$ 5.5 Hz, H-6a), 4.13 (m, 2 H, H-6b and H-5), 2.14, 2.08, 2.02, 1.96 ppm (4 × s, 12H, CH$_3$); ^{1}H NMR (C$_6$D$_6$)^{5a} δ 5.55 (dd, 1H, $J_{3,4}$ = $J_{4,5}$ 10.1 Hz, H-4), 5.41 (dd, 1H, $J_{2,3}$ 3.4 Hz, H-3), 5.25 (dd, $J_{1,2}$ 1.9 Hz, H-2), 4.72 (d, 1H, H-1), 4.27 (dd, 1H, $J_{6a,6b}$ 12.4 Hz, $J_{5,6}$ 5.1 Hz, H-6a), 4.03 (dd, 1 H, $J_{5,6}$ 2.2 Hz, H-6b), 3.84 (m, 1H, H-5), 1.72, 1.69, 1.66, 1.60 ppm (4 × s, 12H, CH$_3$); ^{13}C NMR (CDCl$_3$) δ 170.5, 169.7, 169.6, 169.5 (CO), 87.3 (C-1, $J_{C-1,H-1}$ 170.9 Hz), 70.5 (C-5), 69.0 (C-2), 68.1 (C-3), 65.5 (C-4), 62.0 (C-6), 20.7, 20.6, 20.6, 20.5 ppm (CH$_3$). IR ν_{max} (neat) cm^{-1}: 2959m, 2119s (N$_3$), 1744s, 1369s, 1235s, 1049s, 909s. Electrospray ionization high-resolution mass spectrometry (ESI$^+$-HRMS): [M + Na]$^+$ calcd for C$_{14}$H$_{19}$N$_3$O$_9$Na, 396.1014; found, 396.1007.

ACKNOWLEDGMENTS

This work was supported from Natural Sciences and Engineering Research Council of Canada (NSERC) and a Canadian Research Chair in Therapeutic Chemistry to R.R. Y.M.C. thanks FQRNT (Québec) for a post-doctoral fellowship. The authors are grateful to CFQCU (Conseil Franco-Québecois de Coopération Universitaire) for generous funding on glycosciences. MG and MR thank ANR PCV Glycoasterix.

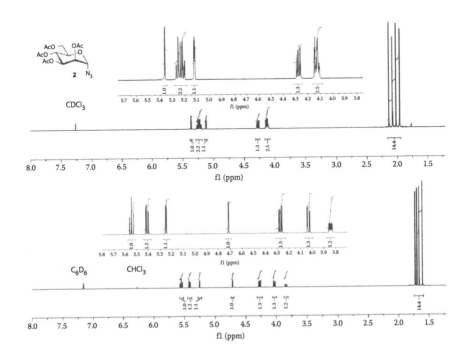

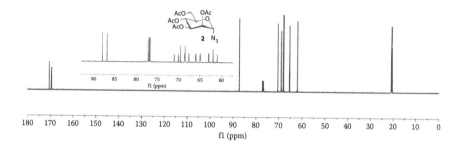

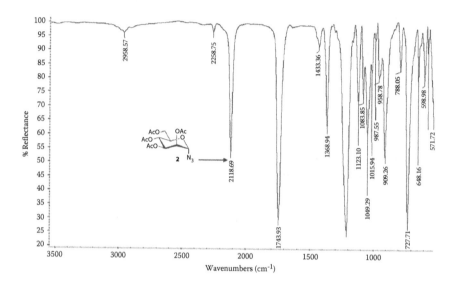

REFERENCES

1. (a) Roy, R. In *Carbohydrate Chemistry*, Ed. G.-J. Boons, Blackie Academic, London, 1998, 243–321; (b) Seitz, O. In *Carbohydrate-Based Drug Discovery*, Ed. C.-H. Wong, Wiley-VCH, New York, 2003, 169–214; (c) Davis, B. G. *Chem. Rev.* 2002, *102*, 579–601; (d) Specker, D.; Wittmann, V. *Top. Curr. Chem.* 2007, *267*, 65–107.

2. (a) Herzner, H.; Reipen, T.; Schultz, M.; Kunz, H. *Chem. Rev.* 2000, *100*, 4495–4537; (b) He, Y; Hinklin, R. J.; Chang, J. Kiessling, L. L. *Org. Lett.* 2004, *6*, 24, 4479–4482; (c) Li, X.; Danishefsky, S. J. *J. Am. Chem. Soc.* 2008, *130*, 16, 5446–5448; (d) Unverzagt, C. *Chem. Eur. J.* 2003, *9*, 6, 1369–1376.

3. (a) Dedola, S.; Nepogodiev, S. A.; Field, R. A. *Org. Biomol. Chem.* 2007, *5*, 1006–1017; (b) Zhang, J.; Garrossian, M.; Gardner, D.; Garrossian, A.; Chang, Y.-T.; Kim, Y. K. Chang, C.-W. T. *Bioorg. Med. Chem. Lett.* 2008, *18*, 1359–1363; (c) Palomo, C.; Aizpurua, J. M.; Balentova, E.; Azcune, I.; Santos, J. I.; Barbero, J. J., Canada, J. Miranda, J. I. *Org. Lett.* 2008, *10*, 11, 2227–2230; (d) Wilkinson, B. L.; Innocenti, A.; Vullo, D.; Supuran, C. T.; Poulsen, S.-A. *J. Med. Chem.* 2008, *51*, 1945–1953.

4. (a) Cosgrove, K. L.; Bernhardt, P. V.; Ross, B. P.; McGeary, R. P. *Aust. J. Chem.* 2006, *59*, 473–476; (b) Zhang, J.; Chen, H.-N.; Chiang, F.-I; Takemoto, J. Y.; Bensaci, M.; Chang, C.-W. T. *J. Comb. Chem.* 2007, *9*, 17–19; (c) Deng, S.; Gangadharmath, U.; Chang, C.-W. T. *J. Org. Chem.* 2006, *71*, 5179–5185; (d) Szarek, W. A.; Achmatowicz Jr., O.;

Plenkiewicz, J.; Radatus, B. K. *Tetrahedron* 1978, *34*, 1427–1433; (e) Ravindranathan Kartha, K. P.; Field, R. A. *Tetrahedron* 1997, *53*, 11753–11766; (f) Maity, S. K.; Dutta, S. K.; Banerjee, A. K.; Achari, B.; Singh, M. *Tetrahedron* 1994, *50*, 6965–6974.

5. Sabesan, S.; Neira, S. *Carbohydr. Res.* 1992, *223*, 169–185.
6. El Meslouti, A.; Beaupère, D.; Demailly, G.; Uzan, R. *Tetrahedron Lett.* 1994, *35*, 23, 3913–3916.
7. (a) Paulsen, H.; Gyorgydeak, Z.; Friedmann, M. *Chem. Ber.* 1974, *107*, 1568–1578; (b) Vicente, V.; Martin, J.; Jimenez-Barbero, J.; Chiara, J. L.; Vicent, C. *Chem. Eur. J.* 2004, *10*, 4240–4251.
8. Block, K.; Pedersen, C. *J. Chem. Soc., Perkin Trans. 2,* 1974, 293–297.

33 Synthesis of Phenyl 2,3,4,6-Tetra-*O*-acetyl-1-thio-β-D-galactopyranoside

Tze Chieh Shiao
Denis Giguère
*Patrick Wisse**
René Roy[†]

CONTENTS

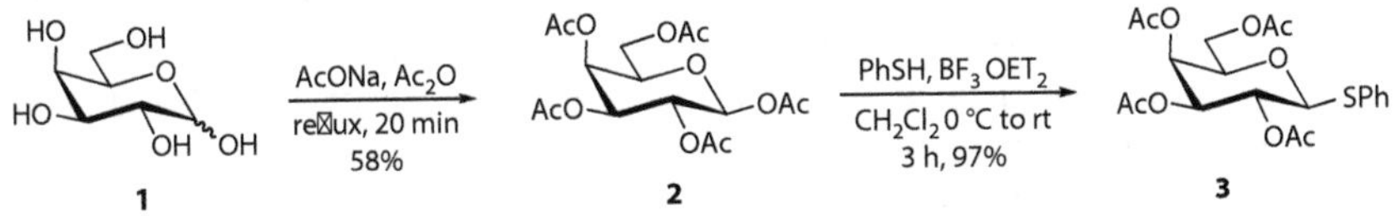

SCHEME 33.1 Synthesis of phenyl 2,3,4,6-tetra-*O*-acetyl-1-thio-β-D-galactopyranoside **3**.

Glycochemistry is an important science, particularly for the synthesis of complex carbohydrates which play key roles in several biological events. Phenyl 1-thio-glycosides in particular and related thioglycosides have been used extensively for the synthesis of complex oligosaccharides.[1] Moreover, aryl 1-thio-β-D-galactopyranosides have been used recently as specific inhibitors of human galectins.[2]

Since Fisher and Delbrück's work,[3] alkyl and aryl 1-thioglycosides of aldoses have been prepared by a large variety of methods: peracylated glycosyl halides by reaction

---

* Checker under supervision of J. Codée: jcodee@chem.leidenuniv.nl.
† Corresponding author: roy.rene@uqam.ca

with thiolate anion;[4] glycosyl halides *via* thiourea intermediates;[5] dithioacetals by partial hydrolysis;[6] unprotected reducing sugars;[7] 1-thioaldose derivatives by reaction with aryldiazonium salts;[8] by decomposition of glycosyl xanthates;[9] glycosyl thiocyanates;[10] 1-thiols by radical addition to alkenes;[11] anhydro sugars;[12] and peracylated aldopyranoses by reaction with a thiol in the presence of Lewis acids.[13–16] The Lewis acids used for this purpose varied greatly from boron trifluoride etherate,[13] TMSOTF,[14a] $SnCl_4$,[14b] $InCl_3$-$TiCl_4$,[14c] AlI_3,[14d] to MoO_2Cl_2.[14e] Mixtures of Fe-I_2[15] and I_2-$NaBH_4$[16] were also used.

We found that the most reliable method for preparation of **3** involves the use of pure β-acetate **2** under boron trifluoride etherate catalysis.[13] We established that **2** reacts faster and more cleanly than the α-acetate, hence explaining the varied results presented in the literature when a mixture of α and β acetates is used.[13] We describe here an efficient synthesis of phenyl 2,3,4,6-tetra-*O*-acetyl-1-thio-β-D-galactopyranoside **3** (Scheme 33.1) from anomerically pure penta-*O*-acetyl-β-D-galactopyranose **2**, obtained by classical acetylation of D-galactose.[17] Importantly, we also observed that the related one-pot reaction starting from D-galactose[13h] was not reproducible in our hands, due mainly to the fact that the acetylation step was cumbersome and incomplete in the time course prescribed.

EXPERIMENTAL

GENERAL METHODS

The reaction was carried under nitrogen atmosphere using freshly distilled solvent (dichloromethane was freshly distilled from P_2O_5). Progress of reactions was monitored by thin-layer chromatography (TLC) using silica gel 60 F_{254} coated plates (E. Merck). Optical rotations were measured with a JASCO P-1010 polarimeter. Melting points were measured on a Fisher Jones apparatus. Nuclear magnetic resonance (NMR) spectra were recorded on Varian Inova AS600 spectrometers. Proton and carbon chemical shifts (δ) are reported in ppm relative to the chemical shift of residual $CHCl_3$, which was set at 7.27 ppm. Coupling constants (*J*) are reported in Hertz (Hz), and the following abbreviations are used for peak multiplicities: singlet (s), doublet (d), doublet of doublets (dd), triplet (t), and multiplet (m). Analyses and assignments were made using COSY (COrrelated SpectroscopY), DEPT (Distortionless Enhancement by Polarization Transfer), and HETCOR (Heteronuclear Chemical Shift Correlation) experiments. High-resolution mass spectra (HRMS) were measured with a LC-MS-TOF (Liquid Chromatography Mass Spectrometry Time Of Flight) instrument (Agilent Technologies) in positive electrospray mode by the analytical platform of UQAM.

PENTA-*O*-ACETYL-β-D-GALACTOPYRANOSE (2)[17]

Using a two-neck, round-bottom flask, a suspension of anhydrous sodium acetate (5.00 g, 61.06 mmol, 1.1 equiv.) in acetic anhydride (70 mL) was stirred and heated at reflux using an oil bath set at 140°C. When reflux began, heating was *immediately* discontinued by removing the flask from the oil bath. D-Galactose **1** (10.00 g, 55.51 mmol, 1.0 equiv.) was added in small portions (1 to 2 g each time) to the mixture. When the mixture became clear, it was stirred under reflux for a further 10 min. The hot solution was poured into ice water (400 mL in a 1 L beaker), and the mixture was stirred vigorously until ice melted. Dichloromethane (120 mL) was added to the solution, and the aqueous layer was separated. The organic phase was washed with ice-cold water (3 × 400 mL), sat. aq. $NaHCO_3$ solution (400 mL), and brine (400 mL). The organic solution was dried over Na_2SO_4, filtered, and concentrated under reduced pressure to give a crude yellow oil whose ^{1}H NMR spectrum showed the presence of four compounds: the peracetylated α,β-furanoses and α,β-pyranoses (15.4:12.4:10.5:61.7%).[*] The yellow oil was dissolved in the minimum amount of Et_2O (~25 mL) and precipitation of **2** was effected by successive addition of petroleum ether (~50 mL) and EtOH (~200 mL). The suspension was kept at 24°C for 2–3 hours and kept at –20°C overnight. The white solid was filtered and washed with petroleum ether to give a product containing ~95% of **2**. Crystallization from a minimum amount of EtOH gave pure (see NMR attached) penta-*O*-acetyl-β-D-galactopyranose **2** (12.57 g, 58%)[†]: mp 142–143°C; [Ref.[18a] mp 142–144°C; Ref.[17] mp 142°C; Ref.[18b] mp 137–139°C (EtOH)]; $R_f = 0.38$, 6:4 hexanes–EtOAc; $R_f = 0.20$, 1:1 Et_2O-petroleum ether (α: $R_f = 0.24$); $[\alpha]_D$ +23.0 (*c* 1.0, $CHCl_3$) [Ref.[18a] $[\alpha]_D$ +23.4 (*c* 1.0, $CHCl_3$)]; ^{1}H NMR (300 MHz, $CDCl_3$) δ 5.69 (d, 1H, $J_{1,2}$ 8.3 Hz, H-1), 5.41 (dd, 1H, $J_{3,4}$ 3.4 Hz, $J_{4,5}$ 1.0 Hz, H-4), 5.32 (dd, 1H, $J_{2,3}$ 10.4 Hz, $J_{1,2}$ 8.3 Hz, H-2), 5.07 (dd, 1H, H-3), 4.17–4.08 (m, 2H, H-6^a and H-6b), 4.05 (ddd, 1H, $J_{5,6}$ 6.6 Hz, H-5), 2.15, 2.11, 2.03, 2.03, 1.98 ppm (5 × s, 15H, 5 × COCH_3); ^{13}C NMR (150 MHz, $CDCl_3$) δ 170.8, 170.6, 170.4, 169.9, 169.5 (*C*OCH$_3$), 92.6 (*C*-1), 72.2 (*C*-5), 71.3 (*C*-3), 68.4 (*C*-2), 67.4 (*C*-4), 61.6 (*C*-6), 20.8, 20.6, 20.6, 20.5 ppm (5 × CO*C*H$_3$). Electrospray ionization high-resolution mass spectrometry (ESI+-HRMS): [M + Na]+ calcd for $C_{16}H_{22}O_{11}Na$, 413.1054; found, 413.1052.

PHENYL 2,3,4,6-TETRA-*O*-ACETYL-1-THIO-β-D-GALACTOPYRANOSIDE (3)

To a solution of penta-*O*-acetyl-β-D-galactopyranose **2** (500 mg, 1.28 mmol) in dry dichloromethane (10 mL, 0.13 M) was added thiophenol[‡] (200 μL, 1.92 mmol, 1.5 equiv.) at 24°C under a nitrogen atmosphere. The solution was cooled to 0°C and $BF_3 \cdot OEt_2$ (237 μL, 1.92 mmol, 1.5 equiv.) was added dropwise. The cooling was removed, and the mixture was stirred for 3 hours at 24°C, when (TLC, 1:1 hexanes–EtOAc) showed that all starting material was consumed. CH_2Cl_2 (10 mL) was added, and the mixture was washed with a sat. aq. $NaHCO_3$ solution (30 mL). The organic solution was dried over Na_2SO_4, filtered, concentrated, and the residue was chromatographed (4:1 → 7:3 hexanes-EtOAc) to give pure phenyl 2,3,4,6-tetra-*O*-acetyl-1-thio-β-D-galactopyranoside **3** as a white solid[§] (545 mg, 97%): $R_f = 0.33$, hexanes/EtOAc 2:1; mp 79–80°C (Et_2O-hexane) [Ref.[14e] 65–67°C; Ref.[13h] 80–81°C (Et_2O-hexane); Ref.[19a] 73–76°C; Ref.[19b] 70.5°C (aqueous ethanol)]; [α]D +4.0 (*c* 1.0, $CHCl_3$) [Ref.[14e] [α]D +4.2 (*c* 1.0, $CHCl_3$); Ref.[13h] [α]D +14 (*c* 2.0, $CHCl_3$); Ref.[19a] [α]D +5 ($CHCl_3$); Ref.[19b] [α]D +4.6 (*c* 1.8, $CHCl_3$); Ref.[19c] $[\alpha]_D$ +11.86 (*c* 1.62, $CHCl_3$)]; ^{1}H NMR (300 MHz, $CDCl_3$) δ 7.53–7.50 (m, 2H, H$_{arom}$), 7.33–7.29 (3H, m, H$_{arom}$),

---

[*] Checker found a ratio of 11.7:12.8:17.0:58.5% which may vary depending on the total time of heating due to the equilibration process.
[†] Yield found by Checker on 12 mmol scale was 45%.
[‡] All operations involving this reagent should be handled in a well-ventilated hood.
[§] Checker obtained a waxy solid.

5.42 (d, 1H, $J_{3,4}$ 3.1 Hz, H-4), 5.24 (t, 1H, $J_{2,3}$ 10.0 Hz, $J_{1,2}$ 10.0 Hz, H-2), 5.05 (dd, 1H, H-3), 4.72 (d, 1H, H-1), 4.19 (dd, 2H, $J_{6a,6b}$ 11.4 Hz, $J_{6a,5}$ 7.0 Hz, H-6a), 4.12 (dd, 2H, $J_{6b,5}$ 6.2 Hz, H-6b), 3.94 (t, 1H, H-5), 2.12, 2.10, 2.04, 1.97 ppm (s, 12H, CH₃); ¹³C NMR (150 MHz, CDCl₃) δ 170.3, 170.1, 170.0, 169.4 (CO), 132.5, 132.4, 128.8, 128.1 (C_{arom}), 86.5 (C-1), 74.3 (C-5), 71.9 (C-3), 67.2 (C-2), 67.1 (C-4), 61.6 (C-6), 20.8, 20.6, 20.6, 20.5 ppm (CH₃). ESI⁺-HRMS: [M + Na]⁺ calcd for $C_{20}H_{24}O_9SNa$, 463.10332; found, 463.10267. The $[\alpha]_D$ and the mp reported herein appear to be consistent with the convergent values observed in the literature for a crystalline compound.

ACKNOWLEDGMENTS

This work was supported by a grant from the Natural Science and Engineering Research Council of Canada (NSERC) to R.R. D.G. is thankful to the FQRNT for

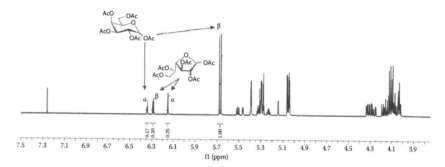

SCHEME 33.2   ¹H NMR of the crude acetylation mixture of D-galactopyranose **1**.

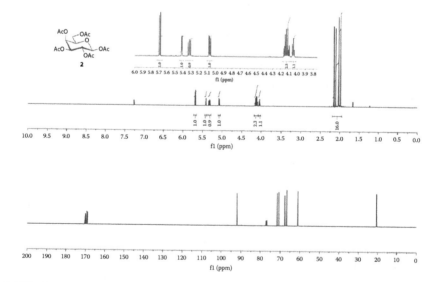

SCHEME 33.3   ¹H and ¹³C NMR of penta-*O*-acetyl-β-D-galactopyranose **2**.

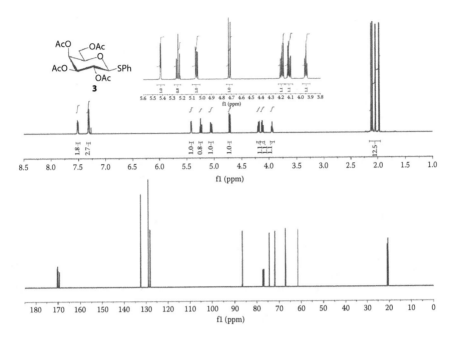

SCHEME 33.4   ¹H and ¹³C NMR of phenyl 2,3,4,6-tetra-*O*-acetyl-1-thio-β-D-galacto-pyranoside **3**.

a post-graduate fellowship. We are also very thankful to the editors G. A. van der Marel and J. D. C. Codée for insightful suggestions.

REFERENCES

1. (a) Norberg, T. In *Modern Methods in Carbohydrate Synthesis*; Khan, S. H., O'Neil, R. A., Eds.; Frontiers in Natural Product Research Series; Harwood Academic, 1995, *Vol. 1*, Chapter 4, pp 82–106; (b) Fügedi, P.; Garegg, P. J.; Lonn, H.; Norberg, T. *Glycoconj. J.* 1987, *4*, 97–108; (c) Garegg, P. J. *Adv. Carbhydr. Chem. Biochem.* 1997, *52*, 179–205; (d) van der Marel, G. A.; van den Bos, L. J.; Overkleeft, H. S.; Litjens, R. E. J. N.; Codée, J. D. C. *ACS Symp. Ser.*, 2007, *960*, 190–208; (e) Codée, J. D. C.; Litjens, R. E. J. N.; van den Bos, L. J.; Overkleeft, H. S.; van der Marel, G. A., *Chem. Soc. Rev.*, 2005, *34*, 769–782; (f) Shiao, T. C.; Roy, R. *Topics Curr. Chem.*, 2011, *301*, 69–108.

2. (a) Giguère, D.; Patnam, R.; Bellefleur, M.-A.; St-Pierre, C.; Sato, S.; Roy, R. *Chem. Commun.* 2006, 2379–2381; (b) Giguère, D.; Sato, S.; St-Pierre, C.; Sirois, S.; Roy, R. *Bioorg. Med. Chem. Lett.* 2006, *16*, 1668–1672; (c) André, S.; Giguère, D.; Dam, T. K.; Brewer, F.; Gabius, H.-J.; Roy, R. *N. J. Chem.* 2010, *34*, 2229–2240; Giguère, D.; André, S.; Bonin, M.-A.; Bellefleur, M.-A.; Provençal, A.; Cloutier, P.; Pucci, B.; Roy, R.; Gabius, H.-J. *Bioorg. Med. Chem.* 2011, *19*, 3280–3287.

3. Fisher, E.; Delbrück, K. *Ber.* 1909, *42*, 1476–1482.

4. (a) Arnott, S.; Scott, W. E.; Rees, D. A.; McNab, C. G. A. *J. Mol. Biol.* 1974, *90*, 253–267; (b) Millane, R. P.; Chandrasekaran, R.; Arnott, S.; Dea, I. C. M. *Carbohydr. Res.* 1988, *182*, 1–17; (d) Roy, R.; Tropper, F. D.; Cao, S.; Kim, J. M. *ACS Symp. Ser.* 1997, *659*, 163–180; (e) Patil, P. R.; Kartha, K. P. R. *Green Chem.*, 2009, *11*, 953–956.

5. Arnott, S.; Mitra, A. K.; Raghunathan, S. *J. Mol. Biol.* 1983, *169*, 861–872.

6. (a) Bluhm, T. L.; Zugenmaier, P. *Carbohydr. Res.* 1979, *68*, 15–21; (b) Winter, W. T.; Arnott, S.; Isaac, D. H.; Atkins, D. T. *J. Mol. Biol.* 1978, *125*, 1–19.

7. Mukhopadhyay, B.; Kartha, K. P. R.; Russell, D. A.; Field, R. A. *J. Org. Chem.* 2004, *69*, 7758–7760.

8. Cerny, M.; Zachystalova, D; Pacak, J. *Coll. Czech. Chem. Commun.* 1961, *26*, 2206–2211.

9. (a) Tropper, F. D.; Andersson, F. O.; Roy, R. *J. Carbohydr. Chem.* 1992, *11*, 741–750; (b) Bogusiak, J.; Wandzik, I.; Szeja, W. *Carbohydr. Lett.* 1996, *1*, 411–416; (c) Sakata, M.; Haga, M.; Tejima, S.; Akagi, M. *Chem. Pharm. Bull.* 1963, *11*, 1081–1082.

10. (a) Fisher, E.; Helferich, B.; Ostman, P. *Ber.* 1920, *53*, 873–886; (b) Ogura, H.; Takahashi, H. *Heterocycles*, 1982, *17*, 87–90; (c) Lindhorst, T. K.; Kieburg, C. *Synthesis*, 1995, 1228–1230; (d) Pakulski, Z.; Pierozynski, D.; Zamojski, A. *Tetrahedron*, 1994, *50*, 2975–2992.

11. Lacombe, J. M.; Rakotomanomana, N.; Pavia, A. A. *Tetrahedron Lett.* 1988, *29*, 4293–4296.

12. (a) Seeberger, P. H.; Eckhardt, M.; Gutteridge, C. E.; Danishesfsky, S. J. *J. Am. Chem. Soc.* 1997, *119*, 10064–10072; (b) Sakairi, N.; Hayashida, M.; Kuzuhara, H. *Tetrahedron Lett.* 1987, *28*, 2871–2874.

13. (a) Chandrasekaran, R.; Lee, E. J.; Thailambal, V. G.; Zevenhuizen, L. P. T. M. *Carbohydr. Res.* 1994, *261*, 279–295; (b) Ogawa, K.; Okamura, K.; Sarko, A. *Int. Biol. Macromol.* 1981, *3*, 31–36; (c) Okuyama, K.; Otsubo, A.; Fukuzawa, Y.; Ozawa, M.; Harada, T.; Kasai, N. *J. Carbohydr. Chem.* 1991, *10*, 645–656; (d) Santra, A.; Sau, A.; Misra, A. K. *J. Carbohydr. Chem.*, 2011, *30*, 85–93; (e) Xu, C.; Liu, H.; Li, X. *Carbohydr. Res.*, 2011, *346*, 1149–1153; (f) Deng, S.; Gangadharmath, U.; Chang, C.-W. T. *J. Org. Chem.*, 2006, *71*, 5179–5185; (g) Janczuk, A. J.; Zhang, W.; Andreana, P. R.; Warrick, J.; Wang, P. G. *Carbohydr. Res.*, 2002, *337*, 1247–1259; (h) Agnihotri, G.; Tiwari, P.; Misra, A. K. *Carbohydr. Res.* 2005, *340*, 1393–1396; (i) Kametani, T.; Kawamura, K.; Honda, T. *J. Am. Chem. Soc.* 1987, *109*, 3010–3017.

14. (a) Li, Z.; Gildersleeve, J. C. *J. Am. Chem. Soc.*, 2006, *128*, 11612–11619; (b) Sau, A.; Misra, A. K. *Synlett*, 2011, 1905–1911; (c) Das, S. K.; Roy, J.; Reddy, K. A.; Abbineni, C. *Carbohydr. Res.*, 2003, *338*, 2237–2240; (d) Weng, S.-S.; Li, C.-L.; Liao, C.-S.; Chen, T.-A.; Huang, C.-C.; Hung, K.-T. *J. Carbohydr. Chem.*, 2010, *29*, 429–440; (e) Weng, S.-S.; Lin, Y.-D.; Chen, C.-T. *Org. Lett.*, 2006, *8*, 5633–5636.

15. Weng, S.-S. *Tetrahedron Lett.*, 2009, *50*, 6414–6417.

16. Valerio, S.; Iadonisi, A.; Adinolfi, M.; Ravida, A. *J. Org. Chem.*, 2007, *72*, 6097–6106.

17. (a) Wolfrom, M. L.; Thompson, A. *Methods in Carbohydrate Chemistry*; Wiley: New York, 1963, *2*, 211–215; (b) Erwig, E.; Koenigs, W. *Ber. Dtsch. Chem. Ges.*, 1889, *22*, 2207–2214; http://gallica.bnf.fr/ark:/12148/bpt6k90718t/f491.image.

18. (a) Pozsgay, V.; Jennings, H. J. *Synthesis,* 1990, 724–726; (b) Wolfrom, M. L.; Thompson, A.; Inatome, M. *J. Am. Chem. Soc.*, 1957, *79*, 3868–3871.

19. (a) Ferrier, R. J.; Furneaux, R. H. *Carbohydr. Res.* 1976, *52*, 63–68; (b) Capon, B.; Collins, P. M.; Levy, A. A.; Overend, W. G. *J. Chem. Soc.* 1964, 3242–3254; (c) Khiar, N.; Martin-Lomas, M. *J. Org. Chem.* 1995, *60*, 7017–7021.

34 An Alternative, Large-Scale Synthesis of 1,2:5,6-Di-*O*-isopropylidene-α-D-*ribo*-hex-3-ulofuranose

Pāvels Ostrovskis
Jevgeņija Mackeviča
Viktors Kumpiņš
Óscar López[*]
Māris Turks[†]

CONTENTS

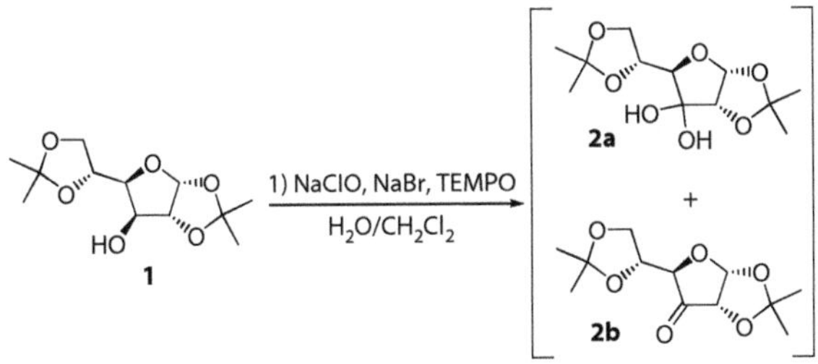

SCHEME 34.1

Carbohydrate-derived ketones are versatile intermediates in organic synthesis. Thus, 1,2:5,6-di-*O*-isopropylidene-α-D-*ribo*-hex-3-ulofuranose (**2b**) has been extensively used.[1-3] Due to steric hindrance of the 1,2-*O*-isopropylidene group and conformational restrictions,[4,5] nucleophiles of choice attack preferentially the β-face (*si*-face) of the

---

[*] Checker: osc-lopez@us.es.
[†] Corresponding author: maris_turks@ktf.rtu.lv.

carbonyl group. This leads to 3-*C*-substituted-D-allose derivatives when *C*-nucleophiles are used.[6] Synthetic approaches that are based on Henry[3,7] and Reformatsky[8] reactions on ketone **2b** have been developed. Logically, also Grignard reagents add to **2b** with good diastereoselectivities.[9–11] Similarly, cyanohydrin formation[12] and reactions with other nucleophiles[13] also proceed well. For example, reduction of ketone **2b** provides an easy access to 1,2:5,6-di-*O*-isopropylidene-α-D-allofuranose.[14] This approach has been used in the straightforward synthesis of kanosamine. Additionally, the Wittig reaction on ketone **2b** opens the door to rich chemistry.[15–17]

Historically, oxidation of diacetone-α-D-glucose (**1**, 1,2:5,6-di-*O*-isopropylidene-α-D-glucofuranose) was realized by ruthenium tetraoxide[2] and various activated DMSO methods.[1,3,18] The latter approach has found industrial application[19] and has been reviewed.[20] Academic laboratories still[13,14] rely mainly on chromium (VI)[21,22] and ruthenium tetraoxide methods.[23] Some of these are rather difficult to scale-up and/or require the use of toxic or expensive reagents.

In 2004, a group at Merck reported a large scale 2,2,6,6-tetramethylpiperidinyl-1-oxyl (TEMPO) catalyzed bleach oxidation of **1**.[24] However, this procedure has not yet become the method of choice for synthesis of 1,2:5,6-di-*O*-isopropylidene-α-D-*ribo*-hex-3-ulofuranose (**2b**). Here, we report on this practical and easily scalable TEMPO-catalyzed oxidation procedure providing **2b** on hundred-gram scale. As shown in Scheme 34.1, diacetone-α-D-glucose (**1**) is oxidized by NaClO in a biphasic (H_2O/CH_2Cl_2) system in the presence of 2,2,6,6-tetramethylpiperidinyl-1-oxyl (TEMPO) and NaBr as a co-catalyst. Pure product is obtained in reproducibly high yield if the concentration and pH of the aqueous NaClO solution are carefully controlled. Additionally, vigorous mechanical stirring and cautious internal temperature control are critical. Besides ketone **2b**, the crude product of oxidation reaction contains the hydrate **2a**. Various methods for azeotropic water removal are available. These include azeotropic distillation with benzene, toluene, and chloroform at ambient or reduced pressure. Sufficiently dry ketone can be obtained by refluxing its chloroform solution in a Soxhlet extractor filled with 3Å or 4Å molecular sieves.[*] Vacuum distillation of the crude product provides dry ketone **2b** suitable for reactions using, e.g., organometallic reagents.

EXPERIMENTAL

GENERAL METHODS

Technical grade, distilled dichloromethane and deionized water were used. The bleach solution used was of technical grade (Stanchem, Poland).[†] 2,2,6,6-Tetramethylpiperidinyl-1-oxyl (TEMPO) and NaBr were purchased from Acros. ^{1}H NMR (300 MHz) and ^{13}C NMR (75 MHz) were recorded on Bruker *Avance 300* spectrometer in $CDCl_3$ (with solvent residual signal as an internal standard).

---

[*] IR spectroscopy can be used as a fast method for qualitative determination of hydrate **2a** in the mixture. The latter shows a characteristic broad absorption band at 3200–3400 cm⁻¹.

[†] The checker used NaOCl solution from Fischer Scientific-Across Organics with a content of hypochlorite not less than 14%.

Optical rotation was measured using a Perkin Elmer automatic polarimeter. The infrared (IR) absorbance was recorded using a Perkin Elmer *Spectrum BX FT-IR* spectrometer.

1,2:5,6-Di-*O*-isopropylidene-α-d-*ribo*-hex-3-ulofuranose-3 (2b)

In an open-mouth stainless steel beaker[*] equipped with mechanical stirrer, a solution of NaBr (4.2 g, 0.04 mol, 0.1 equiv.) in water (20 mL) was added to a solution of diacetone-α-d-glucose (**1**) (104 g, 0.4 mol, 1.0 equiv.) in CH_2Cl_2 (400 mL), followed by TEMPO (0.30 g, 1.9 mmol, 0.5 mol-%). The resulting mixture was cooled to –5 to –10°C (internal temperature) and vigorously stirred, while aqueous solution of NaClO (~1.6 mol/L, pH 9.5, 360 mL, 0.58 mol, 1.45 equiv.)[†] was added dropwise in 30 minutes while keeping the internal temperature in the range of –10–0°C.[25] After addition of the bleach solution, the resulting reaction mixture was stirred for 5 minutes.[‡,§] The organic layer was separated and washed successively with solution of KI (1.6 g, 0.01 mol, 2.5 mol-%) in 0.5 M aqueous hydrochloric acid (100 mL), 10% aqueous solution of $Na_2S_2O_3$ (100 mL), saturated aqueous solution of $NaHCO_3$ (100 mL), and brine (100 mL). After drying (Na_2SO_4), filtration, and concentration under reduced pressure, the crude product (93 g, ~90%) contained a mixture of ketone hydrate **2a**[26] and ketone **2b**[27] in various proportions with the latter being the major component.

The foregoing mixture was diluted with CH_2Cl_2 (10–20 mL) and transferred into a flask suitable for vacuum distillation,[¶] and the residual CH_2Cl_2 was distilled at 1–5 torr (bath temperature 40–50°C). Then, the vacuum was adjusted to 0.2–0.4 torr, and ketone **2b** was distilled at 100–105°C (bath temperature 125–135°C).[**] Fraction I: 2 g (2%, purity by NMR 90%); fraction II: 72 g (70%, purity by NMR 96%); fraction

---

[*] Reaction vessels made of other materials can be used (e.g., glass, PP, etc.). Nevertheless, steel provides the best heat conductivity and thus is most convenient for efficient cooling when an external cooling bath is used.

[†] Commercial bleach with NaClO content of 17-21% was used. The exact molarity of NaClO in this solution was determined to be 2.3 mol/L by iodometric titration. This commercially available 2.3 M aqueous solution of NaClO was adjusted to pH 9.5 (controlled by pH-meter) by addition of saturated aqueous $NaHCO_3$ solution (~110 mL). Thus, the final concentration of the freshly prepared NaClO solution with pH 9.5 was ~1.6 mol/L.

[‡] At this scale, the oxidation process is almost instantaneous. The completion of the reaction is indicated by absence of the exothermic reaction upon addition of a fresh portion of bleach. The reaction mixture in its "active state" should be reddish-brown. The appearance of yellow color indicates the deactivation of the catalytic system.

[§] Progress of the reaction can be monitored by ^1H NMR, as TLC does not appear to be reliable due to ketone–ketone hydrate equilibrium on the TLC plate. Thus, anomeric protons of all three species (**1**, **2a**, **2b**) are well separated (300 MHz, $CDCl_3$): 6.14 ppm (d, $J = 4.5$ Hz, H-C(1) of **2b**), 5.95 ppm (d, $J = 3.6$ Hz, H-C(1) of **1**), 5.85 ppm (d, $J = 3.6$ Hz, H-C(1) of **2a**) (Figure 34.1).

[¶] Magnetic stirring is not appropriate due to relatively high viscosity of **2b**. As a consequence, the distillation system should contain a capillary tube or pipette with a small internal diameter for generation of boiling centers.

[**] Fractions I through III were collected by separating a small amount during the initial and final stages of the distillation process.

III: 7 g (6%, purity by NMR: 85%). The product from fraction II is considered to be sufficiently pure and is used for full characterization. Yield 72 g (70%). $[\alpha]_D^{25}$ +139 (c 3.2, CHCl$_3$)*; IR (film) υ, cm^{-1}: 2989, 2939, 2896, 1773, 1457, 1376, 1261, 1216, 1156, 1079, 1023. ^{1}H and ^{13}C NMR data (300 MHz, CDCl$_3$) were in agreement with those published.[26,27] Anal. Calcd for $C_{12}H_{18}O_6$ (MW 258.27): C, 55.81; H, 7.02. Found: C, 55.86; H, 7.05.

ACKNOWLEDGMENTS

The authors thank the Latvian Council of Science and ERDF project "The development of international cooperation, projects and capacities in science and technology at Riga Technical University," Nr.2DP/2.1.1.2.0/10/APIA/VIAA/003, for financial support. JSC Olainfarm (Latvia) is acknowledged for a scholarship to P.O. and the kind donation of diacetone-α-D-glucose, and the Latvian Academy of Sciences for a scholarship to J.M.

---

* Literature[27]: $[a]_D^{25}$ +131.5 (c 1.3, CH$_2$Cl$_2$).

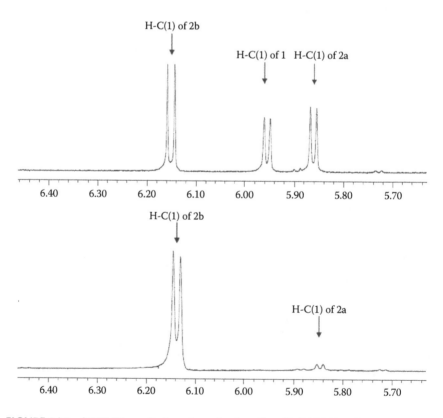

FIGURE 34.1 ^{1}H NMR monitoring of reaction **1** → **2a** + **2b** (300 MHz, CDCl$_3$).

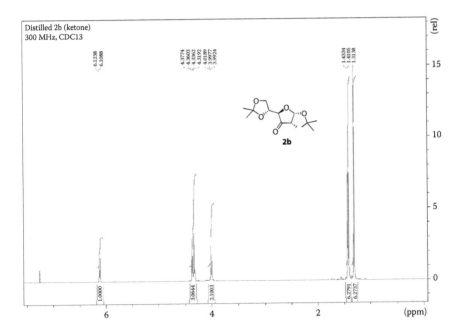

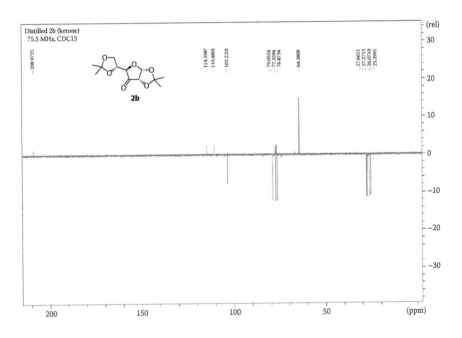

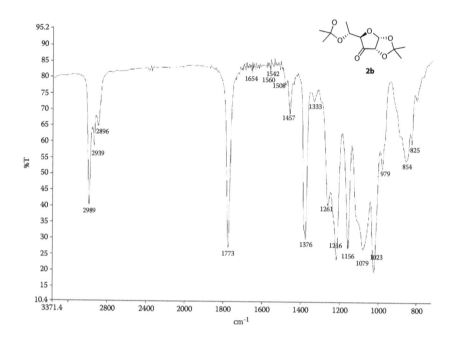

REFERENCES

1. Onodera, K.; Hirano, S.; Kashimura, N. *J. Am. Chem. Soc.* 1965, *87*, 4651–4652.
2. Lawton, B. T.; Szarek, W. A. Jones, J. K. N. *Carbohydr. Res.* 1969, *10*, 456–458.
3. Albrecht, H. P.; Moffatt, J. G. *Tetrahedron Lett.* 1970, *11*, 1063–1066.
4. De Bruyn, A.; Danneels, D.; Anteunis, M.; Saman, E. *J. Carbohydrates, Nucl. Nucl.* 1975, *2*, 227–240.
5. Peciar, C.; Alföldi, J.; Palovcík, R.; Kováč, P. *Chem. Zvesti* 1973, *27*, 90–93.
6. Stütz, E. A. In *Topics in Current Chemistry*; Springer: Heidelberg, 2001; pp. 1–345.
7. Filichev, V. V.; Brandt, M.; Pedersen, E. B. *Carbohydr. Res.* 2001, *333*, 115–122.
8. Gurjar, M. K.; Reddy, D. S.; Bhadbhade, M. M.; Gonnade, R. G. *Tetrahedron* 2004, *60*, 10269–10275.
9. Prashad, M.; Fraser-Reid, B. *J. Org. Chem.* 1985, *50*, 1564–1566.
10. Patra, R.; Bar, N. C.; Roy, A.; Achari, B.; Ghoshal, N.; Mandal, S. B. *Tetrahedron* 1996, *52*, 11265–11272.
11. Mane, R. S.; Ghosh, S.; Chopade, B. A.; Reiser, O.; Dhavale, D. D. *J. Org. Chem.* 2011, *76*, 2892–2895.
12. Gasch, C.; Illangua, J. M.; Merino-Montiel, P.; Fuentes, J. *Tetrahedron* 2009, *65*, 4149–4155.
13. Gasch, C.; Merino-Montiel, P.; López, Ó.; Fernández-Bolaños, J. G.; Fuentes, J. *Tetrahedron* 2010, *66*, 9964–9973.
14. Guo, J.; Frost, J. W. *J. Am. Chem. Soc.* 2002, *124*, 10642–10643.
15. Filichev, V. V.; Pedersen, E. B. *Tetrahedron* 2001, *57*, 9163–9168.
16. Ay, K.; Çetin, F.; Yüceer, L. *Carbohydr. Res.* 2007, *342*, 1091–1095.
17. Xavier, N. M.; Rauter, A. P. *Org. Lett.* 2007, *9*, 3339–3341.
18. Rosenthal, A.; Ong, K.-S. *Tetrahedron Lett.* 1969, *10*, 3981–3983.
19. Christensen, S. M.; Hansen, H. F.; Koch, T. *Org. Proc. Res. Dev.* 2004, *8*, 777–780.

20. Caron, S.; Dugger, R. W.; Ruggeri, S. G.; Ragan, J. A.; Brown Ripin D. H. *Chem. Rev.* 2006, *106*, 2943–2989.
21. Garegg, P. J.; Samuelsson, B. *Carbohydr. Res.* 1978, *67*, 267–270.
22. Herscovici, J.; Antonakis, K. *J. Chem. Soc., Chem. Commun.* 1980, 561–562.
23. Baker, D. C.; Horton, D.; Tindall, C. G. *Meth. Carbohydr. Chem.* 1976, *7*, 3–6.
24. Bio, M. M.; Xu, F.; Waters, M.; Williams, J. M.; Savary, K. A.; Cowden, C. J.; Yang, C.; Buck, E.; Song, Z. J.; Tschaen, D. M.; Volante, R. P.; Reamer, R. A.; Grabowski, E. J. J. *J. Org. Chem.* 2004, *69*, 6257–6266.
25. Anelli, P. L.; Montanari, F.; Quici, S.; Nonoshita, K.; Yamamoto, H. *Org. Synth.* 1990, *69*, 212–217; *Org. Synth. Coll. Vol.* 1993, *8*, 367–372.
26. ¹H-NMR data of 2a: Shing, T. K. M.; Wong, C.-H.; Yip, T. *Tetrahedron: Asymm.* 1996, *7*, 1323–1340.
27. ¹H-NMR data of 2b: Soler, T.; Bachki, A.; Falvello, L. R.; Foubelo, F.; Yus, M. *Tetrahedron: Asymm.* 2000, *11*, 493–517.

Index

Milton Keynes UK
Ingram Content Group UK Ltd.
UKHW021625071024
449327UK00020BA/1191